Probability, Statistics, and Decision Making in the Atmospheric Sciences

Probability, Statistics, and Decision Making in the Atmospheric Sciences

edited by Allan H. Murphy and Richard W. Katz

Westview Press / Boulder and London

Copyright © 1985 by Westview Press, Inc.

Published in 1985 in the United States of America by Westview Press, Inc.,
5500 Central Avenue, Boulder, Colorado 80301; Frederick A. Praeger,
Publisher

Library of Congress Catalog Card Number: 84-52287
ISBN: 0-86531-152-8
ISBN: 0-86531-153-6 (pbk.)

Composition for this book was provided by the editors
Printed and bound in the United States of America

10 9 8 7 6 5 4 3 2 1

Contents

Preface

Methodology drawn from the fields of probability, statistics, and decision making plays an increasingly important role in the atmospheric sciences, both in basic and applied research and in experimental and operational studies. Applications of such methodology can be found in almost every facet of the discipline, from the most theoretical and global (e.g., atmospheric predictability, global climate modeling) to the most practical and local (e.g., crop-weather modeling, forecast evaluation). Almost every issue of the multitude of journals published by the atmospheric sciences community now contains one or more papers involving applications of concepts and/or methodology from the fields of probability and statistics. Despite the increasingly pervasive nature of such applications, very few book-length treatments of probabilistic and statistical topics of particular interest to atmospheric scientists have appeared (especially in English) since the publication of the pioneering works of Brooks and Carruthers (Handbook of Statistical Methods in Meteorology) in 1953 and Panofsky and Brier (Some Applications of Statistics to Meteorology) in 1958. As a result, many relatively recent developments in probability and statistics are not well known to atmospheric scientists and recent work in active areas of meteorological research involving significant applications of probabilistic and statistical methods are not familiar to the meteorological community as a whole.

No single volume could be expected to cover completely the many probabilistic, statistical, and decision-making topics and applications of interest and importance to atmospheric scientists, and this book is no exception. It has been our intention here to place primary emphasis on two types of contributions: (1) surveys of relatively new developments in probability and statistics of potential interest to atmospheric scientists; and (2) reviews of selected areas within the atmospheric sciences, in which probability and statistics have played and continue to play important roles. As in all volumes consisting of contributions from many individuals, the scope and level of treatment varies from chapter to chapter. However, we believe that the treatment in the various chapters is generally appropriate for the intended audience - students in atmospheric sciences who have taken at least one course in probability and statistics and researchers and others in meteorology familiar through their work with basic concepts and methods in these fields. Moreover, some topics that undoubtedly will be new to many readers (exploratory data analysis, decision analysis) are treated at an introductory level. In all chapters, an effort has been made to provide readers with extensive lists of references to the relevant literature for more comprehensive treatments and/or real-world applications. It is our hope that everyone who examines the book will find at least one chapter that is of interest and useful.

Although each chapter can be considered to be an individual unit and read in isolation, many are closely related and can be profitably studied together. Chapter 1 can be viewed as an introduction to the other chapters, since it treats exploratory analysis, the first of two stages of data analysis. Most of the other chapters are concerned primarily with confirmatory analysis, the second stage of data analysis. Multiple regression, undoubtedly the statistical technique most frequently employed in applications to the atmospheric sciences, is the topic of Chapter 2. From the extensive literature on multiple regression, this chapter focuses on some new developments and alternatives. Chapters 3 and 4, on multivariate analysis of a single batch and several batches of data, respectively, can be regarded as extensions of Chapter 1, since an exploratory approach to data analysis is again emphasized.

Chapters 5, 6, and 7 are all concerned with the problem of modeling the dependencies over time that atmospheric measurements typically exhibit. The two related methods of performing time series analysis, the frequency domain approach and the time domain approach, are treated in Chapters 5 and 6, respectively. Probabilistic models, which include the parametric time series models considered in Chapter 6 as a special case, are the subject of Chapter 7.

Chapters 8, 9, and 10 examine certain aspects of weather forecasting. Statistical weather forecasts are routinely produced on an operational basis in many countries, and Chapter 8 reviews various methodological techniques, such as regression and discriminant analysis, used to produce such forecasts. Formulating the forecasts in terms of probabilities, a means of specifying the uncertainty inherent in weather forecasts, is treated in Chapter 9. The problem of evaluating weather forecasts, including forecasts of probabilistic and nonprobabilistic form, is addressed in Chapter 10.

Considerable research in statistics has been devoted to the design and evaluation of experiments to investigate the effectiveness of weather modification. Chapter 11 discusses such experimental designs and evaluation procedures, stressing the need for randomized double-blind experiments and emphasizing the use of distribution-free techniques. Bayesian inference, an approach to statistical inference that is particularly appropriate in situations involving imperfect information and expert judgments, is treated in Chapter 12. Finally, the ultimate aim of many applications of probability and statistics to the atmospheric sciences is to aid in decision making, and Chapter 13 introduces decision analysis, a formal method for analyzing and modeling decision-making problems.

Many individuals and organizations have contributed significantly to the completion of this project and the publication of the book. First we would like to acknowledge the authors of the chapters themselves, without whose contributions this book would not have been possible. In addition to the editors, several individuals read

various chapters and offered comments in an effort to ensure
consistency of terminology and notation within and among chapters.
We would especially like to thank Barbara G. Brown for her efforts in
this regard. Sheri Kellogg typed most of the chapters and Michelle
Holcomb completed this task - they are due special thanks for their
conscientious and careful work. The assistance of several members of
the staff of Westview Press, including Alice Levine, Alice Trembour,
and Edythe Porpa, is gratefully acknowledged. The Department of
Atmospheric Sciences at Oregon State University generously provided
facilities and other assistance that greatly aided in the completion
of the book. The work of the editors on this project was supported
in part by the National Science Foundation (Division of Atmospheric
Sciences) under grants ATM-8004680 and ATM-8209713.

Allan H. Murphy
Corvallis, Oregon

Richard W. Katz
Boulder, Colorado

1
Exploratory Analysis of Atmospheric Data

Thomas E. Graedel and Beat Kleiner

1. INTRODUCTION

Atmospheric data are seldom apposite for the use of the traditional tools of statistics. In addition to the more dramatic effects
that occur on short time scales, the data abound with diurnal, seasonal, annual, and multi-year patterns over a broad range of magni-
tudes. Replicate experiments are often impossible or unfeasible.
These properties suggest that innovative treatment of the data (termed
exploratory data analysis or EDA) may bring substantial rewards. In
this chapter we hope to demonstrate that such a statistical approach
is well within the feasibility of modern computational systems and
that considerable useful information can result from taking such an
approach to the analysis of atmospheric data.

Exploratory data analysis is designed to find out "what the data
are telling us." Its basic intent is to search for interesting relationships and structures in a body of data and to exhibit the results
in such a way as to make them recognizable. This process involves
summarization, perhaps in the form of a few simple statistics (e.g.,
mean and variance of a set of data) or perhaps in the form of a simple
plot (such as a scatterplot). It also involves exposure; that is, the
presentation of the data so as to allow one to see both anticipated
and unexpected characteristics of the data. In general, finding the
unexpected turns out to be much more rewarding than merely confirming
the suspected.

Although EDA is by no means a new subject, it is only in the
recent past that it has become an important part of statistical analysis. A major factor in this development has been the rapidly
increasing availability of computer hardware and software, together
with the challenge of larger bodies of data in many fields and the
accelerating emphasis on quantification in a growing variety of disciplines. Despite its utility, however, EDA has yet to be adequately
incorporated into formal statistical theory. Perhaps the best suc-
cinct introduction to the subject that can currently be given is to
list a few of the important characteristics of EDA techniques:

1

(a) EDA is an iterative process. It proceeds by trial and error and the insights gained from each step are used to guide the next ones.

(b) While doing EDA it is very helpful to form some idea of a model, but this working model should not be taken too seriously. One should try to keep an open mind concerning the relationships that may be revealed.

(c) It is advantageous to look at the data in a variety of different ways. Most of these will, in the end, turn out to be inconsequential, uninteresting, or of no operational value. One should anticipate such a result, however, and not terminate the EDA too quickly.

(d) EDA makes extensive use of robust/resistant statistics and relies heavily on graphical techniques.

In general, the absence of prior information about the data does not allow one to initially assume that the data are normally distributed. On the contrary, it is much more likely that a given set of data is not normal. This may be due to outliers (which inevitably creep into every moderate to large data set, no matter how carefully collected), to non-normality of the process being monitored, or to some other reason. Therefore, simple computation of parameters that describe a set of data completely only if the data are normal will not generally be satisfactory, and other ways have to be used to summarize and expose these data. The most useful techniques to do so are transformation of the data (to approximate normality or at least to approximate symmetry), robust statistical methods, and graphical techniques.

In this chapter it is only possible to discuss selected portions of the entire topic of data exploration. Those that we present were chosen with the atmospheric sciences in mind, and are illustrated by atmospheric examples. Readers desiring to explore further can begin by consulting the books by Tukey (1977), Mosteller and Tukey (1977), and Erickson and Nosanchuk (1977). Two books which contain listings of computer programs for several exploratory data analysis techniques are McNeil (1977) (FORTRAN and APL) and Velleman and Hoaglin (1981) (FORTRAN and BASIC).

In the next section, we describe several methods for summarizing and exposing sets of one-dimensional data. In Section 3 we discuss techniques for comparing two data sets, and Section 4 presents techniques for the comparison of several sets of data. Section 5 contains the summary and conclusions.

2. SUMMARIZATION AND EXPOSURE OF SETS OF ONE-DIMENSIONAL DATA

The analysis of a collection of numbers which can usefully be regarded as a set of one-dimensional data is probably the most frequently encountered problem in data analysis. Examples of such sets of numbers are:

(a) The amount of snowfall in January for several years at a specific weather station.

(b) Frequency distributions of interstroke intervals in lightning flash data.

(c) The average number of heating degree days in all cities within the U.S. whose population exceeds 100,000.

(d) The 15-minute average readings of ozone concentrations in the ambient air at several sites during a summer.

Sets of one-dimensional data do not necessarily have to be independent samples from a univariate distribution. They might be part of a multi-dimensional data set as illustrated by examples (c) and (d), or there might exist non-zero correlations within the set as illustrated by example (d). In these cases, it is often useful to regard the data initially as though they constituted a one-dimensional sample, ignoring for the time being the underlying associated structure.

In analyzing sets of numbers one first wants to get a "feel" for the data set at hand and ask such questions as:

"What are the smallest and largest values?"

"What might be a good single representative number for this set of data?"

"What is the amount of variation or spread?"

"Are the data clustered around one or more values or are they spread uniformly over some interval?"

"Can they be considered symmetric?"

There are innumerable (and often better) ways to investigate a one-dimensional data set besides simply computing its mean as an estimate for the center of the data set (the location) and its standard deviation as an estimate of its variation or spread (the scale). In this chapter we will describe some of the more useful ones.

2.1 One-Dimensional Scatterplots

The one-dimensional scatterplot depicts a set of numbers on a (usually horizontal) axis labelled with the numerical values of the data plotted. The plot provides, at a glance, the answers to some of the questions asked above. Its main advantages are that it is very easy to construct, takes little space, and utilizes the original data, which can be recovered directly from the plot if the data are not too dense. Its major disadvantage is the fact that inferences about the sample distribution of the data cannot be drawn easily from the presentation.

Figure 1 shows a one-dimensional scatterplot of the wind speed measured at 1 a.m. at Newark Airport during December 1974. From the figure we can see that the 31 measurements gave only 11 distinct values, we can very quickly find the extremes of the data set, and we can conjecture that the data are not uniformly distributed. However, it is generally rather difficult to derive further insight into the distribution of the data, especially if many of the data points have the same numerical values, or if the number of data points is so large that the plotting symbols overlap.

One-dimensional scatterplots are most useful with small sets of data (well under 100 data points) and when a large number of data sets are to be compared. Due to their compactness they are also very useful in depicting the individual distribution of each variable in an ordinary (two-dimensional) scatterplot.

2.2 Histograms

The histogram is one of the oldest and most frequently used tools of the data analyst for the investigation of the overall distribution of a sample. For a given set of data, the histogram is constructed by dividing the horizontal axis into segments (also called groups or classes) which together cover the range of the data. Over each segment a rectangle is constructed whose area is proportional to the number of data points in that group. For example, Figure 2 (which depicts the same data as Figure 1) clearly shows that the data are not uniformly distributed, but peak between 10 and 15 km/h and are positively skewed (i.e., skewed toward the higher data values). About one-third of the data lies above 15 km/h. An alternative method of plotting this histogram is to use the fraction of data points in each interval, rather than their number, as the area.

The information derived from a histogram depends in large part upon three choices made prior to its construction: the number of classes, the lengths of these classes, and the breakpoints between them. These decisions determine the usefulness of a histogram as a visualization of a data set. For example, if too few classes are taken, the result will often be a rather featureless and not very illuminating picture; whereas if too many classes are taken (especially with small sample sets), very few observations will fall into each class and the histogram may exhibit a rather irregular pattern.

Usually class lengths are determined first (often on a very subjective basis) and then the number of classes is computed by dividing the data range by class length. The alternative method of first determining the number of classes and then computing class length from

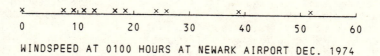

WINDSPEED AT 0100 HOURS AT NEWARK AIRPORT DEC. 1974

Figure 1.   One-dimensional scatterplot of measurements of wind speed
(km/h) at Newark Airport at 0100 hours during December
1974 (data from National Climatic Center, 1974-1975).

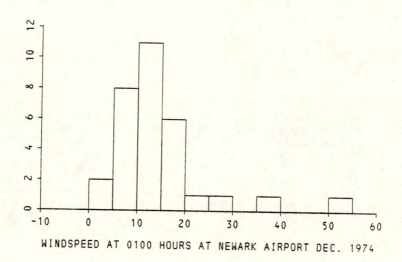

WINDSPEED AT 0100 HOURS AT NEWARK AIRPORT DEC. 1974

Figure 2.   Histogram of wind speed values shown in Figure 1.

the data range has the distinct disadvantage that it is easily influenced by even a single outlying value.

Generally, equal sized classes are used. If different sized classes are employed, one has to make sure that the area (and not the height) of each bar is proportional to the number of data points in the corresponding class, because the histogram is a representation of the density function (where the frequency between any two points in the distribution is represented by the area under the curve between these points). One also has to be careful about the breakpoints between the classes of a histogram, because choosing breakpoints arbitrarily just to give "nice" numbers could interact with the two other choices to give biased and misleading pictures.

Advice on the choice of the number of classes to be used varies widely throughout the literature. It ranges from "we generally advise using 5 to 20 classes" (Mendenhall and Ott, 1972, p. 14) to "the number of occupied groups should be roughly equal to the square root of the number of data points" (Davies and Goldsmith, 1972, p. 11). Doane (1976) suggests that the rule of Davies and Goldsmith be used for symmetric data only and that for asymmetric data one should add a number of groups which increases for increasing skewness of the data. Unfortunately his method is very vulnerable to outliers.

Diaconis and Freedman (1981) suggest that one should first determine an approximate class length h by

$$h = 1.349s[(\ln n)/n]^{1/3},$$

where s is a robust estimate of the standard deviation and n denotes sample size (robust estimates will be introduced and discussed in Section 2.7). This choice for class length will approximately minimize the maximum absolute deviation between the resulting histogram and the underlying density function. Then a tentative number of classes K is computed by dividing the range of the data by h, and approximately K+1 "pretty" breakpoints are determined by a rule similar to the one described in Doane (1976). If more than one third of the data should lie exactly on the breakpoints, all breakpoints are shifted by half the class length. If more than one third of the data should lie on the new breakpoints, they are shifted again, this time by a quarter of the class length.

In Figure 2 the class length h was computed according to the rule of Diaconis and Freedman (1981); with n = 31 and s = 5.5 (taken as the interquartile range/1.349; see Section 2.4), h = 3.6, which resulted in a tentative value of K = 15 for the number of classes. Choosing breakpoints at convenient round numbers resulted in h = 5, 11 classes, and breakpoints at 0, 5, 10, ..., 50, and 55. This rule for determining classes is based on asymptotic considerations and should only be regarded as a reasonable starting point; in each practical situation the numbers of classes or the breakpoints might have to be subjectively adapted to the actual circumstances.

The principal utility of the histogram is that it shows the relative class frequencies in the data and therefore provides information on the data density function. A major disadvantage is due to the grouping of the data into classes, which removes the final product somewhat from the original data. Histograms have proved to be most useful for large data sets with more than 300-500 data points and in cases where the data come already grouped into classes.

2.3 Stem-and-Leaf Displays

Stem-and-leaf displays (Tukey, 1977) are adaptations of the histogram, where information on the data itself is retained while the grouped information is simultaneously displayed. The displays are best described by an illustration. Figure 3 shows a stem-and-leaf diagram of 31 daily maximum temperature readings at Newark Airport in January 1975. Instead of merely making the length of the bars proportional to the number of data points in the class, one represents each data point by its most significant digit within the class. This results in "bars" whose length is proportional to the number of data points in each class, but whose entries allow the analyst to see how the data are distributed within each class.

The preparation of a stem-and-leaf diagram is straightforward. If necessary, each data point is truncated or rounded to an integer multiple of the units used. The integer multiples are separated into a stem (generally all leading digits) and a leaf (the last digit). One then determines from the data the largest and smallest stem and enters these stems and all intervening stems in a vertical column. A vertical line keeps stems and leaves separated. Next one proceeds through the set of data and writes down each leaf on the line corresponding to its stem. Finally, the leaves within each stem are sorted from smallest to largest. Figure 4 shows several steps in the construction of the stem-and-leaf display in Figure 3. The reason for plotting stem-and-leaf displays vertically rather than horizontally is one of convenience. The digits on a vertical stem-and-leaf display are much easier to print and read.

Perhaps the greatest attribute of the stem-and-leaf diagram is that the distributional information for the entire data set is retained while, in addition, it is possible to investigate whether the data are uniformly distributed within classes and whether any special values are present. These features are illustrated in Figure 5, which is a stem and-leaf display of wind speed at Newark Airport (the same data as were used to construct Figures 1 and 2). The display indicates that certain values (like 9.2 or 11.1) occur very often while others (like 10.0 or 13.0) do not occur at all, suggesting that some very special rounding process must have occurred. This turns out to be a correct assumption; the data have been converted from knots to km/h by multiplying by (1.15)(1.609). Therefore, 9.2 corresponds to 5 knots, 11.1 to 6 knots, etc. A histogram illustrating this same effect has been presented by Lockhart (1979); Reiss and Eversole (1978) have discussed missing numbers in visibility data.

DAILY MAX. TEMPERATURE IN NEWARK

31 VALUES, UNITS: 2|8 = 28°F

MEDIAN = 44   QUARTILES = 38, 50

```
2 | 8
3 | 0114
3 | 688
4 | 000122244
4 | 567777
5 | 00014
5 | 7
6 | 1
6 | 6
```

Figure 3. Stem-and-leaf display of daily maximum temperatures (°F)
at Newark Airport in January 1975 (data from National
Climatic Center, 1974-1975). The display is to be read as
follows: 2|8 = 28°F; 3|0114 = 30, 31, 31, 34°F.

DAILY MAX. TEMPERATURE IN NEWARK

31 VALUES, UNITS: 2|8 = 28°F

STEP 0	STEP 1		STEP 5		STEP 31	
2	2		2		2	8
3	3		3		3	0141
3	3		3	8	3	886
4	4	4	4	40	4	402201204
4	4		4		4	767577
5	5		5	0	5	01400
5	5		5		5	7
6	6		6	1	6	1
6	6		6		6	6

Figure 4. Four steps in the process of constructing a stem-and-leaf
display. The data values through step 5 are 44, 50, 61,
40, and 38°F. Sorting the results of step 31 produces the
display shown in Figure 3.

Some data sets require modifications of the basic stem-and-leaf structure. If a few extreme values are well removed from the rest of the data, one would have to add several new stems with no leaves just to accommodate these few data points. To prevent this, one either creates a special stem (usually labelled "HI" or "LO"), which can accommodate all extremely large (small) values, or changes the range each stem encompasses. In each case, one has to make sure that the viewer is aware of the fact that the scale has been changed.

Other adjustments involve cases with too many or too few leaves per stem. If there are too many leaves per stem, a "stretched" stem-and-leaf (Tukey, 1977) simply repeats each stem using the first line for the leaves 0-4 and the second line for leaves 5-9; that is,

$$0\,|\,1112344556777899$$

would become

$$0\,|\,1112344$$
$$0\,|\,556777899.$$

Figure 3 is a stretched stem-and-leaf; Figure 5 shows special stems to accommodate extreme values. If a stretched stem-and-leaf still contains too many leaves, or if the standard stem-and-leaf display has too few leaves, one can use what Tukey (1977) calls a "squeezed" stem-and-leaf, which used 5 lines for each stem. The above example then would become

$$0\,|\,111$$
$$0\,|\,23$$
$$0\,|\,4455$$
$$0\,|\,6777$$
$$0\,|\,899.$$

For further variations of stem-and-leaf displays, we refer the reader to the book by Tukey (1977, Chapter 1).

Stem-and-leaf displays show more details than histograms without involving much extra construction work. In addition to providing the analyst with a "feel" for the general characteristics of a set of data, they also enable the analyst to detect certain unexpected or unusual features in the data (such as unexpectedly frequent or infrequent values). However, stem-and-leaf displays are only useful for moderate size data sets (100-200 data points) and cannot be used if the data are already grouped when they reach the analyst.

WIND SPEED AT 0100 HOURS IN NEWARK

31 VALUES, UNITS: $9 | 2 = 9.2$ KM/H

MEDIAN = 11.1, QUARTILES = 9.25, 16.7

LO: 0.0, 0.0

```
 7 | 444
 8 |
 9 | 22222
10 |
11 | 1111111
12 | 9999
13 |
14 |
15 |
16 | 6666
17 |
18 | 55
19 |
20 |
21 |
22 |
23 |
24 | 0
25 | 9
```

HI: 38.8, 51.8

Figure 5. Stem-and-leaf display of wind speed (km/h) at Newark
Airport at 0100 hours during December 1974. The display
is read as follows: 0|22 = 0.2, 0.2 km/h; 7|444 = 7.4,
7.4, 7.4 km/h.

2.4 Box Plots

Although details concerning a data set are very important, one often wishes to summarize a set of data in terms of a few easily obtained and understood numbers. Such a summary should include information about the range of the data set, its location and scale, and its skewness characteristics.

The range of the data is represented by the extremes; that is, its smallest and largest values. Information about location, scale, and skewness can be obtained from a small set of sample quantiles. A sample quantile q associated with a probability p ($0 < p < 1$) is a number such that, loosely speaking, p(100%) of the data lie below q. The quantile associated with p = 0.5 is called the median (which lies in "the middle" of the data in the sense that half the data lie below the median, half above). The quantiles associated with p = 0.25 and p = 0.75 are the lower and upper quartiles, and quantiles for p = (0.1)k (k= 1, 2, ..., 9) are called deciles.

The median, defined for n odd as the single middle value and for n even as the average of the two middle values, is a good indicator of location. Although the median is not the best possible estimator under the assumption of normality, it remains stable under a wide variety of circumstances. That is, it is robust (Section 2.7 will treat robust estimates in more detail).

Information about scale and skewness is provided by the lower and upper quartiles (p = 0.25 and 0.75). The difference between the upper and lower quartiles (the interquartile range) is a robust estimator of scale (for normally distributed data the interquartile range is equal to 1.349 times the standard deviation), and the differences between the quartiles and the median give us some information about the skewness of the data.

The five values for the extremes, the quartiles, and the median thus summarize a set of one-dimensional data quite well, and can be used to depict a data set graphically by the box plot technique (Tukey, 1977). As shown in Figure 6, a box plot is generated by first drawing a box from quartile to quartile with a bar crossing it at the median. Then one draws a vertical dashed line from each end of the box to the corresponding extreme, marking each extreme with a short horizontal line. Figure 7 shows on the left a box plot of the wind speed data of Figure 5. The data are strongly positively skewed, as can be seen from the distances of the median from the quartiles. Furthermore, the upper extreme is much further away from the upper quartile than the lower extreme is from the lower quartile.

The basic box plot in the form outlined here emphasizes the part of the data set where the middle 50% of the data lie (i.e., the box), but provides no information on that part of the data which lie in the usually large intervals between the quartiles and the extremes. For

12

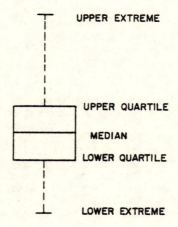

Figure 6. Configuration of a box plot.

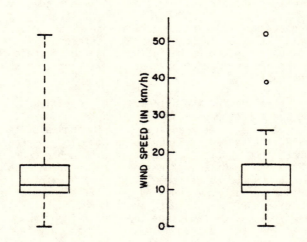

Figure 7. Box plot (left) and schematic plot (right) of wind speed
data shown in Figure 5.

example, in Figure 7 we have no idea if the extreme of 51.8 is a single outlying value or if other days had similarly high wind speeds.

One attempt to add information to the box plot and draw attention to special points outside the box is Tukey's (1977) schematic plot. Here the box is still drawn from quartile to quartile with a line across it at the median, but the vertical lines are not necessarily drawn out to the extremes. In a schematic plot the vertical lines stop at the extremes within the interval (lower quartile minus 1.5 times the interquartile range, upper quartile plus 1.5 times the interquartile range). All observations outside this interval are plotted individually and often specially labelled. The factor 1.5 for the interquartile range has been chosen so that for a normal distribution 95% of the data will lie within the range of the dotted line. Schematic plots therefore show how many (and which) data points are contributing to a long tail. Figure 7 shows on the right a schematic plot of the Newark wind-speed data, in which it is clear that the long upper tail is due to only two values.

Often it is advisable to use a box plot in conjunction with a one-dimensional scatterplot. The scatterplot shows the raw data, while the box plot summarizes their main characteristics. In addition to the fact that they describe the main aspects of a data set very well, box plots and schematic plots are very compact; they are therefore very useful in comparing the distributions of several sets of data (see Section 4.2).

2.5 Probability Plots

The graphical techniques described so far provide a reasonably good idea about the shape of the distribution of the data under investigation, but do not determine how well a data set conforms to a given theoretical distribution. A goodness-of-fit test could be used to decide if the data are significantly different from a given theoretical distribution. However, such a test would not tell us where and why the data differ from that distribution. A probability plot (Gnanadesikan, 1977, Chapter 6), on the other hand, not only indicates how well an empirical distribution fits a given distribution overall, but also shows at a glance how the distributions differ.

There are two basic types of probability plots: P - P plots and Q - Q plots. Both can be used to compare two distributions with each other. The basic principles remain the same if one wants to compare two theoretical distributions, an empirical (or sample) distribution with a theoretical distribution, or two empirical distributions. P - P plots and Q - Q plots can best be explained by Figure 8, where two smooth cumulative distribution functions F_x and F_y have been

drawn. For a P - P plot one starts with a common quantile q and computes the probabilities $p_x(q)$ and $p_y(q)$ (shown on the y-axis of

14

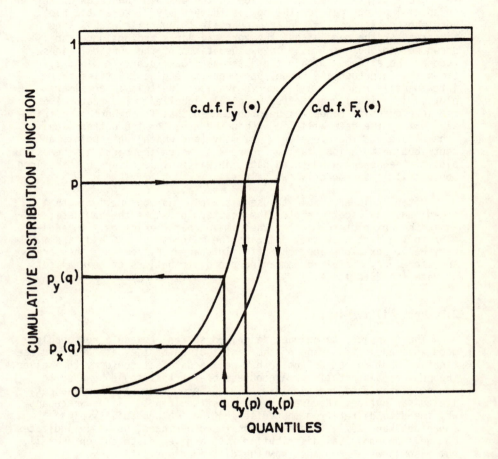

Figure 8.   Illustration of probability plots (from Gnanadesikan, 1977, with permission of John Wiley and Sons, Inc.).

Figure 8) for the sampling distributions of x and y. Thus, the P - P plot consists of points whose coordinates are the probabilities $[p_x(q), p_y(q)]$ for different values of the quantile q.

To construct a Q - Q plot one proceeds in analogous fashion by starting with a common probability p and computing the quantiles $q_x(p)$ and $q_y(p)$ (shown on the x axis of Figure 8). The Q - Q plot is thus a plot of points whose coordinates are the quantiles $[q_x(p), q_y(p)]$ for different values of the probability p.

If the two density functions are roughly bell-shaped (which leads to distributions of the form shown in Figure 8), P - P plots are more sensitive to differences in the middle of the distributions, whereas Q - Q plots are more sensitive to differences in the tails. If $F_x = F_y$, both P - P and Q - Q plots result in a straight line with intercept 0 and slope 1. If the random variables X and Y are related by Y = aX + b, the two distributions are the same except for a linear transformation. In this case, Q - Q plots of q_y versus q_x will still show a straight line (now with intercept b and slope a), because for each fixed p and positive a

$$p = Pr\{X \leq q_x(p)\} = Pr\{aX + b \leq aq_x(p) + b\} = Pr\{Y \leq q_y(p)\},$$

so that

$$q_y(p) = aq_x(p) + b.$$

This linear invariance property does not hold for P - P plots, which probably accounts for their much lower popularity.

The most widely used form of Q - Q plots compares the empirical cumulative distribution function of a data set of size n with a hypothesized, standardized (to have location zero and scale one) theoretical distribution function $F(t,\theta)$, where the parameters θ (which do not include location and scale) have specified values. F is usually determined by some external information or by a tentative model one might have in mind. If we denote by $y_{(1)} \leq y_{(2)} \leq \cdots \leq y_{(n)}$ the n ordered observations of our data set, the Q - Q plot consists of the points $[q_i, y_{(i)}]$ where q_i is the quantile of the distribution function F corresponding to a cumulative probability $p_i = (i - 0.5)/n$. That is, q_i is defined by $F(q_i, \theta) = p_i$.

The hypothesized theoretical distribution with which most data sets are compared initially is the normal distribution. The comparison can be performed either by using normal probability paper or by automated computing and plotting of the quantiles. Milton and Hotchkiss (1969) have, for instance, published algorithms to compute quantiles for the normal distribution. Probability paper and computer programs for other distributions (such as the F-distribution) exist as well. Note that Q - Q plots and plots on the well-known probability paper are one and the same thing. The quantiles for the probability paper are precomputed and incorporated in the grid of lines, while the quantiles for a Q - Q plot are generally determined as needed by computer.

Figure 9 shows a Q - Q plot of the amount of snowfall at Newark Airport versus the quantiles of the normal distribution. The $y_{(i)}$ values are the ordered snowfall measurements, and the q_i values on the x-axis are the quantiles of the standardized normal distribution corresponding to probabilities $p_i = (i - 0.5)/n$, where n is the number of observations. It is immediately clear that the data are not normally distributed, because the points $[q_i, y_{(i)}]$ do not lie on a straight line. We can also see, because of the upward curvature of the data in the upper part of the plot, that the data have a much longer right tail than does the normal distribution. Physically, this indicates that more heavy snowfalls are experienced than would be expected if the monthly amounts of snowfall were normally distributed.

2.6 Power Transformations

Most classical statistical methods rely (implicitly or explicitly) on the assumption that the data are approximately normally distributed (or are, at the very least, symmetric). Therefore, one would like the data to have a distribution which is as close to the normal as possible. However, experience shows that such is seldom the case for data in the atmospheric sciences. For example, Larsen (1969) has demonstrated that measurements of air pollution data are often lognormally distributed.

A convenient procedure to make such one-dimensional data sets more nearly normally (or at least symmetrically) distributed is the application of a power transformation of the form

$$y^{(p)} = \begin{cases} x^p & \text{if } p > 0, \\ \ln x & \text{if } p = 0, \\ -x^p & \text{if } p < 0. \end{cases} \tag{1}$$

Although more detailed methods have been developed whose use would be appropriate if it were known that some transformation of this type would make the transformed data exactly normal, in exploratory data analysis it generally suffices to compute the transformed data for a few chosen values of p and then judge on the basis of stem-and-leaf diagrams, box plots, or Q - Q plots of the original and the

NORMAL PROBABILITY PLOT, 99 VALUES

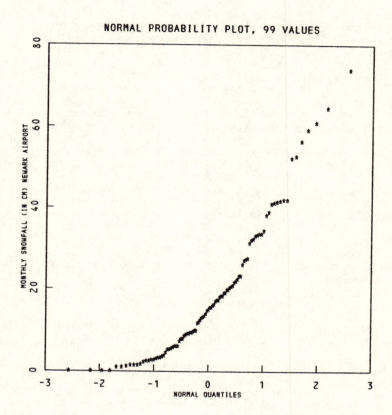

Figure 9. Q - Q plot of monthly snowfall (cm) at Newark Airport
versus quantiles of the normal distribution for the months
of December, January, and February from 1944/45 to 1976/77
(data from National Climatic Center, 1978).

transformed data which power p should be chosen. This method also has the advantage that our choice of p is not overly influenced by a few outliers.

How is p chosen? Generally, one selects p-values from the ladder of expressions ..., -3, -2, -1, -1/2, -1/4, 0, 1/4, 1/2, 1, 2, 3, ... As can be seen from (1) and Figure 10, the above set of re-expressions for x is monotonic. How does one move up and down the ladder for p and where does one start? It is clear that p = 1 will not change the data, that p > 1 will result in larger values getting bigger much faster than smaller ones, and that p < 1 will make larger values decrease relative to smaller values. Therefore, data with a longer right tail should be transformed with a p < 1, data with a longer left tail with p > 1. Many of these power transformations are only defined for positive data and it might be necessary to shift the data by adding a constant before transforming them. Much data in the atmospheric sciences appear to have larger right tails and experience shows that the first p's to be tried often lie between zero and one.

A reasonable initial estimate of p can be found quickly by a method of Hinkley (1977). He suggests transforming the data for dif-ferent values of p among the ladder ..., -3, -2, -1, -1/2, -1/4, 0, 1/4, 1/2, 1, 2, 3, ... and computing for each transformation the mean, median, and an estimate of scale (either the sample standard deviation or one of the robust estimates discussed in Section 2.7.2). One then chooses the value of p which minimizes the measure of asymmetry of the transformed data

$$d_p = (mean - median)/scale.$$

Since this estimate is sensitive to even a single outlier, plotting the transformed data to verify the choice of p is advisable.

Example. Figures 11-13 show the stem-and-leaf displays, box plots, and normal Q - Q plots of the monthly amounts of snowfall dis-cussed above. The figures on the left-hand side show the original data (p = 1) to be skewed to the right; the figures in the middle show the logarithms of the original data (p = 0) which are skewed to the left; and the figures on the right-hand side show the effects of choosing square roots (p = 0.5), a transformation between the original data and the logarithms. The result is a transformed data set that is approximately symmetrically distributed and, as the Q - Q plot shows, quite normal (with the exception of four months with practically no snowfall at all). If a first look at the data reveals points which are different from the rest of the data for a specific reason (such as the four months with practically no snow in the example above), it is often advisable to omit these data before more detailed analyses are performed.

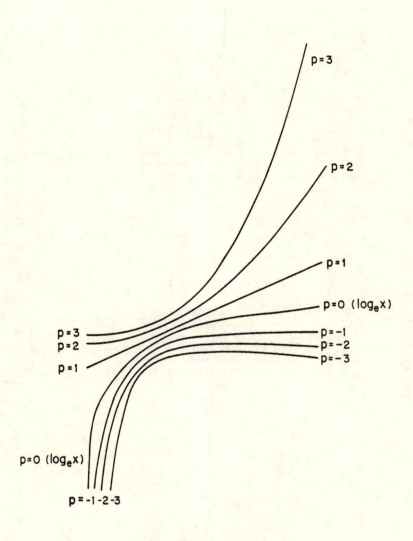

Figure 10.   Shapes of curves $y = x^p$ for p = -3, -2, -1, 0, 1, 2, 3.

MONTHLY SNOWFALL AT NEWARK AIRPORT

99 VALUES

UNITS: 1|5 = 15 CM
MEDIAN = 15.0
QUARTILES = 5.59, 27.2

```
0  00001112222233333333344
0  55566666788999999
1  0002223334
1  55556677788899
2  0001122233
2  6777
3  12233444
3  89
4  111222
4
5  23
5  69
6  1
6  5
7  4
```

P = 1

UNITS: 1|5 = 1 5 LOG CM
MEDIAN = 2.71
QUARTILES = 1.72, 3.30

```
-2  1111
-1
-1
 0  002444
 0  6899
 1  0012233
 1  5777888
 2  01122222333
 2  55566667777888899999
 3  000001111133334
 3  5555556777777777
 4  0001123
```

P = 0

UNITS: 1|5 = 1.5 SQRT CM
MEDIAN = 3.87
QUARTILES = 2.36, 5.21

```
0  4444
1  00122235677789
2  0133445557889
3  00011114555557789999
4  00122233445555667788
5  12226778889
6  22444555
7  23578
8  06
```

P = 0.5

Figure 11.   Stem-and-leaf displays for snowfall data shown in Figure 9. These displays are for the original data (left), logarithms (center), and square roots (right).

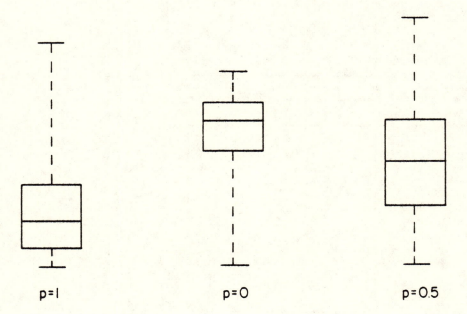

Figure 12. Box plots of original, "logged," and "square-rooted" snowfall data shown in Figure 11.

22

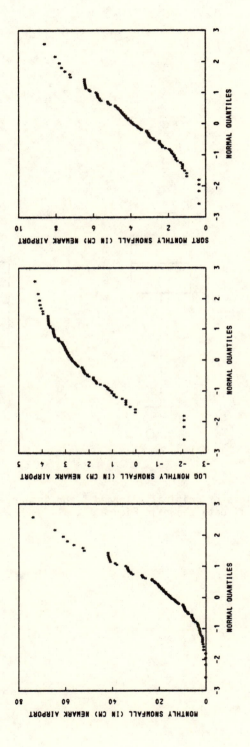

Figure 13. Q - Q plots of original, logged, and square-rooted snowfall data of Figure 11 versus normal quantiles.

2.7 Robust/Resistant Estimates

The utility of describing the characteristics of a data set by a few readily understood descriptors is of such great value that it is often forgotten that the two well known statistics

$$\bar{x} = \frac{1}{n} \sum_{i=1}^{n} x_i \qquad \text{(sample mean)}$$

and

$$s = [\frac{1}{n-1} \sum_{i=1}^{n} (x_i - \bar{x})^2]^{1/2} \qquad \text{(sample standard deviation)}$$

characterize a sample x_1, x_2, ..., x_n completely only if it is drawn from a normal distribution. For normal data, the sample mean and variance are, furthermore, the best estimators of location and $(\text{scale})^2$ of the underlying distribution (best in the sense of having the smallest variance among all estimators which have no bias).

However, most physical data sets are not normally distributed (even after transformation), because the assumption of an underlying normal distribution is a mathematical idealization which is never met exactly in practice and because large data sets inevitably contain outliers. The effect of such problems can best be illustrated by a simple example. Consider the artificial data set {1, 2, 3, 4, 50, 6, 7, 8}. This data set has a sample mean of 81/8 = 10.1 and a sample standard deviation of 19.2, clearly not very representative values for the data at hand. What are needed, therefore, are estimates which may not be optimal for any given distribution, but remain quite good under a variety of circumstances; such estimates are called robust. Estimates which are not unduly influenced by a small number of outliers are called resistant. Although the robust/resistant techniques which will be discussed in the remainder of this section will be insensitive to outliers, by no means does it follow that all outliers should therefore be disregarded. It may very well be that the outliers are the most interesting and important bits of data one has.

2.7.1 <u>Robust Estimates for Location</u>. One of the simplest ways of making the arithmetic mean less sensitive to outliers is to drop or "trim" a proportion of the data from each end of the ordered data set and then to compute the mean of the remaining values. These trimmed means form a family of estimators that can be indexed by α, the fraction of observations trimmed at each end of the sample. Formally, if $x_{(1)} \leq x_{(2)} \leq \cdots \leq x_{(n)}$ are ordered data of a random sample of size n, the α-trimmed mean is of the form

$$\overline{x}_\alpha = \frac{1}{n(1-2\alpha)} \left\{ \sum_{i=k+2}^{n-k-1} x_{(i)} + (1+k-n\alpha)[x_{(k+1)} + x_{(n-k)}] \right\},$$

if $k+2 \le n-k-1$. Otherwise,

$$\overline{x} = \frac{1}{2} [x_{(k+1)} + x_{(n-k)}].$$

Here k is the largest integer less than or equal to $n\alpha$ and $0 < \alpha < 0.5$.

Trimmed means include as special cases the ordinary arithmetic mean ($\alpha = 0$), the midmean ($\alpha = 0.25$), and the median ($\alpha = 0.5$). The choice of α depends on the actual situation, in which the non-normality of the data and the amount of protection desired against outliers play important roles. Frequent choices of values for α include 0, 0.05, 0.1, 0.25, and 0.5.

Applying a 25% trimmed mean to our data set with $\{x_{(1)}, x_{(2)}, \ldots, x_{(8)}\} = \{1, 2, 3, 4, 6, 7, 8, 50\}$ results in

$$\overline{x}_{0.25} = \frac{1}{8(1 - 0.5)} \{x_{(4)} + x_{(5)} + (1 + 2 - 2)[x_{(3)} + x_{(6)}]\}$$

$$= \frac{1}{4} \sum_{i=3}^{6} x_{(i)} = 0.25(3 + 4 + 6 + 7) = 5.$$

Other robust estimators for location are the winsorized mean where the $n\alpha$ extreme values are not trimmed but replaced by the (k+1)st and (n-k)th ordered values, and the Hodges-Lehmann estimator, which is determined by first computing the averages over all $n(n-1)/2$ possible pairs of the n data points and then taking the median of these averages. For a Monte Carlo study on the characteristics of a large number of different robust location estimators, see Andrews et al. (1972).

2.7.2 Robust Estimates of Scale. The estimation of higher order moments (such as the variance) involves resolution of a conflict of interest, since one would like to protect the estimates from outliers, but recognizes that the estimation of higher order moments relies more heavily on information in the tails of the distribution. Thus, a certain amount of caution in using robust estimates of the variance is indicated. With small data sets, the analyst may be well off by studying the samples very carefully and only omitting clearly indicated outliers. With larger data sets, however, the need for semi-automatic treatment of the data demands the use of formal robust/resistant methods.

Trimmed Variances. The ideas underlying trimmed means can also be applied to the estimation of the variance. The trimmed variance is defined as

$$s_\alpha^2 = \frac{c_\alpha}{n(1-2\alpha)} (\sum_{i=k+2}^{n-k-1}[x_{(i)} - \bar{x}_\beta]^2$$

$$+ (1+k-n\alpha)\{[x_{(k+1)} - \bar{x}_\beta]^2 + [x_{(n-k)} - \bar{x}_\beta]^2\}),$$

where $\bar{x}_\beta$ is a robust estimate of location (generally a β-trimmed mean with $\beta = \alpha$), c_α is a normalizing constant to make the trimmed variance an unbiased estimate of the variance under certain conditions, and k is the largest integer less than or equal to $n\alpha$.

The c_α's are based on moments of order statistics of the normal distribution, and implicitly assume that the "middle" of the data is reasonably normal. Tables of values of c_α for small samples ($n \leq 15$) can be found in Johnson and Leone (1964, p. 173). For large samples, c_α can be approximated by a certain integral over the normal distribution (e.g., when $\alpha = 0.05$ and n is large, $c_\alpha = 1.604949$). Because of the reasons just outlined, it is usually advisable to use somewhat smaller trimming proportions for variance estimation than for location.

Median Absolute Deviation (MAD). The scale estimate that is the counterpart of the median as the most resistant estimator for location is the median absolute deviation (MAD) from the median; that is,

$$MAD = \underset{1\leq j\leq n}{median}|x_j - \underset{1\leq i\leq n}{median}(x_i)|.$$

MAD has to be divided by a normalizing constant c_{MAD} to make it a bias-free estimate of the standard deviation and the problems discussed for the normalizing constants of trimmed variances apply here as well. In the normal case, the asymptotic value of c_{MAD} for n tending to infinity is 0.6745. Therefore we define the scale estimate $s_{MAD} = MAD/c_{MAD} = MAD/0.6745$.

The asymptotic efficiency of the MAD-estimate for normal data is very low (only about 40%). On the other hand, it is very easy to compute. MAD can thus be regarded as a crude, but simple and safe, scale estimate. It can be used, when no higher accuracy is required, as a simple check on more complicated computations or as a starting point for iterative procedures. Note that for symmetric data MAD is equal to half the interquartile range (discussed next).

The Interquartile Range (IQ). Another robust measure of scale, which is equivalent to MAD as far as simplicity of computation and low efficiencies under the near normal assumption, is the interquartile range

$$IQ = \text{upper quartile - lower quartile.}$$

For symmetric data, MAD = IQ/2. For non-symmetric data, working with the interquartile range has the advantage that, once the quartiles and the median are computed, the differences (upper quartile - median) and (median - lower quartile) give us some clues about the symmetry of the data. Furthermore, the IQ has a very nice concrete interpretation (it is the range within which the middle 50% of the data lie) and can be directly related to the box plot (the IQ is the length of the box).

The asymptotic value for the normalizing constant c_{IQ} under the assumption of normality in the middle of the data is 1.349, so that

$$s_{IQ} = IQ/c_{IQ} = IQ/1.349.$$

The actual value of these normalizing constants is, of course, unimportant if the scales of several sets of data are compared, because then only the relative sizes are of consequence.

2.7.3 Example. The wind speeds at the Philadelphia airport at 0100 hours for the first 15 days of December 1974 were recorded as follows (in km/h): 22.2, 61.1, 13.0, 27.8, 22.2, 7.4, 7.4, 7.4, 20.4, 20.4, 20.4, 11.1, 13.0, 7.4, 14.8. Clearly, the 61.1 km/h measured on December 2 does not conform to the rest of the data and therefore can be labelled as an outlier (although the measurement seems to be correct, other observing stations on the East Coast reporting similarly high values during a storm in the early hours of December 2). If we compute the ordinary sample mean and standard deviation with all values included the results are $\bar{x}$ = 18.4 and s = 13.5. One could argue that the value 61.1 km/h is clearly atypical and should therefore be excluded in computing averages. If that is done, the computations give $\bar{x}$ = 15.3 and s = 6.8. Trimmed means and variances (adjusted by the asymptotic values for c_α) are

$\alpha(=\beta)$	$\bar{x}_\alpha$	s_α
0	18.4	13.5
0.05	16.6	11.7
0.1	15.8	9.5
0.2	15.8	11.1
0.25	16.0	11.8

$$\text{Median} = 14.8, \qquad \text{MAD} = 7.4, \qquad \text{IQ} = 14.8,$$

$$s_{MAD} = \text{MAD}/0.6745 = 11.0, \qquad s_{IQ} = \text{IQ}/1.349 = 11.0,$$

$$\text{upper quartile - median} = 7.4, \qquad \text{median - lower quartile} = 7.4.$$

Therefore, all robust estimates for location and scale are considerably lower than the ordinary mean and standard deviation, which, if we did not already know, would point us to some irregularity in the data.

Checking data in this way has become one of the major applications of robust/resistant methods, especially in cases where computer time is more plentiful than data analyst time. The procedure is to compute estimates (e.g., for location, scale, correlation, and regression coefficients) both in the classical way and by means of a robust/resistant method. If the results of the two techniques agree closely, they suggest that the data are reasonably well behaved. If the results do not agree, the data should be investigated more thoroughly until the reasons for the discrepancy are understood.

3. ANALYZING TWO SETS OF DATA

The techniques available for the concurrent analysis of two sets of one-dimensional data depend on the relationship between the two data sets. In this section two kinds of relationships are distinguished:

(a) One variable y_k is (or may be) dependent upon another variable x_k that can possess any value. This situation is often investigated by scatterplots (Section 3.1).

(b) Two sets of data $\{y_k, x_k\}$ are to be compared without assuming anything about the dependence between them. Tools for this purpose are the scatterplot (perhaps of transformed variables) accompanied by robust estimates of correlation (Section 3.1) and empirical Q - Q plots (Section 3.2).

3.1 Scatterplots and Robust Correlation

The scatterplot of y_k vs. x_k (k = 1, 2, ..., n) is one of the most powerful tools available to investigate the relationship between two variables x and y. It not only allows the analyst to detect relationships and to get an initial idea about their nature (whether, for example, they are linear), but also demonstrates the strength of a relationship and the presence or absence of outliers. Very often, it pays to transform the data before making a scatterplot. For example, if the data are heavily skewed to the right, transformation to approximate symmetry will prevent the analyst's impressions from being overly influenced by a few extremely large values. The perception of relationships between the variables can be enhanced for noisy or sparse data by the superposition of smoothed curves as described by Cleveland and Kleiner (1975).

If a relationship between data sets x and y is detected, it is desirable to assess it quantitatively. A good measure of the strength of a linear relationship between y and x is the sample correlation coefficient r, defined by

$$\frac{\sum\limits_{i=1}^{n} (x_i - \bar{x})(y_i - \bar{y})}{[\sum\limits_{i=1}^{n} (x_i - \bar{x})^2 \sum\limits_{i=1}^{n} (y_i - \bar{y})^2]^{1/2}} \ .$$

The quantity r is also known as the Pearson product-moment coefficient of correlation. It satisfies the condition $-1 \leq r \leq 1$, where $r = 0$ indicates the absence of any linear relationship, $r = -1$ indicates a perfect negative relationship, and $r = 1$ indicates a perfect positive relationship. The square of r (i.e., r^2) can also be interpreted as the fraction of the total variation of y which is described by fitting a line of the form $y = ax + b$ to the data.

Since correlation coefficients can be greatly affected by outliers, it is advisable to use robust correlation coefficients. There are a variety of different ways to estimate correlation coefficients robustly (for a good overview see Gnanadesikan and Kettenring, 1972). One version which can easily be derived from the robust estimates for location and scale described in Section 2 is computed as follows:

(a) Set

$$\tilde{x}_k = \frac{x_k}{s*(x)}, \quad \tilde{y}_k = \frac{y_k}{s*(y)}, \quad k = 1, 2, \ldots, n,$$

where $s*(x)$ equals the square root of the trimmed variance of x. This means that all x and y values are divided by a robust estimate of their standard deviation.

(b) Then

$$r* = \frac{s*^2(\tilde{x} + \tilde{y}) - s*^2(\tilde{x} - \tilde{y})}{s*^2(\tilde{x} + \tilde{y}) + s*^2(\tilde{x} - \tilde{y})}.$$

This method is called SSD (or Standardized Sums and Differences).

The amount of trimming in computing r* depends on the percentage of bad data points; 5% trimming is probably used most extensively. We note that the bias correction constants c_α, necessary in computing trimmed variances (Section 2.7.2), do not appear here due to their occurrence in both the numerator and the denominator of r*, and that for $\alpha = 0$, r* = r.

The main advantage of the SSD estimate over its competitors is its simplicity. Its disadvantage is that it trims at all extremes, and therefore may delete extreme but perfectly valid data.

Example. Figure 14 shows a scatterplot of the windspeed at 0100 hours for all days in December 1974 at Newark and Philadelphia airports. It is immediately clear that there is one extreme outlier (December 2, which was discussed earlier), but it is not clear how much of a linear relationship there is after the outlying value has been removed. The ordinary regression coefficient is quite high (0.79), but drops considerably if we use a robust estimate for the correlation (to 0.59 with 5% trimming, to 0.39 with 25% trimming). If we recompute the correlation coefficient with no trimming but without the point at (52, 61) we get a correlation of 0.54. Such statistics thus demonstrate the principal relationships in the data, but require other EDA techniques to alert the analyst to the possible significance of extreme values in the data set.

3.2 Empirical Quantile - Quantile Plots

The empirical quantile - quantile (EQQ) plot is a graphical technique for comparing distributions of two unmatched samples x_i (i = 1, 2, ..., m) and y_k (k = 1, 2, ..., n). Unlike most classical techniques (such as a t-test on the difference of the means or a comparison of the variances), EQQ permits the comparison of the two distributions in their entirety. The EQQ plot is a special case of the probability plots discussed in Section 2.5, in which both distributions are empirical (sample) distributions. It is constructed as follows:

(a) Sort x_i (i = 1, 2, ..., m), and y_k (k = 1, 2, ..., n) in ascending order. Denote the ordered values by $x_{(i)}$ and $y_{(k)}$.

(b) Fix a set of probabilities. This is usually done in one of the following two ways:

   (i) Use a predetermined set of probabilities, such as all deciles.

WINDSPEED (KM/H) AT 0100 HOURS DEC. 1974

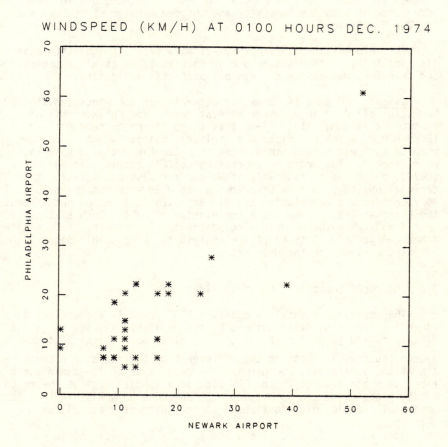

Figure 14.   Scatterplot of wind speeds (km/h) at 0100 hours for December 1974 at Newark and Philadelphia airports (data from National Climatic Center, 1974-1975).

(ii)   Compute the probabilities associated with the sorted values
       of the smaller data set as $p_i = (i - 0.5)/\min(m,n)$ [$i = 1$,
       $2, \ldots, \min(m,n)$].

(c)   Compute empirical quantiles x_i, y_i associated with the above

      percentiles by linear interpolation from the ordered data. If
      the p_i have been determined according to (ii) of (b), only the

      larger data set has to be interpolated.

(d)   Plot the points (x_i, y_i).

      If x and y have the same underlying distribution, their sample
      values will lie around the line y = x on the EQQ plot. If the y
      values are larger, the points will lie above the line y = x. If
      the x values are larger, they will lie below. If one of the two
      distributions is a linear transformation of the other (i.e.,
      y = ax + b), the points will lie around a straight line with
      slope a and intercept b.

      Example. Figure 15 shows an EQQ plot of the daily amount of
solar radiation in Central Park, New York City during May - August
1972 and 1973. It compares the amount of solar radiation on Monday -
Friday (on the x-axis) with the amount on Sundays (on the y-axis).
The extreme values of solar radiation seem to be about the same on
workdays and on Sundays, whereas the middle quantiles are definitely
higher on Sundays (apparently as a result of lower particulate concentrations; see Cleveland et al., 1974). This is a good example of the
main advantage of EQQ plots; although a t-test would also have indicated that there is more sunshine on Sundays (on the average) than on
workdays, it would not have alerted us to the peculiar behavior at the
extremes of the two distributions.

4.   COMPARING SEVERAL SETS OF DATA

4.1   EQQ Plots

      When more than two data sets are to be compared, all the tech-
niques described so far can still be used for pairwise comparisons.
However, the number of pairwise comparisons for n sets is $n(n-1)/2$,
and is therefore growing quite rapidly for increasing n. It is also
relatively difficult to get a good overall impression of the data even
when all pairwise comparisons are available. Thus the use of all
pairwise comparisons is generally not recommended except for a small
number of data sets.

      An alternative to comparing all pairs of data sets is to define a
representative data set and to compare all n data sets with the representative set, a process requiring only n comparisons. The choice of

32

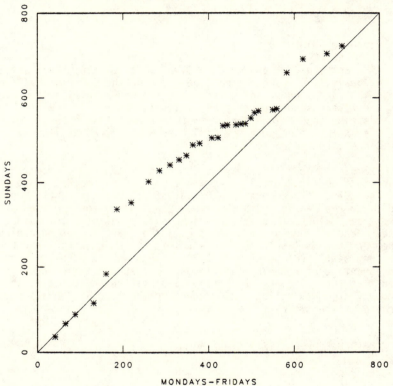

SOLAR RADIATION, MANHATTAN, MAY–AUG 1972,73

Figure 15. Empirical quantile-quantile plot of daily solar radiation
(langleys) in Central Park, New York City, on Sundays
versus workdays (data from New York Department of Environmental Conservation, 1974).

the representative set will depend on the actual circumstances involved. It might, for example, consist of the combined observations from all n data sets or of some standard or target values known from external information.

Figure 16 shows EQQ plots of the monthly average temperatures at six different locations versus a representative data set consisting of the combined observations from all locations. The six locations are all in the United States and have approximately the same latitude, the most extreme latitudinal difference between them being 12' (approximately 24 km). The sites are Eureka, California; Winnemucca, Nevada; Salt Lake City, Utah; Lincoln, Nebraska; Burlington, Iowa; and Newark, New Jersey. While the averages of these temperature data over the 10-year interval at the six locations are similar (ranging from 48.7 F in Winnemucca to 54.4 F in Newark), their distributions are definitely not similar.

From Figure 16, it is clear that Eureka's temperatures behave very differently than does the set consisting of all observations combined; for example, its range is much smaller (from 44° to 60°F, whereas the corresponding values of the combined data range from 20° to 80°F). However, its distribution seems to be about the same as that of the combined set (apart from the different scales) because the points on the EQQ plot form a reasonably straight line. Winnemucca seems to be colder than the representative set except for the very coldest months, where it does not get quite as cold as elsewhere. Salt Lake City's temperature behavior is most similar to the average behavior of these six stations. Lincoln and Burlington, both in the Midwest, behave very similarly. They are both colder than the combined data in the winter months and warmer in the summer, with the exception of the two very coldest months in Lincoln and the hottest months in Burlington. The temperatures are somewhat lower in Burlington, as borne out by the medians over the 10 years of monthly data (Burlington: 52.6°F, Lincoln: 53.6°F) and the overall means (Burlington: 50.4°F, Lincoln: 51.4°F). Newark is clearly warmer in the coldest winter months and during most of the summer months, but not in the extremely hot months and between the lower quartile and the median.

4.2 Box Plots and Schematic Plots

As the number of data sets increases, the pairwise comparisons and comparisons with a yardstick (such as the combined sample) become more and more cumbersome. In such cases, simpler ways of comparing several sample distributions have to be used.

Probably the simplest graphical tools for comparing several sets of data are the box plot and the schematic plot introduced in Section 2.4. Box plots show the range, location, scale, and skewness of the data and therefore allow the viewer to quickly assess trends (i.e.,

34

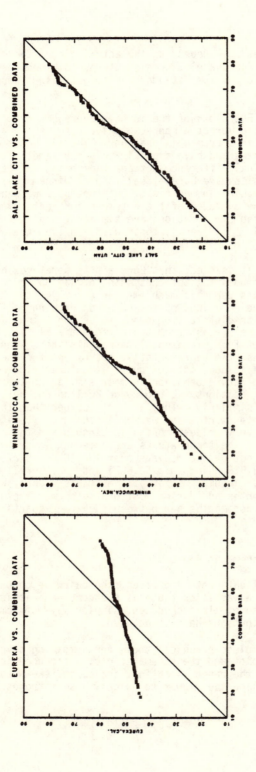

Figure 16. Empirical quantile-quantile plots of monthly average
temperature (°F) from 1964 to 1973 at six different
locations versus the combined observations from all
locations (data from National Climatic Center, 1978).
(a) Eureka, California; Winnemucca, Nevada; Salt Lake
City, Utah.

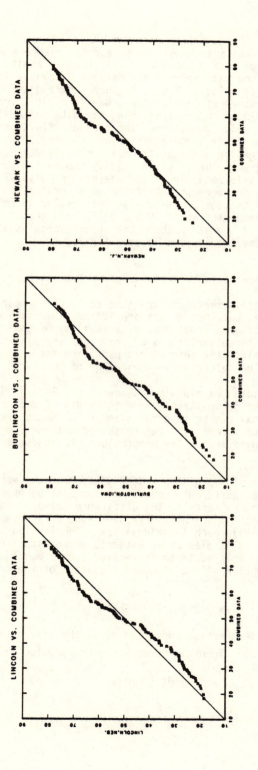

Figure 16. Same as previous page. (b) Lincoln, Nebraska; Burlington, Iowa; Newark, New Jersey.

changes in location) and any changes in spread and skewness when comparing several sets of data. Schematic plots show in addition all data points outside the main body of the data, a display which allows one to spot particular values. Figure 17 shows box plots for the daily maximum temperature at Central Park, New York City, from January 1972 to September 1973, each box containing the data for a single month. The plot not only demonstrates the seasonal fluctuations, but also that the maximum daily temperature never reached 100°F in the months under investigation and that the maximum daily temperature fell as low as 15°F in January 1972. It is readily apparent that temperatures varied most in both Januaries, whereas in July 1973 the maximum temperatures for more than half the days were within 5°F (i.e., the length of the box). It is also interesting to note the very high upper extremes for April 1972 and 1973 and that the medians of March and June 1973 are at or above the upper quartiles of March and June 1972, whereas the reverse is true for May.

4.3 Variations of Box Plots

Box plots and schematic plots are good tools for comparing several sets of data, but they do not provide any information about the relative sizes of each group or the variability of their medians. For these purposes, three variations of box plots are available: the variable width box plot, the notched box plot, and the variable width notched box plot. The latter is a combination of the first two.

The variable width box plot is a means of displaying additional information about the group sizes. Here the width of each box is chosen to be proportional either to group size or to the square root of group size. The square root is often chosen because many measures of scale, such as standard error, are proportional to the square root of the group size.

The notched box plot provides a good idea about the variabilities of the medians of the different groups. It can also be used to make approximate significance tests on the difference between medians of different boxes. Notched box plots are constructed by drawing notches around the median within each individual box. The lengths of these notches are taken as multiples of an estimate of the standard deviation of the median. This estimate is derived from a Gaussian-based asymptotic approximation of the standard deviation of the median and is given by

$$s_i = 1.25 IQ_i / (1.349 n_i^{1/2}),$$

where IQ_i is the interquartile range and n_i is the size of group i (McGill et al., 1978). Then the notches can be computed as

$$median_i \pm c\ s_i,$$

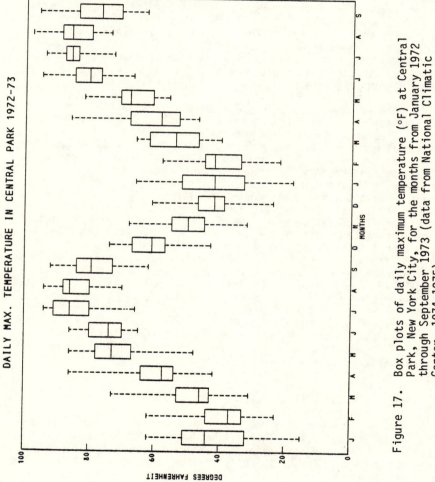

Figure 17. Box plots of daily maximum temperature (°F) at Central
Park, New York City, for the months from January 1972
through September 1973 (data from National Climatic
Center, 1974-1975).

where c is a constant. For c = 1.96, an approximate 95% confidence
interval around each median is obtained.

The notches can also be used to test the significance of the difference between any two medians at approximately the 95% significance
level in the following way. Draw notches with c = 1.7 around the two
medians one wants to assess. If the notches overlap, the two medians
are not significantly different from each other (at an approximately
95% significance level). If the notches do not overlap, the medians
are significantly different. The value of c = 1.7 is a compromise.
Setting c = 1.96 would be appropriate for a 95% significance level if
the sizes of the two groups, whose medians are compared, are vastly
different. If the group sizes were nearly equal, c = 1.386 would be
the right number [for more details see McGill et al. (1978)].

Figure 18 shows schematic plots of pH values in integrated
monthly samples of rainwater at six sites in the continental United
States. Measurements were not necessarily available for every month.
The actual number of measurements was 30 for New York City, New York;
33 for Chester, New Jersey; 35 for Seattle, Washington; 25 for Woods
Hole, Massachusetts; 24 for Argonne National Laboratory, Illinois; 12
for Lawrence Livermore Laboratory, California; and 20 for Beaverton,
Oregon. The sites have been ranked according to their median level.
However, it is not clear if the differences between the medians should
be regarded as significant or not. Figure 19 shows variable width
notched schematic plots of the same data, with the widths of the boxes
proportional to the square root of the group size. It becomes clear
that the difference between the medians of Livermore and Beaverton is
not significant, but that the difference between Beaverton and Argonne
(and therefore all sites to the left of Argonne) is significant. The
strange looking box for Argonne is due to the fact that the lower part
of the notch falls outside the box. Also note the suspiciously low
extremes for Seattle and Woods Hole. The value for Seattle has been
designated by Toonkel (1980) as probably in error. The value for
Woods Hole comes from an extremely small sample of rainwater and may
be unreliable as well.

5. SUMMARY AND CONCLUSIONS

In this paper, we have described several tools for the exploration of statistical data. These techniques can be divided into two
strongly interrelated categories: graphical techniques and robust/
resistant methods. The graphical techniques include histograms, stem-
and-leaf diagrams, box plots, probability plots, scatterplots (and
their enhancement), and empirical quantile-quantile plots. The
robust/resistant methods include robust estimation of location and
scale (in the univariate case) and robust correlation.

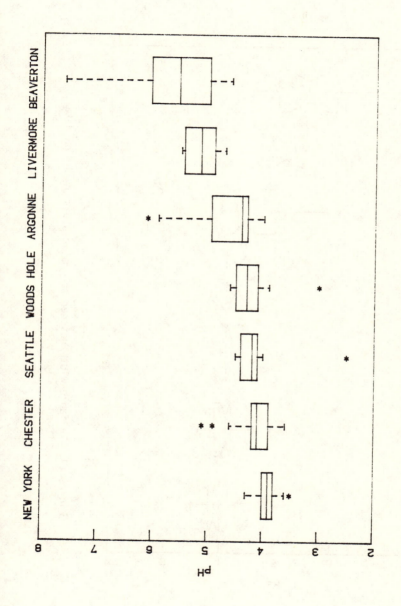

Figure 18. Schematic plots of pH values in integrated monthly samples of rainwater between June 1976 and June 1979 at six different locations (data from Toonkel, 1980).

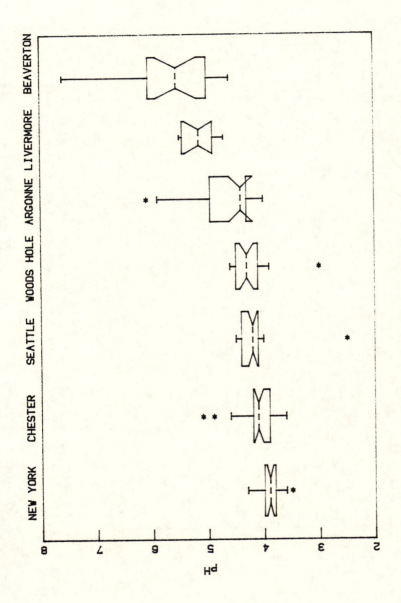

Figure 19. Variable-width notched schematic plots of the pH data shown in Figure 18.

The tools provided here are by no means exhaustive. For example, we have not touched at all on the analysis of data of more than two dimensions (for which an excellent book has been written by Gnanadesikan, 1977), or the regression problem (which is the subject of the entire second part of the book by Mosteller and Tukey, 1977).

Due to the preliminary and evolving nature of EDA, no rules can yet be given on exactly when and how to use a certain technique. We therefore restrict ourselves to reiterating the two fundamental principles of EDA: (a) EDA is an iterative technique where the output from one stage will guide the actions to be taken in the following stages; and (b) the data generally have to be looked at in a variety of ways (most of which will end up in the waste basket). The final results are often unexpected and enlightening, however, thus fully justifying the old Roman philosophy PER ASPERA AD ASTRA.

REFERENCES

Andrews, D. F., P. J. Bickel, F. R. Hampel, P. J. Huber, W. N. Rogers, and J. W. Tukey, 1972: Robust Estimates of Location - Survey and Advances. Princeton University Press.

Box, G. E. P., and D. R. Cox, 1964: An analysis of transformations. Journal of the Royal Statistical Society, Series B, 26, 211-252.

Cleveland, W. S., T. E. Graedel, B. Kleiner, and J. L. Warner, 1974: Sunday and workday variations in photochemical air pollutants in New Jersey and New York. Science, 186, 1037-1038.

Cleveland, W. S., and B. Kleiner, 1975: Enhancing scatterplots with moving statistics. Technometrics, 17, 447-454.

Davies, O. L., and P. L. Goldsmith, 1972: Statistical Methods in Research and Production, Fourth Edition. New York, Hafner Publishing Co.

Diaconis, P., and D. Freedman, 1981: On the maximum deviation between the histogram and the underlying density. Zeitschrift für Wahrscheinlichkeitstheorie und Verwandte Gebiete, in press.

Doane, D. P., 1976: Aesthetic frequency classifications. American Statistician, 30, 181-183.

Erickson, B. H., and T. A. Nosanchuk, 1977: Understanding Data. Toronto, McGraw-Hill Ryerson Ltd.

Gnanadesikan, R., and J. R. Kettenring, 1972: Robust estimates, residuals, and outlier detection with multi-response data. Biometrics, 28, 81-124.

Gnanadesikan, R., 1977: Methods for Statistical Data Analysis of Multivariate Observations. New York, Wiley.

Hinkley, D., 1977: On quick choice of power transformation. Applied Statistics, 26, 67-69.

Johnson, N. L., and F. C. Leone, 1974: Statistics and Experimental Design in Engineering and the Physical Sciences, Volume 1. New York, Wiley.

Larsen, R. I., 1969: A new mathematical model of air pollutant concentration averaging time and frequency. Journal of the Air Pollution Control Association, 19, 24-30.

Lockhart, T. J., 1979: Climate without 19 mph. Bulletin of the American Meteorological Society, 60, 660-661.

McGill, R., J. W. Tukey, and W. A. Larsen, 1978: Variations of box plots. American Statistician, 32, 12-16.

McNeil, D. R., 1977: Interactive Data Analysis. New York, Wiley.

Mendenhall, W., and L. Ott, 1972: Understanding Statistics. Belmont, Calif., Duxbury Press.

Milton, R. C., and R. Hotchkiss, 1969: Computer evaluation of the normal and inverse normal distribution functions. Technometrics, 11, 817-822.

Mosteller, F., and J. W. Tukey, 1977: Data Analysis and Regression. Reading, Mass., Addison-Wesley.

National Climatic Center, 1974-1975: Local Climatological Data, Newark, N.J., and Philadelphia, Pa. Asheville, N.C., National Oceanic and Atmospheric Administration.

National Climatic Center, 1978: Local Climatological Data - Annual Summaries for 1977. Asheville, N.C., National Oceanic and Atmospheric Administration.

New York Department of Environmental Conservation, 1974: Air Quality Data for 1972-1973. Albany, N.Y.

Reiss, N. M., and R. A. Eversole, 1978: Rectification of prevailing visibility statistics. Atmospheric Environment, 12, 945-950.

Toonkel, L. E., 1980: Environmental Quarterly, EML-370. New York, U.S. Department of Energy, Environmental Measurements Laboratory.

Tukey, J. W., 1977: <u>Exploratory</u> Data <u>Analysis</u>. Reading, Mass.,
    Addison-Wesley.

Velleman, P. F., and D. C. Hoaglin, 1981: <u>Applications</u>, <u>Basics</u> <u>and</u>
    <u>Computing</u> <u>of</u> <u>Exploratory</u> <u>Data</u> <u>Analysis</u>. Boston, Duxbury Press.

2
Developing Empirical Models with Multiple Regression: Biased Estimation Techniques

Donald W. Marquardt and Ronald D. Snee

1. INTRODUCTION

Multiple regression analysis is the statistical technique used most widely in data analysis and the development of empirical models. A major contributing factor has been the growth in speed and accessibility of computers. Multiple regression analysis is employed, for example, by the National Weather Service (NWS) in converting the output of numerical models into operational weather forecasts (see Chapter 8). During the past twenty years, there have been a variety of regression techniques proposed in the literature. These techniques will be reviewed briefly and then a detailed discussion of ridge regression and generalized inverse regression will be presented. We focus on these biased estimation procedures because these techniques are most effective in dealing with a serious deficiency of least squares estimation, the inability to produce useful regression models when the predictor variables are highly correlated. It is important to note that many meteorological predictor variables, including some of those used by the NWS in developing weather forecasting equations based on multiple regression analysis, are necessarily highly correlated because of inherent physical relationships.

The objective of regression analysis is to develop a model of the form

$$E(y) = \beta_0 + \beta_1 x_1 + \beta_2 x_2 + \ldots + \beta_p x_p,$$

where y is the response (dependent variable) of interest, x_i is the ith predictor variable, β_0, β_1, $\ldots$, β_p are regression coefficients to be estimated from the data, and the notation "$E(\ )$" refers to the predicted average value of the response. There are several steps in a regression analysis. The following list gives a broad view of what a regression analysis involves:

(a) Care is taken to properly formulate the model (i.e., selection of the proper response and predictor variables).

45

(b) Scatter plots of y versus each x_i as well as plots of each x_i versus all other x_j's are made to aid in selecting the form of the model, check for correlations among the x_i's (i.e., multicollinearity), and to detect data keypunch and recording errors. Histograms of each variable are also helpful in detecting data errors. The need for transformations of the variables is considered.

(c) The postulated model is fit to the data by least squares or another appropriate estimation procedure (e.g., ridge regression).

(d) The model F-ratio is checked to determine the statistical significance of the relationship described by the equation. The related R^2 or R_A^2 (adjusted R^2) statistics are used to describe the proportion of total variation explained by the model.

(e) The regression coefficient variance inflation factors (VIF's) are checked to determine if there are any serious correlations among the x_i's (i.e., multicollinearity) which may render the least squares coefficients useless.

(f) The residuals (i.e., $y - \hat{y}$ = the observed value y minus the predicted value y of the fitted equation) of all observations are analyzed by plotting and other procedures. The purpose is to check for atypical data points (outliers), model inadequacy (lack of fit), and deviations from the assumptions. The key assumptions are that the observation errors are independently and normally distributed and have homogeneous (constant) variance.

(g) The reasonableness of the resulting model is checked by examining the model predictions $\hat{y}$ and coefficients $\hat{\beta}_i$ in the light of previous knowledge. This check should include making predictions for data not used in the model construction process.

(h) When it is concluded that the model is valid, or at least reasonable, the coefficients are interpreted from statistical (F-test or t-test and confidence limits) and practical (direction and magnitude) viewpoints. Particularly when the model contains higher-order terms such as products (i.e., interaction) and squares (X_i^2), contour plots (Myers, 1971) may be helpful in interpreting the equation.

Use least squares to fit the full model. Follow this analysis
with a biased estimation technique such as ridge regression
whenever the least squares coefficient VIF's suggest that the
multicollinearity effects may be large.

It is safe with this approach to eliminate variables with small coefficients, if scientifically appropriate, because the multicollinearity effects are either small (i.e., when least squares is used) or
significantly reduced by the biased estimation procedure.

As with any developing field, there are those who question the
use of the biased estimators (Draper and Van Nostrand, 1979; Smith and
Campbell, 1980). These authors have justly criticized the "automatic
use" of ridge regression. We do not recommend the automatic use of
any statistical technique. These critics also have questioned the
theoretical basis of the biased estimators. Unfortunately, they seem
to have overlooked the fact that the theory underlying the biased
estimation procedures is much better developed than that of the stepwise and best subsets regression procedures. Moreover, the estimator
that emerges from stepwise or best subsets invariably is itself
biased, due to the deletion of one or more predictor variables whose
true coefficients are nonzero. The amount of bias can be greater
than the bias deliberately introduced by a procedure such as ridge
regression. There are many who feel, as we do, that the biased estimators in general, and ridge regression in particular, are an effective data analysis tool that can provide useful results. There have
been successful applications in many fields (Hoerl and Kennard,
1981).

Section 2 of this chapter focuses on the theory of biased estimation; not so much the algebraic details as the concepts involved.
Several illustrative examples help clarify the concepts. While we
emphasize ridge regression, we also discuss its relationship to
generalized inverse regression. Section 3 discusses two larger
examples where one can see ridge regression at work in data analysis
in a realistic setting.

2. THEORY AND ILLUSTRATIVE EXAMPLES

The types of data sets that we are addressing can be messy for a
variety of reasons. The predictor variables may be correlated
because historical data were collected without the aid of an experimental design. Physical and mathematical constraints may induce
correlated predictor variables, even when an experimental design is
used. In particular, many meteorological predictor variables are
correlated as a direct consequence of physical properties. The
presence of gross errors, missing values, correlated errors, split
plotting, nonconstant variance, and other problems can create nonsense results, even when we employ sophisticated regression techniques to deal with the correlation problem. For purposes of this
discussion, we assume that none of these other problems is present.

This list helps us identify where the biased estimation procedures, such as ridge regression, fit into the regression analysis process. Some readers may have encountered some new terminology and concepts in this list. These items will be explained in greater detail at appropriate places later in the chapter. We will focus on the selection of an appropriate estimation procedure [step (c)], the regression coefficient VIF [step (e)], and procedures for checking the validity of the fitted regression model [step (g)]. The other items are well documented in regression textbooks such as those of Draper and Smith (1981), Daniel and Wood (1980), and Montgomery and Peck (1982). We also mention a class of estimation techniques called "robust" regression procedures (Mosteller and Tukey, 1977). These techniques are less sensitive to outliers and nonnormality of residuals and can be considered complementary to the techniques discussed in this chapter.

It is important to begin with some comments on the various regression procedures that have been proposed in the literature. Least squares is the procedure of choice to estimate the regression coefficients provided there are no near multicollinearities and extreme outliers among the predictor variables (Hoerl and Kennard, 1970a, 1970b). To counteract the multicollinearity effects, the following regression procedures have been proposed:

(a) Stepwise regression including forward selection and backward elimination (Draper and Smith, 1981).

(b) Best subsets regression (Daniel and Wood, 1980).

(c) Biased regression, including ridge (Hoerl and Kennard, 1970a, 1970b) generalized inverse (Marquardt, 1970), and latent root regression (Webster et al., 1974).

The stepwise and best subsets procedures use least squares estimation and attempt to reduce the effects of multicollinearity by eliminating some of the variables from the model. Hoerl and Kennard (1970a, 1970b) point out that there is no guarantee that this will work. In some instances, you may be worse off than using the entire set of predictor variables. For this reason and others that will be discussed later, we do not use these procedures. We recognize that some researchers have claimed successful applications of these techniques and refer those interested in these techniques to the books of Draper and Smith (1981) and Daniel and Wood (1980).

Some researchers have claimed that multiple procedures (i.e., least squares and the biased estimation techniques) should be used when analyzing a data set. This Herculean effort may be appropriate in some instances. However, it is not realistic on a routine basis. Our recommendation is very simple:

We emphasize that biased estimation is only one of the tools one uses in the analysis of a set of data. If the VIF's (Marquardt, 1970; Snee, 1973) of the least squares estimates are large, then it is appropriate to consider a biased estimation procedure such as ridge regression in order to minimize the effects of the predictor variable correlations and to develop a set of stable coefficients.

2.1 Comments on Some Common Practices

As we survey the literature and reflect upon the state of the art of regression analysis with large numbers of predictor variables, we have identified a number of practices about which we would like to comment before discussing ridge regression, per se. One common practice we note is failure to remove nonessential ill-conditioning through the use of standardized predictor variables. Standardizing of the predictors is appropriate whenever a constant term is present in the model. The ill-conditioning that results from failure to standardize is all the more insidious because it is not due to any real defect in the data, but only to the arbitrary origins of the scales of which the predictor variables are expressed. In standardizing the predictor variables, the mean is subtracted from each variable ("centering") and then the centered variable is divided by its standard deviation ("scaling"). The standardized variable x_i is thus

$$x_i' = (x_i - \bar{x}_i)/s_i,$$

where $\bar{x}_i$ and s_i are the mean and standard deviation of variable x_i.

Centering removes the nonessential ill-conditioning, thus reducing the variance inflation in the coefficient estimates. In a linear model, centering removes the correlation between the constant term and all linear terms. In addition, in a quadratic model, centering reduces, and in certain situations completely removes, the correlation between the linear and quadratic terms. Scaling expresses the equation in a form that lends itself to more straightforward interpretation and use. In the construction of quadratic and higher-order models, the quadratic and higher-order terms are generated after the basic variables have been standardized. The two-variable quadratic model is

$$E(y) = \beta_0 + \beta_1\left(\frac{x_1 - \bar{x}_1}{s_1}\right) + \beta_2\left(\frac{x_2 - \bar{x}_2}{s_2}\right) + \beta_{12}\left(\frac{x_1 - \bar{x}_1}{s_1}\right)\left(\frac{x_2 - \bar{x}_2}{s_2}\right)$$

$$+ \beta_{11}\left(\frac{x_1 - \bar{x}_1}{s_1}\right)^2 + \beta_{22}\left(\frac{x_2 - \bar{x}_2}{s_2}\right)^2.$$

An alternative form of centering and scaling is

$$x_i' = (x_i - MR_i)/HR_i,$$

$$MR_i = (x_{max} + x_{min})/2,$$

$$HR_i = (x_{max} - x_{min})/2,$$

where MR_i, HR_i, x_{min}, and x_{max} are the midrange, half-range, minimum, and maximum of variable x_i, respectively. This is the same scaling as that used in the analysis of two-level factorial designs (Daniel, 1976) and results in $-1 \leq x_i' \leq 1$. Using this scaling, the results of a regression analysis of a two-level factorial design will be equivalent to that produced by the Yates analysis procedure (Daniel, 1976). Many will find this comparison helpful in understanding the importance of standardization of predictor variables in regression. The reader is referred to Marquardt (1980) for further discussion of standardized predictor variables and their essential role in conducting both least squares and biased regression analyses.

The acetylene data (Himmelblau, 1970; Kunugi et al., 1961) shown in Table 1 is the first example we will discuss. This is a typical set of response surface data for which a full quadratic model in x_1, x_2, and x_3 is an appropriate candidate (Myers, 1971). Two of the predictor variables are plotted against each other in Figure 1. Correlations like this cause inflation of the variance of the estimated coefficients in the least squares model. The variance of the ith regression coefficient estimate $\hat{\beta}_i$ can be written as

$$Var(\hat{\beta}_i) = \frac{\sigma^2 R_{ii}^{-1}}{(n-1)s_i^2},$$

where σ is the observation error standard deviation, R_{ii}^{-1} is the ith diagonal element of the inverse of the correlation matrix R of the predictor variables, n is the number of observations, and s_i is the standard deviation of x_i. This formulation shows clearly the four things that influence the variance of a regression coefficient: $Var(\hat{\beta}_i)$ is an increasing function of the observation error standard deviation σ, an increasing function of the multicollinearity of the predictor variables R_{ii}^{-1}, a decreasing function of the sample size n, and a decreasing function of the variation (i.e., range) of x_i.

Table 1. Acetylene data.

x_1 Reactor temperature (°C)	x_2 Ratio of H_2 to n-heptane (mole ratio)	x_3 Contact time (sec)	y Conversion of n-heptane to acetylene (%)
1300	7.5	0.0120	49.0
1300	9.0	0.0120	50.2
1300	11.0	0.0115	50.5
1300	13.5	0.0130	48.5
1300	17.0	0.0135	47.5
1300	23.0	0.0120	44.5
1200	5.3	0.0400	28.0
1200	7.5	0.0380	31.5
1200	11.0	0.0320	34.5
1200	13.5	0.0260	35.0
1200	17.0	0.0340	38.0
1200	23.0	0.0410	38.5
1100	5.3	0.0840	15.0
1100	7.5	0.0980	17.0
1100	11.0	0.0920	20.5
1100	17.0	0.0860	29.5

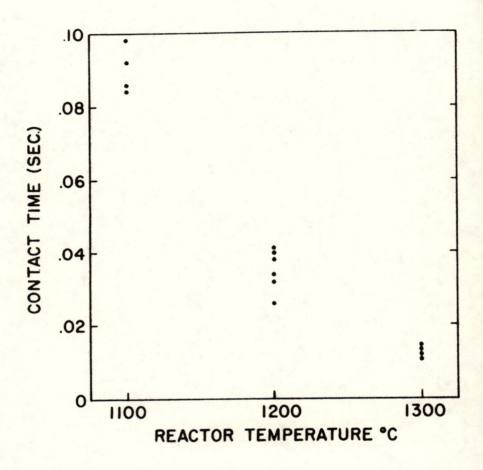

Figure 1. Acetylene data: correlation between reactor temperature (x_1) and contact time (x_3).

We focus here on R_{ii}^{-1}, which Marquardt (1970) called the <u>vari-ance inflation factor</u> (VIF) of the coefficient estimate $\hat{\beta}_i$. The role of R_{ii}^{-1} is apparent when we recall that

$$R_{ii}^{-1} = (1 - R_i^2)^{-1}$$

where R_i is the multiple correlation between x_i and all the other predictor variables (x_j, $j \neq i$) in the model. The VIF (Marquardt, 1970; Snee, 1973) for each term in the model measures the collective impact of these simple correlations on the variance of the coefficient of that term. The VIF's are the diagonal elements of the inverse of the simple correlation matrix. As the multiple correlation of any predictor with the other predictors approaches unity (i.e., when $R_i^2 \simeq 1$), the corresponding VIF becomes infinitely large.

The VIF's for the acetylene data are shown in Table 2. We believe that the maximum VIF is the single most useful measure of the conditioning of the data. For any predictor orthogonal to all other predictors, the VIF equals one. The inflation factors for the stand-ardized model are in the second column. Scaling does not affect the VIF's, but centering does. A maximum factor of six thousand is horrible, but two million is unthinkable and unnecessary. As a general rule, we recommend that the use of ridge regression proce-dures be considered when the maximum VIF is larger than ten.

The Kentucky utility data discussed by Hocking and LaMotte (1973) further illustrate the usefulness of the VIF. The objective of this work was to use weather variables and other information to predict electricity usage y. The VIF's for the dry bulb temperature, wet bulb temperature, dew point, and relative humidity variables are very large, indicating that the associated coefficients would be poorly estimated (see Table 3). This should come as no surprise, since these variables are mathematically related. It is important to note, however, that the pairwise correlations do not reflect the high degree of multicollinearity identified by the VIF (see Table 3). The mathematical relationships among these four variables suggest that only two of the four variables are independent. Hence, because of common usage, one might choose to use dry bulb temperature and humi-dity to model the collective effects of these four variables (maximum VIF = 2.67 from Table 3).

From a correlation viewpoint, we could choose any two of these four variables because the VIF is never greater than six for any of the six possible pair-wise combinations of variables (maximum VIF = 5.95 for the five-variable model containing wet bulb temperature and dew point). The use of three of the four variables is of little

Table 2. Acetylene data regression results: ten-coefficient quadratic model.

Term	Least squares - VIF Unstandardized	Standardized	Correlation basis coefficient Least squares	Ridge k = 0.01	Ridge k = 0.05	Generalized inverse (r = 3.8)
x_1 = Temperature	2,856,748.93	375.25	0.336	0.589	0.522	0.507
x_2 = H$_2$/n-heptane	10,956.14	1.74	0.233	0.216	0.209	0.180
x_3 = Contact time	2,017,162.52	680.28	-0.676	-0.327	-0.379	-0.414
x_1x_2	9,802.90	31.04	-0.480	-0.326	-0.202	-0.095
x_1x_3	1,428,091.88	6,563.35	-2.034	-0.094	-0.061	-0.051
x_2x_3	240.36	35.61	-0.266	-0.083	0.042	0.123
x_1^2	2,501,944.59	1,762.58	-0.835	0.126	0.125	0.165
x_2^2	65.73	3.16	-0.090	-0.054	-0.047	-0.063
x_3^2	12,667.10	1,156.77	-1.001	-0.069	-0.024	-0.053
Maximum VIF			6,563.35	12.38	2.63	0.46
R_A^2			0.994	0.990	0.983	0.973

Table 3. Kentucky utility data: correlation coefficients and VIF's.

Predictor variable	Correlation coefficients						VIF[a]	
	DB temp	WB temp	Dew point	Humidity	Atmospheric pressure	Sky cover	A	B
Dry bulb temperature	1.00						222.38	2.67
Wet bulb temperature	0.84	1.00					184.53	--
Dew point	0.53	0.90	1.00				147.26	--
Humidity	-0.71	-0.23	0.21	1.00			122.52	2.35
Atmospheric pressure	-0.01	-0.11	-0.17	-0.13	1.00		1.21	1.21
Sky cover	0.17	0.33	0.39	0.12	-0.32	1.00	1.28	1.28
Wind speed	0.44	0.43	0.34	-0.23	-0.27	0.21	1.37	1.36

[a] Least squares VIF's: A – all seven variables in the model, B – five variables in the model (i.e., wet bulb temperature and dew point not included).

help, because the maximum VIF is never less than 162. It is also of
interest to note that in the full model the VIF's for atmospheric
pressure, sky cover, and wind speed are all less than 1.4 (see Table
3). These results indicate that the correlation among these three
variables and between these variables and any subset of the other
four variables is of no practical concern.

In the Kentucky utility data just discussed, the multicollinearity demonstrated by the VIF's was removed by using physical considerations to guide the choice of which variables to delete. Such
guidance is seldom so straightforward. A case in point is the acetylene data. The question remains: once we have standardized, how do
we carry out a meaningful analysis of data with a VIF greater than
six thousand?

The classic methodology is least squares. It does not provide
good estimates when the data are ill-conditioned. In achieving optimum fit to the estimation data, least squares often destroys good
prediction of new data (possibly outside the region covered by the
estimation data).

The earliest alternative to the classical methodology is
variable selection as a technique for reducing the degree of ill-
conditioning. Variable selection implies a simplistic two-valued
classification logic wherein any predictor variable must either be
important or unimportant. Large prediction biases can result from
elimination of apparently "nonsignificant" predictors. It is better
to use a little bit of all the variables than all of some variables
and none of the remaining ones. This is what biased estimators do.
We will show how biased estimators can alleviate both of these limitations.

Finally, in our comments on current practices, we observe that
most statisticians restrict themselves to models linear in the parameters. Frequently the known background of the problem suggests a
function nonlinear in the parameters (Marquardt, 1963; Draper and
Smith, 1981), one that may provide a simpler and more natural model.
We will illustrate this with the Laird and Cady corn yield example.

2.2 Formulation of the Problem

In linear estimation one postulates a model of the form

$$Y = X\beta + \varepsilon.$$

The n by p matrix X contains the values of p predictor variables at
each of n data points, Y is the vector of observed values, β is the p
by one vector of population values of the parameters, and ε is an n
by one vector of experimental errors having the properties $E(\varepsilon) = 0$

and $E(\varepsilon\varepsilon') = \sigma^2 I$. Here I is the n by n identity matrix. For convenience, we assume that the predictor variables are scaled so that X'X has the form of a correlation matrix. This scaling will be discussed later in the paper.

The conventional estimator for β is the least squares estimator $\hat{\beta}$, with $\hat{\beta}$ chosen to minimize the sum of squares of residuals

$$\Phi(\hat{\beta}) = (Y - X\hat{\beta})'(Y - X\hat{\beta}).$$

The solution to this minimization problem is

$$\hat{\beta} = (X'X)^{-1}X'Y = (X'X)^{-1}g,$$

where g is the gradient of Φ. The two key properties of $\hat{\beta}$ are that it is unbiased [i.e., $E(\hat{\beta}) = \beta$], and that it has minimum variance among all linear unbiased estimators. The covariance matrix of $\hat{\beta}$ is

$$Var(\hat{\beta}) = \sigma^2(X'X)^{-1}.$$

In their development of ridge regression, Hoerl and Kennard (1970a, 1970b) focus attention on the eigenvalues of X'X. A seriously non-orthogonal (or "ill-conditioned") problem is characterized by the fact that the smallest eigenvalue λ_{min} is very much smaller than unity (the eigenvalues are all unity in the orthogonal case). For example, the smallest eigenvalues of the X'X matrices for the acetylene data (p = 9, p = 5) and the Laird and Cady (p = 33) and GC-ASTM (p = 15) data (to be discussed later) are 0.00010, 0.01005, 0.00207, and 0.00027, respectively. Hoerl and Kennard summarized the dramatic inadequacy of least squares for nonorthogonal problems by noting that the expected squared length of the coefficient vector is

$$E(\hat{\beta}'\hat{\beta}) = \beta'\beta + \sigma^2 Tr(X'X)^{-1}$$

$$> \beta'\beta + \sigma^2/\lambda_{min}.$$

Here "Tr" denotes the trace of a matrix. Thus, the least squares coefficient vector $\hat{\beta}$ is much too long, on the average, for ill-conditioned data, since $\lambda_{min} \ll 1$. The least squares solution yields coefficients whose absolute values typically are too large and whose signs may actually reverse with negligible changes in the data.

It is important to note that the VIF mentioned earlier is a measure of how close the smallest eigenvalues are to zero (Marquardt, 1970; Snee, 1973). For example, an eigen analysis of the Kentucky utility data correlation matrix (Table 3) produced two small eigenvalues (0.0034 and 0.00262), indicating that the seven predictor variables represented, in effect, only five dimensions (Table 4).

Table 4. Kentucky utility data: eigen analysis of predictor variable correlation matrix.

Variable	Eigenvector						
	1	2	3	4	5	6b	7b
Dry bulb temperature	0.50[a]	0.35	-0.02	0.12	-0.17	-0.10	0.76
Wet bulb temperature	0.54	0.01	-0.27	-0.06	-0.16	0.72	-0.29
Dew point	0.46	0.30	-0.42	-0.18	-0.11	-0.65	-0.26
Humidity	-0.20	0.65	-0.33	-0.29	0.10	0.23	0.53
Atmospheric pressure	-0.14	-0.42	-0.65	0.09	0.61	0.01	0.00
Sky cover	0.26	0.42	0.14	0.76	0.39	0.00	0.00
Wind speed	0.35	0.00	0.44	-0.52	0.64	0.00	0.00
Eigenvalue	3.090	1.644	1.024	0.680	0.556	0.00341	0.00262

[a]Tabled values are eigenvector elements.

[b]The elements for atmospheric pressure, sky cover, and wind speed are very small for eigenvectors no. 6 and no. 7. These results indicate that the other four variables are producing the multicollinearity identified by the two small eigenvalues associated with these two eigenvectors.

Examination of the eigenvectors associated with these eigenvalues (Table 4) shows that dry bulb temperature, wet bulb temperature, dew point, and humidity are the variables producing the high degree of multicollinearity. This analysis thus shows mathematically what is known physically; namely, that these four variables represent only two independent dimensions. We will not discuss eigen analysis further here and refer the reader to Marquardt (1970), Snee (1973), and Montgomery and Peck (1982) for further details on this useful method for identifying sources of multicollinearity among the predictor variables in a regression model.

The "fly in the ointment" with least squares is its requirement of unbiasedness. Figure 2, top, illustrates the situation where an estimator $\hat{\beta}$ is unbiased but is plagued by a large variance. Typical confidence limits for this estimator would be nearly half the width of the figure. At the bottom of Figure 2 is the corresponding frequency function for a biased estimator with much smaller variance. Statistical limits for this situation would be perhaps twenty percent of the width of the figure. Thus, it is meaningful to focus on the achievement of small mean squared error as the relevant criterion, if a major reduction in variance can be obtained as a result of allowing a little bias. This is precisely what the ridge and generalized inverse solutions accomplish.

2.3 Ridge Solution

The ridge estimator is obtained by solving

$$(X'X + kI)\hat{\beta}* = g$$

to give

$$\hat{\beta}* = (X'X + kI)^{-1}g,$$

for $k \geq 0$. We note that when $k = 0$, the ridge estimator is simply the least squares estimator $\hat{\beta}$. Values of k between zero and 1.0 should be explored. The upper limit of 1.0 is not a mathematical limit, but a practical value beyond which it is not necessary to explore the results. In general, there is an "optimum" value of k for any problem, but it is desirable to examine the ridge solution for a range of admissible values of k. "Admissible" means having smaller mean squared error in the parameters than the least squares solution. Since regression models are generally employed in meteorology to make predictions, it is important to observe that the mean squared error in future predictions is also reduced correspondingly. Hoerl gave the name ridge regression (Hoerl, 1962) to his procedure, because of the similarity of its mathematics to methods he used earlier (i.e., "ridge analysis," Hoerl, 1959) for graphically depicting the characteristics of second-order response surface equations in many predictor variables.

VARIANCE AND BIAS IN AN ESTIMATOR

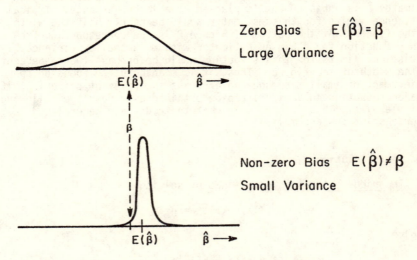

Zero Bias $\quad E(\hat{\beta}) = \beta$

Large Variance

Non-zero Bias $\quad E(\hat{\beta}) \neq \beta$

Small Variance

MSE = Mean Square Error = Variance + (Bias)^2

Figure 2. Schematic illustrating the variance and bias in an estimator.

Key properties applicable to ridge regression are (Hoerl and Kennard, 1970a, 1970b; Marquardt, 1970):

(a) If $\hat{\beta}*$ is the solution of $(X'X + kI)\hat{\beta}* = g$, then $\hat{\beta}*$ minimizes the sum of squares of residuals on the sphere centered at the origin whose radius is the length of $\hat{\beta}*$. The sum of squares of residuals is an increasing function of k.

(b) The length of $\hat{\beta}*$ is a decreasing function of k.

(c) The angle γ between the ridge solution $\hat{\beta}*$ and the gradient vector g is a decreasing function of k.

In Figure 3 we illustrate the geometry of ridge regression for a hypothetical problem involving only two parameters β_1 and β_2. The point $\hat{\beta}$ at the center of the ellipses is the least squares solution. At $\hat{\beta}$ the sum of squares of residuals Φ achieves its minimum. The small ellipse is the locus of points in the (β_1, β_2)-plane where the sum of squares Φ is constant at some value larger than the minimum value. The circle about the origin is tangent to the small ellipse at $\hat{\beta}*$. We note that the ridge estimate $\hat{\beta}*$ is the shortest vector that will give a residual sum of squares as small as the value anywhere on the small ellipse. Thus, the ridge estimate gives the smallest regression coefficients consistent with a given degree of increase in the residual sum of squares. The gradient g is perpendicular to the Φ-contour through the origin. The ridge estimate $\hat{\beta}*$ is always between $\hat{\beta}$ and g, the angle γ getting steadily smaller as the quantity k added to the diagonal increases.

Other key properties of the ridge estimator are:

(a) $\hat{\beta}*$ is a linear transform of $\hat{\beta}$:

$$\hat{\beta}* = Z_k\hat{\beta} = (X'X + kI)^{-1}(X'X)\hat{\beta}$$

Thus, $\hat{\beta}*$ is a biased estimator of β with $E(\hat{\beta}*) = Z_k\beta$.

(b) $Var(\hat{\beta}*) = \sigma^2(X'X + kI)^{-1}(X'X)(X'X + kI)^{-1}$

(c) $ESD = Tr[Var(\hat{\beta}*)] + \beta'(Z_k - I)'(Z_k - I)\beta$

$= Variance + (Bias)^2,$

where ESD denotes the expected squared distance of $\hat{\beta}*$ from β. Two crucially important corollaries are:

(i) The variance term is a decreasing function of k.

(ii) The bias term is an increasing function of k.

GEOMETRY OF RIDGE REGRESSION

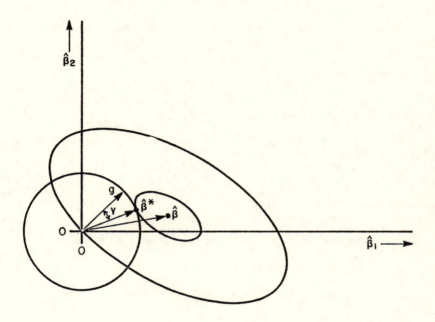

Figure 3. Geometry of ridge regression for a hypothetical problem involving two parameters β_1 and β_2.

(d) If $\beta'\beta$ is bounded, then there exists a $k > 0$ such that $ESD(\hat{\beta}\star) < ESD(\hat{\beta})$.

We note in passing two further points that are frequently emphasized by R. W. Kennard. First, all the theory developed for the regression coefficients $\hat{\beta}\star$ also hold for the model predictions $X\hat{\beta}\star$, because the model predictions are a linear transform of the model coefficients. Second, least squares implies an assumption of an unbounded uniform prior distribution on the coefficient vector. This unboundedness assumption can be used in place of the unbiasedness requirement in deriving least squares estimators. When selecting the amount of bias, we work with predictor variables and response variables both scaled to correlation form. In this scaling, it is exceedingly rare for the population value of any regression coefficient to be larger than three in a real problem. In any case, the regression coefficient vector is surely finite. The ridge estimate is equivalent to placing mild boundedness requirements on the coefficient vector (see Marquardt, 1980, for further discussion of this point).

Theobald (1974) generalized the conditions under which ridge is known to produce a smaller expected squared distance than least squares. It is also known that the expected improvement of ridge over least squares depends on the orientation of the true regression vector relative to the principal axes defined by the eigenvectors of the X'X matrix, the expected improvement being greatest when the orientation of β coincides with the eigenvector associated with the largest eigenvalue of X'X. Other results appear in the papers of Allen (1974), Lindley and Smith (1972), Mallows (1973), Marquardt (1974), and Mayer and Willke (1973).

A complete sequence of corresponding properties of the generalized inverse solution is developed in Marquardt (1970). An illustrative example in that paper demonstrates the close geometric similarity between the generalized inverse solution expressed as a function of the rank r assigned to the matrix X'X, and the ridge solution, expressed as a function of the bias parameter k added to the diagonal elements of X'X. In Marquardt (1970) it is emphasized that for this purpose the assigned rank is best defined as a piecewise continuous variable.

2.4 Analysis of the Acetylene Data

We return now to the acetylene data example. The standardized full quadratic model is

$$E(y) = \beta_0 + \sum_{j=1}^{3} \beta_j x_j + \sum_{1 \leq j < j'}^{3} \beta_{jj'} x_j x_{j'} + \sum_{j=1}^{3} \beta_{jj} x_j^2,$$

where y = % conversion, x_1 = (Temperature - 1212.50)/80.623, x_2 = [H_2/(n-heptane) - 12.44]/5.662, and x_3 = (Contact time - 0.0403)/ 0.03164.

We recommend that a polynomial model be reported to (and used by) the ultimate consumer in this form. We note that each predictor variable is standardized, but the expansion terms (squares and cross-products) are created directly from the standardized linear terms. The model is not standardized with respect to y. Numerical evaluation in this form is accurate, and interpretation of the coefficients is straightforward.

It is necessary, however, for selection of the amount of bias to examine the equation and its fit with all variables scaled to correlation form, including y, the linear predictors, and the expansion predictor variables (see Marquardt, 1980, for further discussion of this point). The model written in "correlation form" is

$$y' = b_1 V_1 + b_2 V_2 + \ldots + b_p V_p,$$

where

$$y' = (y - \overline{y})/SSy, \quad SSy^2 = \Sigma(y - \overline{y})^2,$$

$$V_i = (Z_i - \overline{Z}_i)/SSZ_i, \quad SSZ_i^2 = \Sigma(Z_i - \overline{Z}_i)^2,$$

and Z_i is the ith predictor variable in the model (i.e., Z_i could be ln x_j, x_i, $x_i x_j$, etc.). Written in this form, V_i has a mean of zero and sum of squares of one. Hence, the associated X'X matrix (i.e., V'V) will be the correlation matrix with one's down the diagonal (Draper and Smith, 1981, p. 263). Since V'V = X'X, there is no need to construct a correlation-form model. The coefficients can be calculated directly using the X'X matrix. We refer to the regression coefficients b_i as the "correlation-basis" regression coefficients, because they result from expressing the model in correlation form. Other authors (Ezekiel and Fox, 1959; Chatterjee and Price, 1977) have referred to these coefficients as "beta" coefficients.

In regular use of ridge regression, we display the correlation-basis regression coefficients in tables and graphs for about 25 values of k, spaced approximately logarithmically over the interval [0,1]. Table 2 shows the correlation-basis regression coefficients by least squares, and by ridge with two values of k. Both X'X and X'Y are in correlation form. Hence, the correlation-basis coefficient b_i is related to the regression coefficient $\hat{\beta}_i$ by $\hat{\beta}_i$ = $b_i(Sy/SZ_i)$, where Sy and SZ_i are the standard deviations of y and Z_i, respectively [i.e., $(n-1)Sy^2 = SSy^2$ and $(n-1)SZ_i^2 = SSZ_i$]. The correlation-basis coefficient b_i is therefore the expected change in y, measured in y-standard deviations, given an increase in Z of one standard deviation (e.g., b_i = 0.5 implies that y increases 0.5 standard deviations when Z_i is increased one standard deviation).

In the acetylene data, the correlation-basis coefficients have stabilized by k = 0.05 (see Section 3.1), and the VIF's are also reasonable for this value of k. We note that the large least squares coefficients for the x_1x_3 interaction and for x_3^2 have all but disappeared in the ridge model. Also, x_2x_3 and x_2^2 have small coefficients. Similar results are obtained with the generalized inverse model, shown for r = 3.8, in which case the regression vector length is about the same as the ridge model. Suppose these four terms are eliminated. Variable selection is a safer strategy here, since the bias has removed most of the ill-conditioning. For a nearly orthogonal model in the correlation-basis scaling, all coefficients will have nearly equal variances. Hence, variable selection can be made on the basis of absolute values of the coefficients. The mild bias component of the mean squared errors of the coefficients is ignored for this purpose. Table 5 shows that further biasing of the five-term model doesn't change the coefficients much.

Figure 4 shows the predictions for the least squares nine-term full quadratic, the ridge nine-term model, and the least squares five-term model. The prediction points shown are the extreme points of the data; they define the boundary of the region covered by the data. For practical purposes, all three models predict equally well here.

Figure 5 shows predictions outside the data. Notice that we are not extrapolating beyond the ranges of the individual predictor variables, but only to the corners of the region. Consider the upper right corner. The least squares nine-term model predicts minus 86.2% conversion. This is physically impossible. The other models predict 32.9% and 38.0% conversion, much more realistic predictions. A similar situation holds at the lower left corner.

What do we conclude from this example? Well, first of all, we conclude that the nine-term full quadratic given by least squares is not a good model, even though it is the one that fits the estimation data most closely. Secondly, we find that biased regression, routinely applied, produces a much better model, still using all nine terms. Finally, we have been guided by the ridge results to a subset model that also does a good job. In this case, Snee (1973) arrived at the same subset model by plotting the raw data. We do not claim that the ridge regression estimates of the model coefficients are optimal under all conditions for these data, only that these estimates are much better than the least squares coefficients. The model checks we were able to perform are not sufficient for establishing the adequacy of the model. Rather, these checks are minimal requirements for validation using only the available data.

Table 5. Acetylene data regression results: five-coefficient quadratic model.

Term	VIF Least squares	Correlation basis coefficient			
		Least squares	Ridge		Generalized inverse (r = 4.0)
			k = 0.01	k = 0.05	
x_1 = Temperature	43.11	0.602	0.557	0.514	0.518
x_2 = H$_2$/n-heptane	1.07	0.194	0.192	0.187	0.193
x_3 = Contact time	53.52	-0.323	-0.368	-0.391	-0.417
$x_1 x_2$	1.09	-0.273	-0.270	-0.258	-0.272
x_1^2	4.68	0.173	0.180	0.169	0.197
Maximum VIF		53.52	13.63	1.72	1.15
R_A^2		0.991	0.990	0.989	

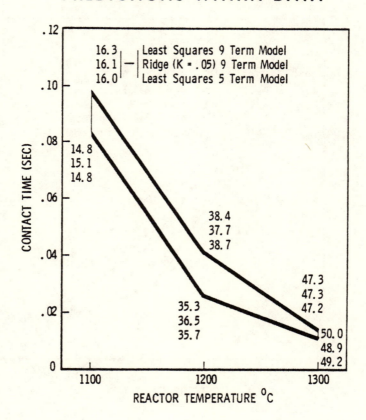

Figure 4. Acetylene data: predictions of the 9-term least squares, 9-term ridge regression (k = 0.05), and 5-term least squares models <u>within</u> the data.

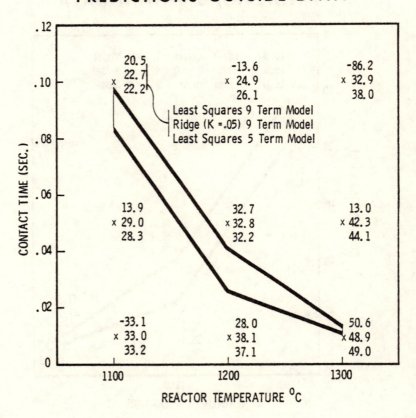

Figure 5. Acetylene data: predictions of the 9-term least squares, 9-term ridge regression (k = 0.05), and 5-term least squares models <u>outside</u> the data.

2.5 When is Variable Selection a Good Strategy?

At this point, it is appropriate to put forth the conditions under which variable selection is, or is not, a good strategy. Variable selection is a good strategy when the candidate variables already are mutually orthogonal, or nearly so. It is also a good strategy when the variables have been made effectively orthogonal, or nearly so, by introducing bias in the estimator. The only other condition under which it is a good strategy is when the selection of the candidate subsets can be strongly guided by knowledge of the background of the problem and the definitions or properties of the variables (e.g., the Kentucky utility data). This condition means information external to the numerical values of the data is available.

Variable selection is a poor strategy when the candidate variables are highly correlated. In essence, the extreme variance inflation completely unstabilizes all the criteria one can compute from the least squares estimates, leading to highly unstable subset selection. It is also a poor strategy when the candidate variables include curvilinear effects (for example, squares) of other candidate variables. This is illustrated in Figure 6.

Consider the two candidate models A and B. Both can represent the quadratic curve that passes smoothly through the solid dot data, with $y = a_0$ at $x = 0$. Both models can also represent the open circle data, with minimum at $x = \overline{x}$, where $y = b_0$. Now, consider what happens if the linear terms are dropped or not selected by the subset procedure. Then if the data are like the open circles, model A is a total disaster, while model B does a great job. The converse holds for data like the solid dots. Thus, the operational behavior of all subset-selection procedures is not invariant with the arbitrary choice of computing origin.

Now, in most situations where quadratic or higher polynomials are applied, the model is really functioning like a Taylor-series expansion of some function in the region of the data. There are normally only two appropriate points for each predictor variable dimension about which the expansion could be natural. These are the mean point of the predictor variable data and the origin of the predictor variable data. This implies that any variable selection procedure should be done multiple times with curvilinear models, computing about the mean point, and computing about the origin for each predictor variable in order to find out which, if either, gives a simple, well-behaved model.

2.6 A Simulation Experiment

The final example of this section is a synthetic one. It illustrates how both ridge and generalized inverse estimators do better

VARIABLE SELECTION WITH CURVILINEAR MODELS

A. $y = a_0 + a_1 x + a_{11} x^2$

B. $y = b_0 + b_1 (x - \bar{x}) + b_{11} (x - \bar{x})^2$

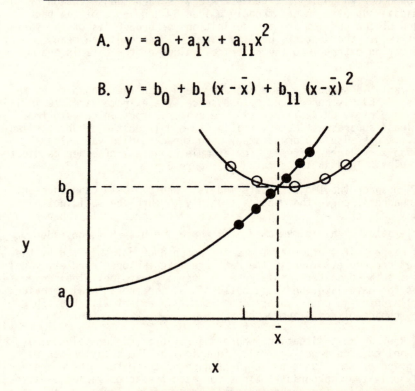

THUS, VARIABLE SELECTION IS NOT INVARIANT WITH THE CHOICE OF COMPUTING ORIGIN.

Figure 6. Schematic of two curvilinear models illustrating the importance of the computing origin ($x = 0$ or $x = \bar{x}$) when using automatic variable selection techniques.

than either least squares or any subset. It also illustrates the mechanics of using the biased estimation procedures.

For this "three-predictor example," the data structure is as shown in Figure 7. There are eight estimation data points shown by the open circles. If the parameter α is zero, the design is a classical 2^3 factorial. As α approaches 1.0, four of the points move in the directions of the arrows. The correlation of x_1 and x_2 becomes progressively larger, with unit correlation as the limit.

We generated data for three values of α. It should be noted that x_3 is orthogonal to x_1 and x_2 for all values of α. The solid dots are prediction points. The true model is $E(y) = x_1 + x_2 + x_3$; that is, all coefficients are unity and the regression constant is zero. We have introduced additive errors selected randomly from a normal distribution with mean zero and standard deviation $\sigma = 0.8$.

The actual data are as follows:

i	x_1	x_2	x_3	$E(y)$	e_i
1	-1	-1	-1	-3	-0.305
2	1	1	-1	1	-0.321
3	-1	-1	1	-1	1.900
4	1	1	1	3	-0.778
5	-1	$(1 - 2\alpha)$	-1	$-1 - 2\alpha$	0.617
6	1	$-(1 - 2\alpha)$	-1	$-1 + 2\alpha$	-1.430
7	$-(1 - 2\alpha)$	1	1	$1 + 2\alpha$	0.267
8	$(1 - 2\alpha)$	-1	1	$1 - 2\alpha$	0.978

The correlation between x_1 and x_2 is $r_{12} = \alpha/(1 - \alpha + \alpha^2)$. Thus $r_{12} = 0.110, 0.667, 0.989$ for $\alpha = 0.1, 0.5, 0.9$, respectively.

The fitted regression model is $y = b_0 + b_1 x_1 + b_2 x_2 + b_3 x_3$. In all cases, we included a constant term on the hypothetical assumption that the data analyst does not know in advance that the true value of the regression constant is zero. The criterion by which we judge the quality of our regression models is the prediction standard error at the eight corners of the cube.

Table 6 shows the ridge regression results for $\sigma = 0.8$ for several values of k and for three value of α. The quantities tabulated are as follows:

(a) S_e = residual standard error from the estimation data.

(b) R_A^2 = adjusted $R^2 = 1 - S_e^2/S_y^2$, where

THREE - PREDICTOR EXAMPLE
DATA STRUCTURE

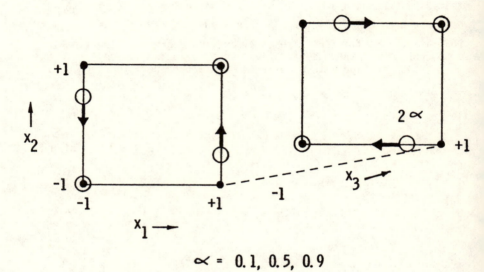

$$\propto \; = \; 0.1, \; 0.5, \; 0.9$$

OPEN CIRCLES ARE ESTIMATION POINTS
SOLID DOTS ARE PREDICTION POINTS

Figure 7. Data structure for the three-predictor simulation experiment.

Table 6. Three-predictor example (σ = 0.80): ridge regression.

α		Ridge k				
		0	0.1	0.2	0.4	0.8
0.1	S_e	0.729	0.759	0.828	0.995	1.286
	R_A^2	0.856	0.843	0.813	0.730	0.550
	Maximum VIF	1.012	0.833	0.698	0.511	0.309
	S_p	0.609	0.598[a]	0.619	0.702	0.878
0.5	S_e	0.591	0.626	0.701	0.876	1.173
	R_A^2	0.906	0.895	0.868	0.794	0.630
	Maximum VIF	1.800	1.155	0.825	0.510	0.309
	S_p	0.713	0.651	0.636[a]	0.674	0.817
0.9	S_e	0.537	0.621	0.699	0.878	1.189
	R_A^2	0.936	0.914	0.891	0.829	0.685
	Maximum VIF	45.751	0.826	0.694	0.510	0.309
	S_p	1.972	0.560	0.536[a]	0.583	0.738
$(VIF)_0$		1.00	0.826	0.694	0.510	0.309

[a]Minimum of tabulated values.

$$S_y = \text{standard deviation of } y,$$

$$y_i = E(y_i) + e_i.$$

(c) Maximum VIF = maximum variance inflation factor.

(d) S_p = prediction standard deviation at the eight prediction points; that is,

$$S_p = \{ \sum_{i=1}^{8} [y_i - E(y_i)]^2/8 \}^{1/2}.$$

Let us start by examining results for $\alpha = 0.9$. We note how the estimation residual error, which is 0.537 when k = 0, increases as the bias k is increased. As a consequence, R_A^2 decreases. However, we note the reduction in maximum VIF as k is increased. Finally, the proof of the pudding is the composite effect of these two opposing trends. This is seen in the prediction residual error S_p, which goes through a minimum at k = 0.2. This tells us that a ridge estimator with bias of k = 0.2 gives predictions at the corners of the cube with a standard error only 0.536 compared with the least squares prediction residual error S_p = 1.972. We note that similar but less dramatic results occur for the smaller values of α.

At the bottom of the table is an extra line of numbers. These are the VIF's for orthogonal predictors, as a function of k. We note that the minimum prediction residual error here occurs for values of k at or just beyond the value where the maximum VIF is about the same size as if the factors were orthogonal.

Table 7 shows the corresponding results using generalized inverse regression. Again focusing first on $\alpha = 0.9$, we note how the regression residual gets larger as the assigned rank of the X'X matrix is decreased. Again R_A^2 decreases, and so does the maximum VIF. Again the prediction residual error S_p goes through a minimum. The minimum occurs at assigned rank 1.5 which, it should be noted, is not an integer. In this example, the minimum prediction residual error S_p by generalized inverse is smaller even than that by ridge regression. We do not interpret this as a general result, but only as an indication that either type of biased estimator can give results substantially better than least squares.

Finally, for completeness, we show in Table 8 the corresponding results for all possible subset models. Focusing again on $\alpha = 0.9$,

Table 7. Three-predictor example ($\sigma = 0.80$): generalized inverse regression.

α		Generalized inverse r				
		3.0	2.5	2.0	1.5	1.0
0.1	S_e	0.729	0.741	0.776	1.247	2.101
	R_A^2	0.856	0.851	0.836	0.577	0.000
	Maximum VIF	1.012	1.000	1.000	0.500	0.450
	S_p	0.609	0.581	0.571	0.527[a]	1.087
0.5	S_e	0.591	0.606	0.649	1.172	2.057
	R_A^2	0.906	0.901	0.887	0.630	0.000
	Maximum VIF	1.800	1.050	1.000	0.500	0.300
	S_p	0.713	0.634	0.605	0.563[a]	1.105
0.9	S_e	0.537	0.553	0.599	1.146	2.042
	R_A^2	0.936	0.932	0.920	0.708	0.072
	Maximum VIF	45.751	23.001	1.000	0.500	0.251
	S_p	1.972	1.100	0.563	0.518[a]	1.083

[a]Minimum of tabulated values.

Table 8. Three-predictor example ($\sigma = 0.80$): all possible subset models.

α		Subset model					
		x_1	x_2	x_3	x_1, x_2	x_1, x_3	x_2, x_3
0.1	S_e	1.942	1.807	1.320	1.864	1.214	0.934
	R_A^2	0.000	0.110	0.525	0.054	0.598	0.763
	Maximum VIF	1.00	1.00	1.00	1.01	1.00	1.00
	S_p	1.46	1.42	1.47	1.11	1.13	1.08[a]
0.5	S_e	1.807	1.693	1.340	1.825	0.933	0.624
	R_A^2	0.121	0.229	0.517	0.104	0.766	0.895
	Maximum VIF	1.00	1.00	1.00	1.80	1.00	1.00
	S_p	1.42	1.43	1.47	1.17	1.07[a]	1.09
0.9	S_e	1.687	1.656	1.643	1.811	0.603	0.492
	R_A^2	0.367	0.389	0.399	0.270	0.919	0.946
	Maximum VIF	1.00	1.00	1.00	45.75	1.00	1.00
	S_p	1.47	1.48	1.47	2.18	1.15[a]	1.16

[a]Minimum of tabulated values.

we note that the prediction residual error varies between a minimum of 1.15 and a maximum of 2.18. The two best subset models are, not unexpectedly, the ones involving x_1 with x_3 and x_2 with x_3. Both do better than the full least squares model, whose residual prediction error was 1.972; but all of these least squares models are much poorer than the ridge or generalized inverse models.

2.7 Model Validation

After we have developed a prediction equation, it is imperative that a measure of the accuracy of the model coefficients and predictions be obtained. One way to accomplish this is a study of the physical nature and theoretical basis of the system being studied. For example, in the acetylene data discussed earlier, negative percent conversion was predicted in some parts of the factor space by the ten-coefficient least squares quadratic model. Negative conversion is physically impossible and clearly indicates the associated model does not give an accurate description of the system which generated the data.

Another method of model validation is to collect additional data and see how well the model predicts the new data. This often is not possible. One way to simulate the collection of new data is to split the data in hand into two subsets. One subset, called the "estimation data," is used to estimate the coefficients in the model. The remaining subset, called the "prediction data," is used to measure the prediction accuracy of the model. When data are ordered with respect to time, some point in time can be used to split the data. For example, the corn yield data (Laird and Cady, 1969), to be discussed later, were collected over a four-year period. Laird and Cady used the first three years of data as the estimation data and the fourth year as the prediction data. The CADEX Procedure (Kennard and Stone, 1969) is another way of splitting the data and will be used in the analysis of the GC-ASTM Data. Snee (1977) discusses several model validation procedures including the use of DUPLEX, a general purpose data-splitting algorithm, developed by R. W. Kennard. The DUPLEX algorithm is similar to CADEX, but provides a more stringent test of the model developed from the estimation data.

3. USE OF BIASED ESTIMATION IN DATA ANALYSIS

In Section 2 we reviewed the theory underlying ridge regression and biased estimation, presented the results of a small simulation experiment, and illustrated the use of ridge regression in developing a model for the acetylene data. The remainder of this chapter will describe two data sets which illustrate the use of biased estimation in problems with a large number of variables. The reader may also wish to review the paper by Katz (1979) that describes the use of ridge regression in the development of a wheat yield model involving meteorological variables.

3.1 Interpreting the Ridge Trace

When the predictor variable correlation matrix contains several large (in absolute value) correlation coefficients, it is difficult to untangle the relationships among the predictor variables by inspection of the simple correlation coefficients. Some automatic procedures, such as stepwise, best subsets, and PRESS regression (Allen, 1971), attempt to untangle the variables by selecting some "best" subset of the predictors. However, these methods do not give insight into the structure of the factor space and the sensitivity of the results to the particular set of data at hand. In the Gorman and Toman ten-factor problem, Hoerl and Kennard (1970b) showed that a "best subsets" procedure does not necessarily reduce predictor variable correlations. The correlations may be greater than those among the original variables.

One of the big advantages of ridge regression is that a graphical display, called the "ridge trace," can help the analyst to see which coefficients are sensitive to the data. Thus, sensitivity analysis is an aim of ridge regression. The ridge trace is a plot of the value of each coefficient versus k. The trace will have one curve, or trace, per coefficient. For clarity we recommend that not more than ten coefficient traces be plotted on a given graph. It was noted earlier that the variance of a coefficient is a decreasing function of k and the bias is an increasing function of k. Thus, as k increases, the coefficient mean squared error (variance plus square bias) decreases to a minimum and then increases. The objective is to find a value of k which gives a set of coefficients with smaller mean squared error than the least squares solution. Of course, as k increases, the residual sum of squares will also increase. This behavior should not be of great concern, because the objective is not to obtain the closest possible fit to the estimation data, but to develop a "stable" set of coefficients that will do a good job of predicting future observations. By stable we mean that the coefficients are not sensitive to small changes in the estimation data. If the predictor variables are highly correlated, the coefficients will change rapidly for small values of k and gradually stabilize (change little) at larger values of k. The value of k at which the coefficients have stabilized gives the desired set of coefficients. If the predictor variables are orthogonal, then the coefficients would change very little (i.e., the coefficients are already stable) indicating the least squares solution is a good set of coefficients.

Many statisticians have expressed concern about the selection of k. It is our experience that this is not a problem in practice. As will be pointed out later in the examples, the plot of prediction standard deviation of new data versus k usually has a flat minimum;

hence, there is a range of k-values that give equivalent results from a practical point of view. We have also observed that the ridge trace is easy for scientists to interpret. They respond quickly to graphical output, and after observing two or three examples they can usually interpret the trace readily.

The k selected by means of the ridge trace is, technically speaking, a random variable. Even though that fact is not a practical concern in selecting the regression estimates, it does complicate the theory of confidence limits and hypothesis tests, due to the introduction of bias. Because of the bias, the mean squared errors are all mildly dependent upon the true coefficient vector β, which is unknown. This is a fertile area for continued research.

On the practical side, it should be noted that models with no constant term ($\beta_0 = 0$) require special consideration. If the constant term is zero because the regression must pass through the origin, then ridge models typically require a smaller value of k (often smaller than 0.01) than models with a constant term ($\beta_0 \neq 0$). If the constant term is zero because the model is a mixture model written in Scheffe canonical form (Snee, 1971), then the ridge regression must be specially modified (Marquardt and Stanley, 1982). Also, models with low R_A^2 statistics usually require larger values of k than models with high R_A^2. Increasing k indefinitely will ultimately drive all coefficients (except the constant term) to zero, but for smaller values of k it is not uncommon to see a coefficient (perhaps after an initial sign change) increase in absolute value as k increases. In this situation, we have often found that good results are obtained by using a value of k about where the coefficient passes through the maximum absolute value. This procedure was used to select the value of k for the GC-ASTM model to be discussed later. Plots of the generalized inverse coefficients can be constructed and interpreted in the same manner as the ridge trace by using the assigned rank r as the abscissa of the trace in place of k.

3.2 Laird and Cady Corn Yield Data

Our first example of a large data set is the corn yield data published by Laird and Cady (1969). Cady and Allen (1972) later used these data to illustrate the PRESS procedure. The response is the corn yield, in tons per hectare, measured at each of four applied nitrogen levels in each of 72 experimental sites. This experiment resulted in 288 data points over a four-year period.

There are 11 predictor variables:

applied nitrogen	N	soil slope	F
soil nitrogen	A	soil texture	G
previous crop	B	hail	H
excess moisture	C	blight	J
drought	D	weeds	L
rooting zone depth	E		

Four of these are measured (N, A, E, F), one is expressed as an index (D), and the remaining variables have been assigned a value on a subjective scale. The non-nitrogen variables will be referred to as "site variables." The model used by Cady and Allen to describe these data was a subset of the full quadratic containing a constant term, 11 linear terms, 18 cross-product or interaction terms, and four squared terms for a total of 33 terms (Table 9). We will note here for later reference that 13 out of the 18 interaction terms involve the two nitrogen variables (N and A).

Cady and Allen chose to divide the data into two sets. The 228 data points collected in the first three years were used to estimate the coefficients in the model. The 60 observations obtained in the fourth year are used to test the predictability of the model. These data sets will be called the "estimation data" (n = 228) and "prediction data" (n = 60), respectively.

Cady and Allen found the residual standard deviation of the fits of the full, stepwise, and PRESS models to be 0.59, 0.62, and 0.65 for the estimation data and 1.03, 0.84, 0.72 for the prediction data. The PRESS model did the best job of predicting. This might be expected since the variables in the full and stepwise models are highly correlated, with maximum VIF of 180 and 122, respectively, while the variables in the PRESS model are less correlated, with maximum VIF = 12.

What does ridge regression do with this problem? We found that the ridge trace computed from the first three years of data stabilized around k = 0.3. The prediction standard deviation for the fourth year of data is plotted in Figure 8 versus the value of k used in developing the ridge model. As shown there, the prediction standard deviation decreases as k increases, reaching a very flat minimum of 0.71 at k = 0.6. A bias of only k = 0.01 reduced the prediction standard deviation from 1.03 to 0.82. At k = 0.3, which we selected from the ridge trace, the prediction standard deviation is 0.72, which is identical to the prediction standard deviation of the PRESS model. The k = 0.3 ridge coefficients ranked in order of absolute value are shown in Table 9.

Table 9. Laird and Cady corn yield data: 33-term model coefficients.

Rank	Variable	Ridge regression (k = 0.3)	Generalized inverse (r = 9.5)	Rank	Variable	Ridge regression	Generalized inverse
1	[a]N	0.249	0.179	18	A^2	0.035	0.040
2	BN	0.188	0.166	19	JN	0.033	-0.009
3	AN	0.182	0.181	20	BL	0.031	0.000
4	[a]J	-0.119	-0.113	21	AL	0.027	0.017
5	[a]AC	-0.112	-0.096	22	A	0.026	0.042
6	AJ	-0.101	-0.108	23	G^2	0.025	0.039
7	AH	-0.099	-0.069	24	BH	-0.017	-0.078
8	[a]DN	-0.096	0.028	25	HN	0.017	0.008
9	BJ	-0.091	-0.109	26	LN	0.015	0.086
10	[a]N^2	0.091	0.175	27	[a]B^2	-0.014	0.011
11	C	-0.082	-0.095	28	D	-0.013	-0.062
12	[a]H	-0.074	-0.069	29	CN	-0.007	-0.006
13	[a]F	-0.072	-0.066	30	L	-0.005	0.006
14	BD	-0.069	-0.057	31	AD	-0.005	-0.053
15	[a]AB	0.056	0.050	32	E	0.003	0.054
16	BC	-0.050	-0.088	33	B	0.002	0.013
17	G	0.043	0.036				
				Vector length		0.239	0.236

[a]Variables selected by PRESS.

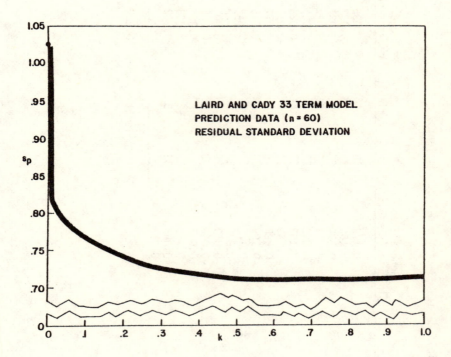

Figure 8.   Laird and Cady 33-term model: plot of the prediction data
(n = 60) residual standard deviation versus the estimation
data ridge regression bias parameter k.

The three largest coefficients are N, BN and AN, all involving applied nitrogen N. The nine terms in the PRESS model, denoted by an "*," rank from the largest to 27th out of 33; eight of them being in the top 15 and the ninth (B^2) ranking 27th. It is even more interesting that out of the top 15 ridge coefficients, nine are interactions and nine involve either applied nitrogen N or soil nitrogen A.

What is this telling us? Applied nitrogen plus soil nitrogen ("nitrogen" for short) is the dominant variable. When nitrogen is absent (i.e., A + N = 0), there cannot be any corn yield. This would suggest a zero intercept with respect to nitrogen. However, the other variables do not have a zero intercept. From this we conjectured that a multiplicative model would be most natural for these data. This is consistent with the dominance of the many interaction terms involving nitrogen in the 33-term model. Our postulated multiplicative model is

$$E(y) = (\text{applied nitrogen} + \text{soil nitrogen}) \times (\text{site variables})$$

$$= Z_1 Z_2,$$

where

$$Z_1 = (N + \beta_2 A) + \beta_{12}(N + \beta_2 A)^2,$$

and

$$Z_2 = \beta_1 + \beta_3 B + \beta_4 C + \dots + \beta_{11} L.$$

Schematically, corn yield is modeled as a product of two factors. The first is applied nitrogen plus soil nitrogen, which we denote by Z_1. The second factor contains the site variables denoted by Z_2. Factor Z_1 is a quadratic function of $N + \beta_2 A$, and Z_2 is a linear function of the non-nitrogen site variables. The β_2 coefficient is needed, because applied nitrogen N and soil nitrogen A are measured in different units. The model contains twelve coefficients, as compared to ten coefficients in the PRESS model.

The ten coefficient PRESS model fits the estimation and prediction data with residual standard deviations of 0.65 and 0.72 respectively (see Table 10). The twelve-coefficient multiplicative model, fitted by nonlinear least squares (Marquardt, 1963), had residual standard deviations of 0.72 and 0.75 for the estimation and prediction data, which are similar to those of the PRESS model.

An examination of the coefficient confidence limits suggested that the three coefficients corresponding to previous crop B, root depth E, and weeds L were not significant. When these variables were deleted, the resulting nine-coefficient multiplicative model fit the

Table 10. Laird and Cady corn yield data: fit of PRESS and multiplicative models.

| Model | Number of coefficients | Residual standard deviation | |
		Estimation (n = 228)	Prediction (n = 60)
PRESS	10	0.65	0.72
Multiplicative	12	0.72	0.75
Multiplicative	9	0.73	0.64

estimation and prediction data with residual standard deviations of
0.73 and 0.64 respectively, giving a somewhat better fit to the prediction data than the PRESS model (see Table 10).

There are other differences between the PRESS and multiplicative
models which should be noted:

(a)   The PRESS model contains a constant term; the multiplica-
      tive model does not.

(b)   Previous crop B is included in the PRESS model, but not in
      the multiplicative.

(c)   Soil texture G is included in the multiplicative model, but
      not in the PRESS model.

To summarize this problem, by keeping all the terms in the model
and reducing the variable correlations with ridge regression, we were
able to:

(a)   Obtain a model that predicts well; and

(b)   Learn more about the roles of all the variables in the
      model.

In this case, the ridge regression results suggested a nonlinear
alternative model. While this may not be the ultimate model, it is
consistent with the physical background of the problem as described
in the Laird and Cady paper, and gives the scientist a different way
to think about the mechanism under study.

3.3   GC-ASTM Model

The second example concerns the relationship between the ASTM
and gas chromatograph, or GC for short, distillation of a gasoline
sample. One of the properties that determines the quality of a
gasoline is volatility as measured by the percent of the blend evaporated at various temperatures (°F). The standard method of
measuring volatility is an ASTM distillation, in which the gasoline
sample is heated and the vapors pass through an ice bath and con-
dense. The cumulative percent evaporated at various temperatures is
recorded. In the ASTM distillation, some of the higher boiling components "hold back" the lower boiling components. The GC distillation is much more accurate and each component "comes off" at its true
boiling point.

ASTM and GC distillation curves are not identical, and gasoline
volatility specifications are written in terms of ASTM. In order for
a refinery to use the GC for on-line control of volatility, a model

is needed to predict the ASTM distillation of a blend from a GC distillation of the blend. This example is typical of those situations where a property of a material is measured by a series of points which form a curve. The use of too many points results in redundant information. In this example, the points on the GC curve were selected by the engineer to give a proper description of the curve. The points on the ASTM curve to be predicted corresponded to specifications of various gasolines.

The GC temperature range was divided into fifteen cuts: 0-15, 15-40, ..., 414-478°F (see Table 9). The fifteen predictor variables in the model, x_1, x_2, ..., x_{15}, are the volume fraction of the blend evaporated in each of the cuts. The responses to be predicted, y_1, y_2, ..., y_{14}, are the cumulative percent of the blend evaporated at each of the 14 ASTM temperatures. While a model was developed for each of the 14 ASTM temperatures, we will concentrate our attention on the three most important specifications: y_4 = ASTM 158, y_6 = ASTM 212, and y_{10} = ASTM 302.

The postulated model is

$$E(y) = \beta_1 x_1 + \beta_2 x_2 + \dots + \beta_{15} x_{15},$$

where y is the cumulative percent evaporated at a given ASTM temperature and x_i is the fraction evaporated in the ith GC cut. The constant term β_0 has been deleted because the x_i's sum to 1.0, and would produce a singular X'X matrix if β_0 were included (see Section 3.4).

The main uses of the model would be prediction of gasoline blend ASTM distillations, for which the prediction standard deviation had to be less than or equal to 1.5%, and as input to linear programming calculations to determine optimum volatility blending procedures. Hence, it was imperative that the ASTM predictions also be responsive to changes to the GC curve in any temperature range, and that the estimated coefficients in the model be "realistic" in light of the available engineering knowledge. Part of the prior history on this problem was that coefficients developed by least squares and stepwise regression were unacceptable from a physical viewpoint.

GC and ASTM distillation data were available on 59 blends. It was felt that it would be advantageous to have an independent estimate of the model prediction standard deviation. We used the CADEX algorithm (Kennard and Stone, 1969) to split the data into two sets, 29 estimation blends and 30 prediction blends. (While we do not claim that this is necessarily the best way to split the data, we are describing the analysis the way it was actually conducted.) As in

the Laird and Cady corn yield example, the coefficients were estimated from the 29 estimation blends. The 30 prediction blends were
used to obtain a measure of the model prediction standard deviation.

The ridge trace for y_4 = ASTM 158 is shown in Figure 9 where
the x-axis is k x 10^{-1}. The traces for the first 10 coefficients are
shown in Figure 9a and the traces for the last five coefficients are
shown in Figure 9b. The system stabilizes around k = 0.005 or 0.01.
We decided to use the coefficients at k = 0.006. Before studying
these coefficients further, let us look at the plot of the prediction
standard deviation of the 30 prediction blends versus k shown in
Figure 10.

For y_4 = ASTM 158, the prediction standard deviation has a
value of 1.28 at the least squares solution, decreases as k
increases, and reaches a minimum near k = 0.006, the value of k we
chose from the ridge trace. The ASTM 212 curve follows a similar
pattern, reaching a minimum around k = 0.003. At the 302 and 375
temperatures, the experimental error is smaller. The prediction
standard deviation increases as k increases, although the ASTM 375
curve is flat until k = 0.006.

We now turn our attention to the coefficients in the ASTM 158
model shown in Table 11. First, we note the standard error of prediction at the bottom of the columns. As noted previously, the ridge
regression model is a better predictor. It has a standard error of
1.01%, compared to a standard error of 1.28% for the least squares
model. Again, these two numbers were computed from the 30 blends not
used in computing the coefficients. An examination of the models
reveals that the coefficients in the least squares model are in
general larger than the coefficients in the ridge regression model.
In the least squares model, the largest coefficients are around 200
in absolute value (in cuts 1, 13 and 14), while the largest coefficients in the ridge regression model are around 100 in absolute
value (in cuts 1 and 3). In addition, the least squares coefficients
are not well behaved with respect to sign. These characteristics can
be seen easily when the models are compared graphically.

In Figure 11, the coefficients are plotted versus GC cut temperature. The top graph shows the least squares coefficients, and the
second graph shows the ridge regression coefficients. We note that
the ridge coefficients 5, 14 and 15 have changed sign compared to
least squares, and coefficients 1, 4, 9 and 13 are considerably
smaller.

Previous knowledge indicated that all the coefficients at the
higher temperatures should be negative. To gain greater insight into
this problem, we designed a theoretical distillation experiment centered around the 59 blends. The amounts of each of the 15 GC cuts
were varied according to a pseudocomponent simplex design for mixtures (Snee, 1971) comprised of 15 pure component and 105 binary

88

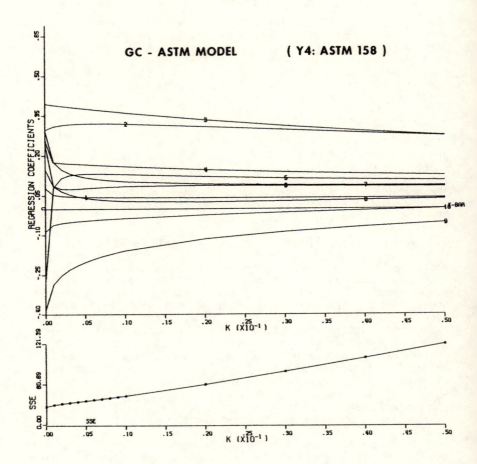

Figure 9a. Ridge trace for coefficients 1-10 in the GC-ASTM model for
y4 = ASTM 158.

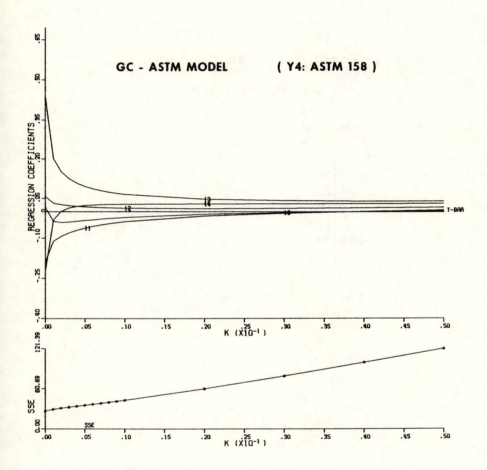

Figure 9b. Ridge trace for coefficients 11-15 in the GC-ASTM model
for y_4 = ASTM 158.

GC — ASTM MODEL
PREDICTION STANDARD DEVIATION
VERSUS k (n=30)

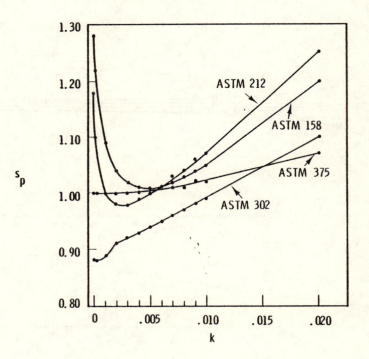

Figure 10. GC-ASTM model: plot of the prediction data (n = 30) residual standard deviation versus the estimation data ridge regression bias parameter k.

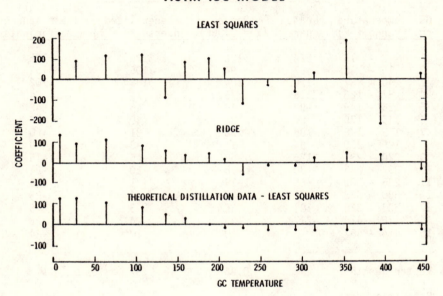

Figure 11. ASTM 158 model: plot of the coefficients in the least squares, ridge regression (k = 0.006), and theoretical distillation data models versus GC temperature.

Table 11. Model for 158° ASTM data.

GC cut	Temperature range (°F)	Coefficients estimated from data			Theoretical distillation least squares coefficients
		Least squares	Ridge (k = 0.006)	Generalized inverse (r = 7)	
1	0 - 15	224	126	120	125
2	15 - 40	87	94	105	132
3	40 - 87	110	104	109	102
4	87 - 126	116	75	74	84
5	126 - 145	-92	45	46	42
6	145 - 175	80	26	45	20
7	175 - 198	96	38	21	0
8	198 - 220	54	14	-8	-9
9	220 - 237	-125	-62	-45	-15
10	237 - 285	-30	-15	-20	-21
11	285 - 300	-65	-19	1	-29
12	300 - 333	21	6	-4	-25
13	333 - 376	181	36	5	-25
14	376 - 414	-217	25	24	-25
15	414 - 487	22	-40	-10	-24
Vector length			0.36	0.37	
Prediction data standard deviation		1.28	1.01	0.96	

blends. The distillations for these 120 blends were calculated using Raoult's Law with activity coefficients of unity and atmospheric pressure. The 15 GC cuts were treated as pure hydrocarbons, having true boiling points at the midpoint of the cut. This theoretical distillation is similar to the ASTM distillation and can provide corollary information concerning the relationship between the model coefficients and GC temperature.

The models developed for the theoretical distillation data confirmed our previous theories. At the bottom of Figure 11, we see that the theoretical distillation coefficients decrease in size with increasing temperature and finally go negative at the higher temperatures. It is immediately obvious from this graph that the ridge coefficients bear a closer resemblance to the theoretical distillation coefficients than do the least squares coefficients. The ridge regression coefficients and theoretical distillation coefficients follow a smooth pattern as the GC temperature increases; however, the least squares coefficients do not follow this nice relationship. Ridge regression gave equally meaningful coefficients in the models at the other ASTM temperatures as evidenced by the coefficients in the ASTM 212, 302, and 375°F models shown in Figure 12. As the scientific background of the problem indicated, the number of large positive coefficients in the model increases as the ASTM temperature increases.

We can summarize this example by saying that the project goals were met and on-line process control of volatility went into operation on schedule. Both the ridge model and the theoretical distillation model worked in practice, whereas the least-squares and stepwise regression models had not. In a subsequent paper, Snee (1977) reanalyzed these data using Kennard's DUPLEX algorithm to split the data. The results of this analysis were in general agreement with those obtained using the CADEX split. A minor difference was that the plots of prediction standard deviation versus the ridge bias k were better behaved (showed definite minima at k $\simeq$ 0.006 for ASTM 302 and ASTM 375) than those for the CADEX data shown in Figure 10.

3.4 Mixture Models

The GC-ASTM problem as we have formulated it is really a mixture model expressed in Scheffe canonical form, due to the fact that the x_i's add to 1.0 (Snee, 1971). Marquardt and Stanley (1982) developed a special procedure for properly structuring ridge regression in view of the null hypothesis implied by the mixture model (Marquardt and Snee, 1974). In this GC-ASTM example, the Marquardt and Stanley procedure makes only a modest improvement in the prediction standard error compared to the results reported here. Indeed, ordinary ridge regression often exhibits remarkable robustness to such misspecification of the implied null hypothesis. Nevertheless,

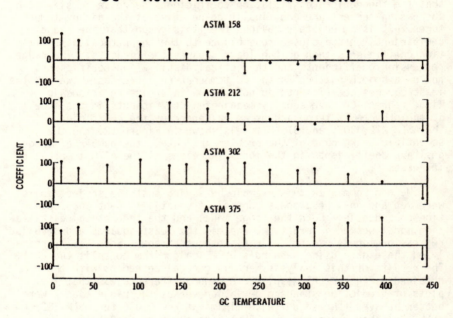

Figure 12. GC-ASTM data: plot of the coefficients in the ridge regression (k = 0.006) models for responses ASTM 158, ASTM 212, ASTM 302, and ASTM 375 versus GC temperature.

researchers who encounter mixture problems should refer to the Marquardt and Stanley procedure for correct application of ridge regression.

3.5 Generalized Inverse Results

In our discussion of the Laird and Cady corn yield data and the GC-ASTM distillation model, we have focused on the ridge estimators. The corresponding results for the generalized inverse estimators are shown simultaneously in Tables 9 and 11. The generalized inverse results shown correspond to a value of r for which the regression vector length is approximately equal to the length for the selected ridge bias k. Detailed examination of the coefficients shows that the ridge and generalized inverse coefficients are remarkable similar. Table 9 shows that the generalized inverse model achieves a reduction of the prediction standard deviation comparable to the ridge model. Table 12 shows the correlation coefficients between ridge regression coefficients and generalized inverse coefficients (chosen to have the same approximate vector length) for eight sets of data, including the three sets discussed in this paper. The correlation coefficients are all very high.

3.6 Computing Ridge Regression and Generalized Inverse Coefficients

One of the advantages of the ridge and generalized inverse estimators is ease of computation. The X'X and X'Y matrices are computed once and scaled to form the correlation matrix. For ridge regression, 10-30 inversions of the X'X + kI matrices, one for each value of k, are usually sufficient to determine where the ridge trace stabilizes. The generalized inverse coefficients are computed from the eigenvalues and eigenvectors of the correlation matrix. Numerical analysis is not a problem, nor is it a problem in the case of the ridge estimates where the addition of k to the diagonal of the correlation matrix reduces the nonorthogonality and thereby improves the numerical analysis. Furthermore, a separate matrix inversion is calculated for each value of k and roundoff errors cannot accumulate as in some stepwise algorithms. With the biased estimators, the same inverse matrix, for a given k or rank r, can be used to calculate the coefficients in the models for all responses. The best subset and stepwise algorithms require a separate computer run for each response.

Some researchers may choose to implement ridge regression by revising a regression program with which they are familiar to perform the necessary calculations. Alternatively, one may wish to use one of the computer program packages in the public domain such as BMDP (Dixon and Brown, 1979; Hill, 1975) and SAS (Helwig and Council, 1979) or the ridge regression programs developed by Enslein (1974), Hoerl (1980), and Hui and Jagpal (1979).

Table 12. Correlation between ridge and generalized inverse regression coefficients.

Data set	Number of coefficients	Ridge		Generalized inverse		
		Bias (k)	Vector length ($\beta'\beta$)	Bias (r)	Vector length ($\beta'\beta$)	Correlation coefficient[a]
Acetylene data	9	0.05	0.524	3.8	0.522	0.98
Gorman and Toman (1966)	10	0.26	0.373	6.6	0.384	0.92
Corn yield data	33	0.30	0.239	9.5	0.236	0.89
GC – ASTM 158	15	0.006	0.358	7.0	0.374	0.96
Steam data[b]	8	0.30	0.329	3.5	0.318	0.96
Rocket engine[b]	13	0.10	0.817	9.0	0.811	0.90
McDonald and Schwing (1973)	15	0.18	0.376	8.5	0.384	0.91
Liver cirrhosis[c]	4	0.30	0.239	1.0	0.262	0.98

[a] Linear correlation coefficient.

[b] Draper and Smith (1981).

[c] Brownlee (1965).

ACKNOWLEDGMENTS

We are grateful to R. W. Kennard for suggestions concerning the presentation of this material. The theoretical distillation results were generated from a model developed by J. B. Jones. M. H. Sarner and R. W. McGill did other computer programming required for this study. Appreciation is expressed to F. B. Cady for sending us the raw data associated with the corn yield example. This exposition was adapted from the version which was published in the February 1975 issue of The American Statistician. The American Statistical Association holds the copyright to the original publication and we thank them for permission to publish this expanded version.

REFERENCES

Allen, D. M., 1971: Mean square error of prediction as a criterion for selecting variables. Technometrics, 13, 469-475.

Allen, D. M., 1974: The relationship between variable selection and data augmentation and a method for prediction. Technometrics, 16, 125-127.

Brownlee, K. A., 1965: Statistical Methodology in Science and Engineering, Second Edition. New York, Wiley.

Cady, F. B., and D. M. Allen, 1972: Combining experiments to predict future yield data. Agronomy Journal, 64, 211-214.

Chatterjee, S, and B. Price, 1977: Regression Analysis by Example. New York, Wiley-Interscience.

Daniel, C., 1976: Applications of Statistics to Industrial Experimentation. New York, Wiley-Interscience.

Daniel, C., and F. S. Wood, 1980: Fitting Equations to Data, Second Edition. New York, Wiley-Interscience.

Dixon, W. J., and M. B. Brown (Editors), 1979: BMDP-79, Biomedical Computer Programs, P-Series. Berkeley, University of California Press.

Draper, N. R., and H. Smith, 1981: Applied Regression Analysis, Second Edition. New York, Wiley.

Draper, N. R., and C. R. Van Nostrand, 1979: Ridge regression and James-Stein estimation: review and comments. Technometrics, 21, 451-466.

Enslein, K., 1974: Ridge regression program. Genesee Statistics Newsletter, V. 2, No. 1. Rochester, N.Y.

Ezekiel, M., and K. A. Fox, 1959: Methods of Correlation and Regression Analysis. New York, Wiley.

Gorman, J. W., and R. J. Toman, 1966: Selection of variables for fitting equations to data. Technometrics, 8, 27-51.

Helwig, J. T., and K. A. Council (Editors), 1979: Statistical analysis system. Cary, N.C., SAS Institute, Inc.

Hill, A. M., 1975: Ridge regression using BMDP2R. BMD Communications No. 3. Los Angeles, Health Sciences Computing Facility, University of California, pp. 1-2.

Himmelblau, D. M., 1970: Process Analysis by Statistical Methods. New York, Wiley.

Hocking, R. R., and L. R. LaMotte, 1973: Using the select program for choosing subset regressions. Proceedings of the University of Kentucky Conference on Regression with a Large Number of Predictor Variables, Lexington, Ky.

Hoerl, A. E., 1959: Optimum solution of many variable equations. Chemical Engineering Progress, 55, 69 ff.

Hoerl, A. E., 1962: Application of ridge analysis to regression problems. Chemical Engineering Progress, 58, 54-59.

Hoerl, A. E., 1980: A full ridge regression program. Newark, Department of Mathematical Sciences, University of Delaware.

Hoerl, A. E., and R. W. Kennard, 1970a: Ridge regression: biased estimation for nonorthogonal problems. Technometrics, 12, 55-67.

Hoerl, A. E., and R. W. Kennard, 1970b: Ridge regression: applications to nonorthogonal problems. Technometrics, 12, 69-82.

Hoerl, A. E., and R. W. Kennard, 1981: Ridge regression - 1980: advances, algorithms, and applications. American Journal of Mathematical and Management Sciences, 1, 5-83.

Hui, B. S., and H. S. Jagpal, 1979: RIDGE: an integrated ridge regression program. Journal of Market Research, 16, 571-572.

Katz, R. W., 1979: Sensitivity analysis of statistical crop-weather models. Agricultural Meteorology, 20, 291-300.

Kennard, R. W., and L. Stone, 1969: Computer aided design of experiments. Technometrics, 11, 137-148.

Kunugi, T., T. Tamura, and T. Naito, 1961: New acetylene process uses hydrogen dilution. Chemical Engineering Progress, 57, 43-49.

Laird, R. J., and F. B. Cady, 1969: Combined analysis of yield data from fertilizer experiments. Agronomy Journal, 61, 829-834.

Lindley, D. V., and A. F. M. Smith, 1972: Bayes estimates for the linear model (with discussion). Journal of the Royal Statistical Society, Series B, 34, 1-41.

Mallows, C. L., 1973: Some comments on C_p. Technometrics, 15, 661-675.

Marquardt, D. W., 1963: An algorithm for least-squares estimation of nonlinear parameters. Journal of the Society for Industrial and Applied Mathematics, 11, 431-441.

Marquardt, D. W., 1970: Generalized inverse, ridge regression, biased linear estimation and nonlinear estimation. Technometrics, 12, 591-612.

Marquardt, D. W., 1974: Discussion of "The fitting of power series, meaning polynomials, illustrated on band-spectroscopic data" by A. E. Beaton and J. W. Tukey. Technometrics, 16, 189-192.

Marquardt, D. W., 1980: You should standardize the predictor variables in your regression models. Discussion of "A critique of some ridge regression methods" by G. Smith and F. Campbell. Journal of the American Statistical Association, 75, 87-91.

Marquardt, D. W., and R. D. Snee, 1974: Test statistics for mixture models. Technometrics, 16, 533-537.

Marquardt, D. W., and R. M. Stanley, 1982: Biased estimators for mixture models and smooth regression: examples of driving the candidate model estimate toward the null model estimate. Submitted for publication.

Mayer, L. S., and T. A. Willke, 1973: On biased estimation in linear models. Technometrics, 15, 497-508.

McDonald, G. C., and R. C. Schwing, 1973: Instabilities of regression estimates relating air pollution to mortality. Technometrics, 15, 463-481.

Montgomery, D. C., and E. A. Peck, 1982: Introduction to Linear Regression Analysis. New York, Wiley-Interscience.

Mosteller, F., and J. W. Tukey, 1977: Data Analysis and Regression. Reading, Mass., Addison-Wesley.

Myers, R. H., 1971: Response Surface Methodology. Boston, Allyn and Bacon.

Smith, G., and F. Campbell, 1980: A critique of some ridge regression methods (with discussion). Journal of the American Statistical Association, 75, 74-103.

Snee, R. D., 1971: Design and analysis of mixture experiments. Journal of Quality Technology, 3, 159-169.

Snee, R. D., 1973: Some aspects of nonorthogonal data analysis, Part I. Developing prediction equations. Journal of Quality Technology, 5, 67-79.

Snee, R. D., 1977: Validation of regression models: methods and examples. Technometrics, 19, 415-428.

Theobald, C. M., 1974: Generalizations of mean square error applied to ridge regression. Journal of the Royal Statistical Society, Series B, 36, 103-106.

Webster, J. T., R. F. Gunst, and R. L. Mason, 1974: Latent root regression analysis. Technometrics, 16, 513-522.

3
Exploratory Multivariate Analysis of a Single Batch of Data

K. Ruben Gabriel

1. INTRODUCTION

Chapters 3 and 4 present a very idiosyncratic view of multivariate analysis and reflect what the author has found useful in statistical practice. They stress exploration, virtually ignore tests of significance, and emphasize a particular graphical technique developed by the author - the biplot display. The author hopes that this material will help readers obtain a better grasp of the structure of multivariate data and the fundamentals of multivariate analysis. He also hopes that the chapters will be useful to readers who wish to pursue multivariate analysis in its more classical form, to which references are given.

The methods employed in these chapters lean heavily on least squares, which is most appropriate for normally distributed data and data that have been transformed to near normality after outliers have been omitted.

2. ONE BATCH OF MULTIVARIATE DATA AND THEIR DESCRIPTIVE STATISTICS

2.1 A Multivariate Data Matrix

The essence of multivariablity is that several variables are observed on each unit. Thus, if the units are days, one might observe maximum and minimum temperatures on each day, as well as surface barometric pressure at 6 a.m., 12 noon, 6 p.m., and midnight. These data would be six-variate observations. It is convenient to think of the data as a nxm matrix Z in which row $\underline{z}_i'$ (i = 1, ..., n) contains the m-variate observations for unit i (out of n units) and column $\underline{z}_{(v)}$ (v = 1, ..., m) contains all n units' observations on the vth variable. Thus, $z_{i,v}$ is unit i's observation on variable v. In this example, element $z_{2,3}$ would be the 6 a.m. pressure on day 2, $\underline{z}_2'$ would be the six-variate observations on day 2 and $\underline{z}_{(3)}$ the n days' observations on the third variable (6 a.m. pressure).

We have adopted the convention of denoting a matrix by a Latin capital letter and any of its elements by the corresponding lower case letter with two indices. The latter indicate, respectively, the row and column in which the element is located. We have denoted both rows and columns of the matrix by the lower case letter underlined and with a single index. If the index is in parentheses, a column is denoted, whereas if no parentheses are shown, a row is indicated.

For a detailed illustration, consider the data of Table 1; namely, mean monthly temperatures for 20 stations during six months of the year 1951. Here, n = 20, m = 6, and $z_{3,1}$ is the third station's mean January temperature, whereas $z_{1,3}$ is the first station's mean May temperature. The location of the 20 stations is shown on the map in Figure 1.

A first glance at such a data matrix is apt to be somewhat confusing. Some idea of the general pattern can be obtained from the means and standard deviations shown at the bottom of Table 1. The temperature averages are seen to be much the same in all six of the months (evidently because the stations are spread out on both sides of the equator). However, there is considerable variation from station to station, as evidenced by the large standard deviations. The latter are around 50 (i.e., 5°C) for each month, though a bit less for March and May.

2.2 Summary Statistics for the Variables' Configuration

The common statistics for the batch of n units are readily obtained from the matrix Z. Thus, the means $\overline{z}_{(1)}$, ..., $\overline{z}_{(m)}$ of all m variables are arrayed in the vector

$$\overline{z}' = (1/n)\ \underline{1}'Z, \tag{1}$$

where $\underline{1}$ is a vector of n ones. Deviations from each variable's mean are given in the matrix

$$Y = Z - \underline{1}\ \overline{z}', \tag{2}$$

which has the typical element

$$y_{i,v} = z_{i,v} - \overline{z}_{(v)}. \tag{3}$$

The means for the temperature illustration were noted in Table 1. An example of a deviation $y_{i,v}$ from the mean is $y_{1,3} = -7.9 = 224.0 - 231.9 = z_{1,3} - \overline{z}_{(3)}$.

Table 1. Mean monthly temperatures (10 x °C) at 20 North and South
American stations in 1951.

Month
v

Station i	1 (Jan)	2 (Mar)	3 (May)	4 (Jul)	5 (Sep)	6 (Nov)
1	260	251	224	196	215	240
2	259	268	249	226	275	268
3	325	226	193	166	184	234
4	204	188	149	129	157	205
5	192	172	156	147	145	172
6	269	275	254	233	247	263
7	273	278	263	260	273	277
8	255	256	253	250	267	271
9	248	251	247	242	254	260
10	259	256	259	236	268	261
11	124	136	132	130	136	127
12	241	240	216	191	186	191
13	247	256	253	252	266	265
14	261	281	277	250	276	278
15	263	269	271	271	269	268
16	242	249	273	284	280	270
17	192	218	266	278	273	215
18	198	204	237	278	284	254
19	132	174	238	283	272	146
20	68	125	228	298	255	106
Means	225.60	228.65	231.90	230.00	239.10	228.55
Standard deviations	59.08	45.93	41.56	51.42	48.04	52.10

104

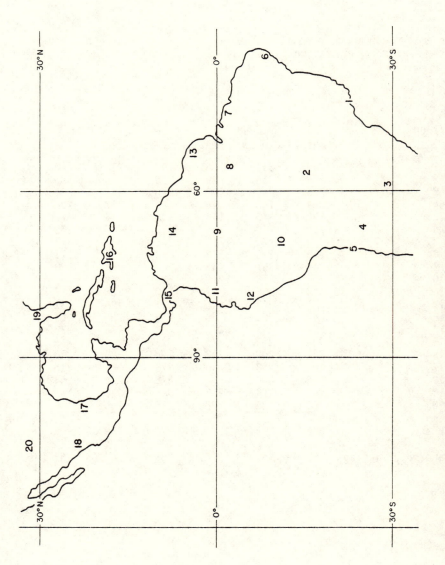

Figure 1. Location of stations for which temperatures are given in Table 1.

From the y's one may compute the <u>variance</u> <u>matrix</u> (often referred to as the <u>variance-covariance</u> <u>matrix</u>)

$$S = \frac{1}{n} Y'Y, \tag{4}$$

the <u>standard</u> <u>deviations</u>

$$s_v = (s_{v,v})^{1/2} \quad (v = 1, \ldots, m), \tag{5}$$

and, defining the diagonal matrix with elements $d_{v,v} = s_v$ as D_s, the <u>correlation</u> <u>matrix</u>

$$R = D^{-1}SD^{-1}, \tag{6}$$

whose elements are the correlations $r_{v,v'}$ between variables.

For the temperature illustration, the standard deviations were shown in Table 1, and the variance and correlation matrices are given in Table 2 and 3, respectively.

The configuration of correlations in an mxm matrix R may not be easy to grasp at first, especially if m is 10 or more. In the present illustration with m = 6, one may begin to study Table 3 by concentrating on the highest correlations. One notices two distinct sheaves of months. Specifically, November, January, and March are highly intercorrelated and so are, even more strongly, May, July, and September. The correlations between months not belonging to the same sheaf (or season) are seen to be much lower.

Such a perusal of correlations is not always easy, especially if the variables do not group neatly into highly intercorrelated sheaves. It is sometimes helpful also to consider the inverse of the variance matrix, where

$$S^{-1} = (\frac{1}{n} Y'Y)^{-1}, \tag{7}$$

because its elements $s^{v,v'}$ have the following interpretation in terms of the multiple regression coefficients of the vth variable on <u>all</u> <u>other</u> variables. Take the vth row of S^{-1}, divide each off-diagonal element by the diagonal element and change sign; then

$$b_{v,v'} = -s^{v,v'}/s^{v,v} \tag{8}$$

is the coefficient of variable v' in the regression of variable v on the m-1 other variables. Furthermore, using diagonal terms from both S and S^{-1},

$$r^2 = 1 - 1/(s_{v,v}s^{v,v}) \tag{9}$$

gives the square of the multiple correlation of variable v on all m-1 other variables.

Table 2. Variance matrix of mean temperatures ($^{\circ}C^2$).

Month v	1 (Jan)	2 (Mar)	3 (May)	Month v' 4 (Jul)	5 (Sep)	6 (Nov)
1 (Jan)	3490.7	2376.4	895.86	-277.15	463.09	2635.4
2 (Mar)	2376.4	2109.5	1324.7	612.2	1117.8	2221.4
3 (May)	895.86	1324.7	1726.8	1843.3	1873.6	1388.2
4 (Jul)	-277.15	612.2	1843.3	2644.5	2300.4	714.15
5 (Sep)	463.09	1117.8	1873.6	2300.4	2307.5	1338.1
6 (Nov)	2635.4	2221.4	1388.2	714.15	1338.1	2714.1

Table 3. Correlation matrix of mean temperatures.

Month v	1 (Jan)	2 (Mar)	3 (May)	Month v' 4 (Jul)	5 (Sep)	6 (Nov)
1 (Jan)	1	0.87573	0.36489	-0.091219	0.16317	0.8562
2 (Mar)	0.87573	1	0.69405	0.2592	0.50664	0.92836
3 (May)	0.36489	0.69405	1	0.86259	0.93859	0.64123
4 (Jul)	-0.091219	0.2592	0.86259	1	0.93124	0.26656
5 (Sep)	0.16317	0.50664	0.93859	0.93124	1	0.53471
6 (Nov)	0.8562	0.92836	0.64123	0.26656	0.53471	1

For the temperature illustration, the inverse S^{-1} and the multiple correlation and regression coefficients are given in Tables 4 and 5, respectively. Each month's temperature is seen to be quite highly correlated with the temperatures of all other months. Of course, the multiple correlations are all higher than the correlations with individual variables; that is,

$$r_v \geq r_{v,v'} \quad (v' \neq v). \tag{10}$$

It is interesting to see the pattern of regression coefficients. Each month's coefficients with adjacent months are positive, but with months about half a year away (2 or 3 variables away in the circular order ...123456123...) the coefficients are negative. This result makes good sense in terms of consistent seasonal patterns.

2.3 Distances in the Units' Scatter

The above statistics describe the <u>configuration of the variables</u> (monthly temperatures) for the entire batch of units - no attention being paid to the individual units (stations). If one is interested in the individual units, in particular their similarities and differences, a description of the <u>scatter of the units</u> is needed.

This scatter may be described by the units' nxn standardized inner-products matrix

$$U = YS^{-1}Y', \tag{11}$$

which can be interpreted in terms of <u>standardized distances</u> in the following manner. The diagonal elements of U,

$$u_{i,i} = \underline{y}_i' \, S^{-1} \, \underline{y}_i,$$
$$= (\underline{z}_i - \overline{\underline{z}})' S^{-1} (\underline{z}_i - \overline{\underline{z}}), \tag{12}$$

are squares of standardized distances of units i from the <u>centroid</u>; that is, from the multivariate mean of the batch. The tetrad differences

$$d_{i,e} = (u_{i,i} - u_{i,e} - u_{e,i} + u_{e,e}),$$

$$= (\underline{y}_i - \underline{y}_e)' S^{-1} (\underline{y}_i - \underline{y}_e), \tag{13}$$

$$= (\underline{z}_i - \underline{z}_e)' S^{-1} (\underline{z}_i - \underline{z}_e),$$

are squares of standardized distances between units i and e. Such distances should be understood as measuring statistical differences simultaneously on all m variables. They are equal to zero if and

Table 4. Inverse of variance matrix S for mean temperature data.

Variable

Variable	1	2	3	4	5	6
1	0.003122	-0.0019469	-0.00089546	0.000063023	0.0021828	-0.0020728
2	-0.0019469	0.019515	-0.024644	0.011706	0.002688	-0.0058826
3	-0.00089546	-0.024644	0.042981	-0.019968	-0.0075252	0.0080195
4	0.000063023	0.011706	-0.019968	0.015857	-0.0044671	-0.0013988
5	0.0021828	0.002688	-0.0075252	-0.0044671	0.012391	-0.005404
6	-0.0020728	-0.0058826	0.0080195	-0.0013988	-0.005404	0.0061264

Table 5. Multiple correlation and regressions of each mean temperature variable on all other variables (correlations on diagonal, regression coefficients on off-diagonal).

Onto variable

Regression of Variable	1	2	3	4	5	6
1	0.9530	0.6236	0.2868	-0.0202	-0.6992	0.6639
2	0.0998	0.9878	1.2628	-0.5998	-0.1377	0.3014
3	0.0208	0.5734	0.9932	0.4646	0.1751	-0.1866
4	-0.0040	-0.7382	1.2593	0.9880	0.2817	0.0882
5	-0.1762	-0.2169	0.6073	0.3605	0.9824	0.4361
6	0.3383	0.9602	-1.3090	0.2283	0.8821	0.9695

only if the units compared have equal observations on all variables, and they increase when the differences in any one or more variables becomes larger.

The matrix of standardized distances $d_{i,e}^{1/2}$ between each pair of stations on the six month's temperature data is given in Table 6. All distances are positive except the "self-distances" in the diagonal which are identically zero. The matrix is symmetric; that is, the i-to-e distance $d_{i,e}^{1/2}$ equals the e-to-i distance $d_{e,i}^{1/2}$. Small distances, such as $d_{8,9} = 0.6$, indicate that stations 8 and 9 have very similar mean monthly temperatures; whereas large distances, such as $d_{2,3} = 5.3$, show that very considerable differences in mean monthly temperatures exist between stations 2 and 3. The reader can verify these statements by examining Table 1.

It is difficult to inspect a table of distances of this magnitude (not to speak of distance matrices for a hundred or more units). We shall therefore require methods of disentangling the pattern of distances of a scatter of units and of making some sense of such a distance matrix. These topics will be discussed below in Section 5.

2.4 Some Further Remarks on Standardized Statistical Distances

These distances are based on differences standardized by standard deviations. Thus, the same difference on a variable v with large variability s_v will contribute less to these distances than if it occurred on a variable v' with lesser variability $s_{v'}$ ($< s_v$). Thus, on the vth variable alone the standardized difference is

$$d_{i,e}^{1/2}(v) = |y_{i,v} - y_{e,v}|/s_v. \tag{14}$$

Similarly, for the linear combination of variables (LCV) with coefficients $\underline{a} = (a_1, \ldots, a_m)'$ the standardized difference is

$$d_{i,e}^{1/2}(\underline{a}) = |\underline{a}'\underline{y}_i - \underline{a}'\underline{y}_e|/(\underline{a}'S\underline{a})^{1/2}, \tag{15}$$

since the variance of that LCV is $\underline{a}'S\underline{a}$. A generalized i-to-e difference, or distance, for all variables and LCVs together can then reasonably be defined as the maximum of all such LCVs' differences. But it can be proved that this maximum over all LCVs (i.e., all possible sets of coefficients $\underline{a}$) is precisely the square root of $d_{i,e}$ in (13). Thus, the proposed generalized i-to-e distance of (13) can be regarded as a maximum difference over all variables and LCVs.

Table 6. Inter-station standardized distances for mean temperature data.

Station

Station	1	2	3	4	5	6	7	8	9	10	11	12	13	14	15	16	17	18	19	20
1	0.0	3.0	4.2	2.7	2.3	1.1	2.4	1.7	1.4	2.6	2.9	2.3	1.7	2.1	2.1	2.8	2.8	4.0	3.7	3.9
2	3.0	0.0	5.3	3.6	4.7	3.3	3.4	3.0	3.3	2.9	4.1	4.7	3.0	3.3	4.2	4.7	4.3	4.4	3.4	5.1
3	4.2	5.3	0.0	5.1	4.4	4.9	5.0	4.4	4.5	4.5	5.6	5.0	4.7	5.4	4.7	4.9	5.0	5.2	5.2	5.6
4	2.7	3.6	5.1	0.0	2.2	3.2	3.7	2.7	2.7	3.8	1.8	4.3	2.8	3.8	3.8	3.9	4.2	3.5	4.6	4.7
5	2.3	4.7	4.4	2.2	0.0	2.9	3.7	2.8	2.5	3.9	2.1	3.3	2.9	3.7	3.0	3.0	3.4	4.0	4.6	3.8
6	1.1	3.3	4.9	3.2	2.9	0.0	1.7	1.7	1.3	3.3	3.6	2.2	1.4	2.4	1.4	2.7	3.1	4.0	3.8	4.0
7	2.4	3.4	5.0	3.7	3.7	1.7	0.0	1.9	2.0	4.3	4.3	3.1	1.5	3.7	1.8	3.3	4.3	3.4	3.5	4.4
8	1.7	3.0	4.4	2.7	2.8	1.7	1.9	0.0	0.6	2.8	3.5	3.6	0.6	2.4	1.7	2.0	2.8	2.5	3.6	3.7
9	1.4	3.3	4.5	2.7	2.5	1.3	2.0	0.6	0.0	2.8	3.3	3.2	0.6	2.3	1.4	1.8	2.6	2.9	3.7	3.6
10	2.6	2.9	4.5	3.8	3.9	3.3	4.3	2.8	2.8	0.0	4.0	4.5	3.1	1.8	3.8	3.3	2.1	4.5	4.2	4.4
11	2.9	4.1	5.6	1.8	2.1	3.6	4.3	3.5	3.3	4.0	0.0	4.0	3.4	3.8	4.2	4.3	3.9	4.4	4.2	3.9
12	2.3	4.7	5.0	4.3	3.3	2.2	3.1	3.6	3.2	4.5	4.0	0.0	3.3	3.9	2.8	4.2	4.1	5.7	4.1	4.3
13	1.7	3.0	4.7	2.8	2.9	1.4	1.5	0.6	0.6	3.1	3.4	3.3	0.0	2.6	1.5	2.2	3.1	2.7	3.5	3.7
14	2.1	3.3	5.4	3.8	3.7	2.4	3.7	2.4	2.3	1.8	3.8	3.9	2.6	0.0	3.0	2.8	2.0	4.6	4.5	4.4
15	2.1	4.2	4.7	3.8	3.0	1.4	1.8	1.7	1.4	3.8	4.2	2.8	1.5	3.0	0.0	1.8	3.1	3.5	4.0	3.7
16	2.8	4.7	4.9	3.9	3.0	2.7	3.3	2.0	1.8	3.3	4.3	4.2	2.2	2.8	1.8	0.0	2.2	3.1	4.7	3.6
17	2.8	4.3	5.0	4.2	3.4	3.1	4.3	2.8	2.6	2.1	3.9	4.1	3.1	2.0	3.1	2.2	0.0	4.3	4.1	2.9
18	4.0	4.4	5.2	3.5	4.0	4.0	3.4	2.5	2.9	4.5	4.4	5.7	2.7	4.6	3.5	3.1	4.3	0.0	4.4	4.2
19	3.7	3.4	5.2	4.6	4.6	3.8	3.5	3.6	3.7	4.2	4.2	4.1	3.5	4.5	4.0	4.7	4.1	4.4	0.0	2.8
20	3.9	5.1	5.6	4.7	3.8	4.0	4.4	3.7	3.6	4.4	3.9	4.3	3.7	4.4	3.7	3.6	2.9	4.2	2.8	0.0

A similar explanation can be given for the structure of the standardized distance $u_{i,i}^{1/2}$ to the centroid.

2.5 What Are the Units and What Are the Variables?

In the analyses discussed above, the treatment of the rows and columns of data matrix Z is asymmetric - columns are correlated with <u>equal</u> <u>weight</u> attached to each row (station); rows are compared by distances standardized with respect to the different variables (months). In that sense, the rows - units - are regarded as independent replications of equal weight and comparable size, whereas the columns - variables - are allowed to have unequal means and variances and to be correlated.

It is not always obvious which of the classifications of data one wants to regard as units and which as variables. In the example mentioned above, time has appeared as a variable - but in a long series of successive observations on a few stations or measurements, time points might be taken to be the units. In any particular application, the decision as to what to regard as units and what to consider as variables will determine what will be weighted equally and what will be standardized.

Statistics textbooks often treat only the description of the variables' configurations and ignore that of the units' scatter. This situation presumably arises because statisticians have been concerned mostly with random samples in which the individual units are not of interest in themselves. In practical data analysis, however, the units are often of real interest and their description is as relevant as that of the variables.

3. THE GEOMETRY AND DISPLAY OF A BATCH OF MULTIVARIATE DATA

3.1 The Configuration of Variables

We begin by considering the configuration of a batch, in terms of its means, standard deviations, and correlations. The reader may be bewildered by the magnitude of a correlation (or variance) matrix and may need guidance to make any sense of the, say, $\binom{20}{2} = 190$ correlations from a 20-variate data batch. We will provide a method of representing such a configuration and show an example of interpreting it. For brevity, we will illustrate this method on the above six-variate data of mean monthly temperatures.

Geometry is most useful in grasping the structure and patterns of multivariate data. One may think of the m variables as m vectors

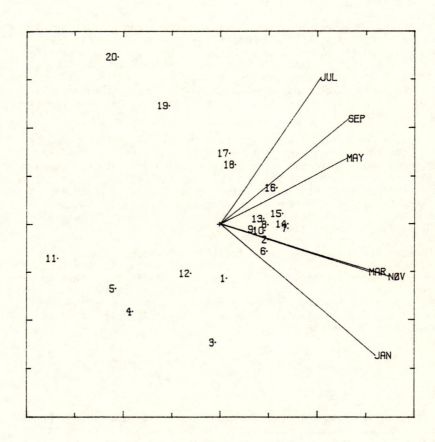

Figure 2. Biplot of temperature data (Table 1).

emanating from one origin. The vectors are chosen so that (a) the length of each vector is proportional to the standard deviation of the corresponding variable and (b) the cosine of the angle between any two vectors is the correlation between the corresponding two variables. In fact, it follows from (4) and (5) that the Euclidean length of $\underline{y}_{(v)}$ is

$$\| \underline{y}_{(v)} \| = (n)^{1/2} s_v, \tag{16}$$

and from (6) that

$$\cos(\underline{y}_{(v)}, \underline{y}_{(v')}) = r_{v,v'}. \tag{17}$$

Long vectors thus correspond to high variability, whereas short vectors represent low variability. Tight sheaves of vectors correspond to highly correlated variables, vectors at right angles to one another represent uncorrelated variables, and vectors in opposite directions correspond to negatively correlated variables.

3.2 Approximation of the Variables' Configuration in the Plane

Exact representation of standard deviations and correlations by such vectors may require high dimensional hyper-space, a conceptualization that may not be to everyone's taste. However, an approximate representation in the plane, or in three dimensions, is often quite useful in revealing many of the features of such a configuration. To illustrate, consider the variances of Table 2 and the approximate representation of their configuration by the arrows of Figure 2. This display is called a biplot. The method of approximation will be discussed later (see Section 3.3). For the moment, it is sufficient to say that the goodness of fit of this planar display is 96.7% for the temperature illustration, so that little of interest could have been lost by reducing this configuration to the plane. (The dots on Figure 2 represent the stations; see Section 3.6).

The configuration of the arrows in the biplot of Figure 2 is particularly simple. The lengths of the arrows are quite similar, indicating similar variabilities of all months. The arrows for March and May are the shortest, since the standard deviations for these months are smallest (see Table 1). All arrows are within the quadrant formed by those for January and July. The angle separating the latter two arrows is close to 90°. This angle indicates virtually zero correlation between these two months (Table 3 shows this correlation to be -0.09, which is indeed negligible). All other months, with smaller angles, lie in between these two extremes, indicating positive correlation. One may describe the entire configuration roughly by two sheaves of arrows; namely, a Fall-Winter sheaf of arrows separated by small angles (i.e., highly correlated, with March and November being particularly highly correlated), and a

Spring-Summer sheaf with slightly greater angles (i.e., less highly correlated). This description will be noted to accord completely with that obtained from Table 3.

The practical usefulness of the biplot is even more evident when the number of variables is larger. In that case it is much easier to see patterns and sheaves on the biplot than by inspection of the matrix of correlations. It is a matter of not seeing the wood (configuration) for the trees (correlations) because there are so many of the latter. An important function of multivariate data analysis is to provide simple descriptive tools such as the biplot, since these tools allow the investigator to make sense out of the mass of correlations and other data spewed out by modern computers.

3.3 Computation of Planar Approximations

It is good that readers understand the principles involved in these approximations even if they do not check the derivations but apply the procedures by means of a standard program (e.g., BIPLOT, which is available from the author).

The method of obtaining the planar approximation of the variables' configuration is to solve

$$Y'Y\underline{q}_\alpha = \lambda_\alpha^2 \underline{q}_\alpha \tag{18}$$

for the largest two eigenvalues λ_1^2 and λ_2^2 ($\lambda_1^2 \geq \lambda_2^2$) and their associated eigenvectors $\underline{q}_1$ and $\underline{q}_2$ (normalized to length one). One then forms the mx2 matrix

$$H = (\lambda_1 \underline{q}_1, \lambda_2 \underline{q}_2), \tag{19}$$

whose rows $\underline{h}_1'$, ..., $\underline{h}_m'$ are plotted as arrows emanating from a common origin. This method is equivalent to least squares fitting and its goodness of fit can be gauged by the coefficient

$$\lambda_{[2]}^{(4)} = (\lambda_1^4 + \lambda_2^4)/\text{tr}(Y'YY'Y)$$
$$= 1 - \|Y'Y - HH'\|^2/\|Y'Y\|^2. \tag{20}$$

It provides the approximations

$$\|\underline{h}_v\| \simeq (n)^{1/2} s_v \tag{21}$$

and

$$\cos(\underline{h}_v, \underline{h}_{v'}) \simeq r_{v,v'} \tag{22}$$

corresponding to (16) and (17) above.

3.4 Approximation of the Units' Scatter in the Plane

The foregoing calculations are equivalent to those of the first two principal components, a fact that will be commented on later. They also lead to a useful representation of the units in terms of their statistical scatter. One forms the mx2 matrix

$$F = (\lambda_1^{-1} g_1, \lambda_2^{-1} g_2) \tag{23}$$

and computes the nx2 matrix

$$G = Y F, \tag{24}$$

whose rows g_1', ..., g_n' are plotted as points. The distances between the plotted points then provide an approximate representation of the standardized statistical distances between the corresponding units. That is,

$$\| g_i - g_e \| \simeq (d_{i,e}/n)^{1/2} \tag{25}$$

and

$$\| g_i \| \simeq (u_{i,i}/n)^{1/2}. \tag{26}$$

The coefficients obtained by performing these calculations on the temperature data are shown in Table 7.

The interpretation of such a g-scatter is obvious. Distant points represent units which are dissimilar in their y values; points close together represent units with similar y's; clusters of points represent groups of similar units; and sequences of points ordered across the biplot represent units whose y's can be ordered systematically.

Each station i is represented by its appropriate g_i (row of G) vector on the biplot of Figure 2. (Note that different scales can be used for the g's and the h's, although for each of them the horizontal and vertical scales must be the same.)

Figure 2 shows that the scatter of g points mimics to some extent the geographical spread of the stations - see map in Figure 1. Thus, the northernmost stations appear on top of the biplot with a clear diagonal trend associated with latitude. Stations on or near the northern coast of South America form a tight cluster (high degree of statistical similarity), whereas stations farther south and west in South America trail out towards the lower left of the biplot. If we split the stations into four geographically contiguous groups, we should expect greater homogeneity of temperature profiles within each group and considerable inter-group differences. Table 8 groups the

Table 7. Biplot (and bimodel) coordinates for temperature data.

$\underline{h}_1'$	198.05	-169.37	-29.23	$\underline{f}_1'$	0.0010	-0.0017	-0.0056
$\underline{h}_2'$	190.76	-62.32	-16.19	$\underline{f}_2'$	0.0010	-0.0006	-0.0031
$\underline{h}_3'$	162.02	85.17	-25.32	$\underline{f}_3'$	0.0008	0.0009	-0.0049
$\underline{h}_4'$	128.44	188.45	-13.20	$\underline{f}_4'$	0.0007	0.0019	-0.0025
$\underline{h}_5'$	163.86	135.40	17.81	$\underline{f}_5'$	0.0009	0.0014	0.0034
$\underline{h}_6'$	215.63	-68.44	54.51	$\underline{f}_6'$	0.0011	-0.0007	0.0104

$\underline{g}_1'$	0.0206	-0.1865	-0.1004	$\lambda_1 = 437.841$	$\lambda_1^2 = 191704$
$\underline{g}_2'$	0.1605	-0.0533	0.1526		
$\underline{g}_3'$	-0.0167	-0.4055	-0.3292	$\lambda_2 = 313.613$	$\lambda_2^2 = 98353$
$\underline{g}_4'$	-0.2971	-0.2990	0.3781	$\lambda_3 = 72.247$	$\lambda_3^2 = 5220$
$\underline{g}_5'$	-0.3549	-0.2212	0.0300		
$\underline{g}_6'$	0.1572	-0.0923	-0.1148		
$\underline{g}_7'$	0.2279	-0.0155	-0.0235		
$\underline{g}_8'$	0.1604	-0.0025	0.1362		
$\underline{g}_9'$	0.1143	-0.0180	0.0810		
$\underline{g}_{10}'$	0.1498	-0.0227	0.0190		
$\underline{g}_{11}'$	-0.5509	-0.1157	0.1811		
$\underline{g}_{12}'$	-0.1000	-0.1692	-0.5192		
$\underline{g}_{13}'$	0.1459	0.0179	0.1099		
$\underline{g}_{14}'$	0.2273	-0.0004	0.0124		
$\underline{g}_{15}'$	0.2092	0.0362	-0.1138		
$\underline{g}_{16}'$	0.1897	0.1254	0.0817		
$\underline{g}_{17}'$	0.0294	0.2422	-0.0913		
$\underline{g}_{18}'$	0.0504	0.2036	0.5040		
$\underline{g}_{19}'$	-0.1752	0.4054	-0.2196		
$\underline{g}_{20}'$	-0.3480	0.5711	-0.1744		

Table 8. Median distances within and between groups of stations.

		Group			
Group	I	II	III	IV	Stations
I	3.3	3.6	4.1	4.3	1,3,4,5,11,12
II	3.6	2.4	3.0	3.8	2,6,7,8,9,10,13,14,15
III	4.1	3.0	2.0	1.9	16,17,18
IV	4.3	3.8	1.9	2.8	19,20

inter-station distances of Table 6 accordingly, and this display confirms that the biplot clustering does produce relatively homogeneous groups.

Clearly, geographical proximity is associated with similarity in annual temperature profiles. But this association is not perfect, as witness to the fact that the west coast stations 11 and 12 are more similar to southern stations 1, 3, 4, and 5 than to stations 9 and 15 which are much closer geographically.

3.5 Plotting of Extra Data Points

The biplot has been constructed to display a data matrix Z about its centroid $\bar{z}'$; that is, it represents $\bar{z}'$ at its origin and displays deviations Y. At times one may wish to display further data points that were not fitted in obtaining the biplot. Thus, one may have an additional unit with m-variate observations z_0'. This unit is to be centered as deviation $y_0' = z_0' - \bar{z}'$ and its biplot coordinates then calculated, analogously to (24), as

$$g_0' = y_0' F, \qquad (27)$$

or, equivalently, as

$$g_0' = (z_0' - \bar{z}')F. \qquad (28)$$

To illustrate, one might consider a hypothetical station whose temperatures for each month were exactly one standard deviation above the mean. That is,

$$z_0' = (284.68, 274.58, 273.46, 281.42, 287.14, 280.65).$$

Calculation of (27) or (28) yields biplot coordinates $g_0' = (0.2748, 0.0379)$. Such a point would appear in the biplot - Figure 2 - slightly to the right of g_{15}. Indeed, the temperatures for station 15 are similar, but slightly smaller, than those of this hypothetical station.

3.6 Joint Variables and Units Approximation - The Biplot

The biplot (Gabriel, 1971) displays both the configuration of the months (variables - columns of data matrix Z) and the scatter of stations (units - rows of Z). Because it displays them jointly this diagram is called a biplot. Such a simultaneous representation allows more insight into the data than could be obtained from the separate inspections of variables (Section 3.3) and of rows (Section 3.4), which have been illustrated above.

The biplot displays the actual deviations $y_{i,v} = z_{i,v} - \bar{z}_{(v)}$ by inner products

$$y_{i,v} \simeq \underline{g}_i' \underline{h}_v, \qquad (29)$$

with goodness of fit

$$\lambda_{[2]}^{(2)} = (\lambda_1^2 + \lambda_2^2)/tr(Y'Y)$$

$$= 1 - \|Y - GH'\|^2/\|Y\|^2. \qquad (30)$$

In other words, the deviation for station i on variable v can be visualized as the length of $\underline{h}_v$ times the length of the projection of $\underline{g}_i$ (considered as a vector from the origin) onto $\underline{h}_v$. The sign of the lengths' product is positive or negative according to whether $\underline{g}_i$'s projection onto $\underline{h}_v$ is in the same or opposite direction to $\underline{h}_v$ itself. Clearly, then, a station i whose $\underline{g}_i$ is far out in the direction of (opposite to) the vector $\underline{h}_v$ has large positive (negative) deviation $y_{i,v}$ on variable v. When the $\underline{g}_i$'s are less far out in the $\underline{h}_v$ direction (or opposite it) the deviations are smaller.

Figure 3 displays the January arrow $\underline{h}_1$ and the station 1 and station 20 points $\underline{g}_1$ and $\underline{g}_{20}$ of the biplot of Figure 2. It also shows the orthogonal projection of these two points onto the line through $\underline{h}_1$. The projection of $\underline{g}_1$ is seen to be of length 0.132 in the direction of $\underline{h}_1$, whose length is 256. Hence the biplot approximation of $y_{1,1} = 34.4$ is $\underline{g}_1'\underline{h}_1 = 0.132 \times 256 = 33.8$. Similarly, the projection of $\underline{g}_{20}$ onto the line through $\underline{h}_1$ is of length 0.625 in the direction opposite $\underline{h}_1$. Hence the biplot approximation of $y_{20,1} = -157.6$ is $\underline{g}_{20}'\underline{h}_1 = -0.625 \times 256 = -160$ (the minus sign being attached because the projection is <u>opposite</u> the vector projected upon).

This relation between $\underline{g}_i$ points and $\underline{h}_v$ arrows is useful in interpreting the scatter of $\underline{g}$ points. Thus, one may identify the variables (months) on which a cluster of units (stations) is particularly large or small. As an example, we note that the northernmost stations in Figure 2 are aligned in a direction opposite the Fall-Winter sheaf. Evidently, the farther north the station, the

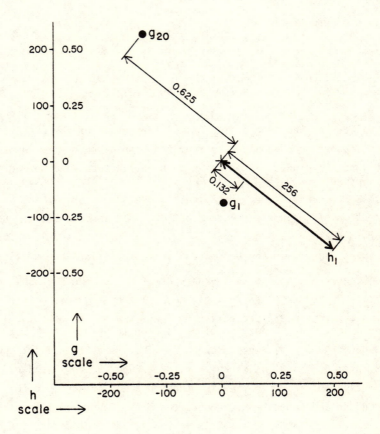

Figure 3. Biplot reconstruction of two observations.

lower its Fall-Winter temperatures. On the other hand, the difference between the second and third clusters of stations is associated with the direction of the Spring-Summer temperatures. That is, the north coast stations have higher Spring-Summer temperatures than the western and southern South American stations.

At this stage it is well to realize that one can inspect the biplot also for linear combinations of variables (LCVs) beyond the actual variables of the data displayed. One may do this inspection by vector addition of $\underline{h}$ arrows. Thus, for example, a March plus May sum would be represented by the vector $\underline{h}_2 + \underline{h}_3$ which is readily constructed on the biplot - the dashed line on Figure 4 - and found to be roughly horizontal. Also, a Spring-Summer sum $(\underline{h}_3 + \underline{h}_4 + \underline{h}_5)$ - dashed and dotted line on Figure 4 - slants up at roughly $45°$, whereas a Spring-Summer versus Fall-Winter difference $(\underline{h}_3 + \underline{h}_4 + \underline{h}_5) - (\underline{h}_1 + \underline{h}_2 + \underline{h}_6)$ - dashed and double dotted line on Figure 4 - is quite close to vertical.

Such LCVs are often useful. For example, we note that the Northern Hemisphere station points are mostly above the biplot origin and the Southern Hemisphere station points are below. This _vertical_ difference evidently is one of Spring-Summer versus Fall-Winter temperatures. The latter agrees with the very well known fact that temperatures in the Northern (Southern) Hemisphere are maximal in the Spring-Summer (Fall-Winter). Similarly, the north coastal South American station points are farthest to the right of the biplot, indicating that _average_ temperatures are highest in that region (again, a well-known fact).

These features of the biplot are of considerable importance for data analysis. They allow one to go beyond _separate_ descriptions of variables and of units and actually account for clusters and patterns of units in terms of the variables that determine them.

3.7 Joint Approximation in Three Dimensions - The Bimodel

The biplot displays the rank 2 least squares approximation of Y by GH'. One could similarly obtain a rank 3 approximation by solving (18) and (23) also for $\alpha = 3$ and by adding a further column to H, to F, and to G. The resulting _bimodel_ could be constructed in three-dimensional space, since each $\underline{g}_i$ and $\underline{h}_v$ now has three coordinates. Higher dimensional approximations can also be calculated by solving (18) and (23) for further α's, but these approximations cannot be constructed physically.

It is, however, feasible to inspect the three or higher dimensional approximations by displaying various projections on a CRT.

122

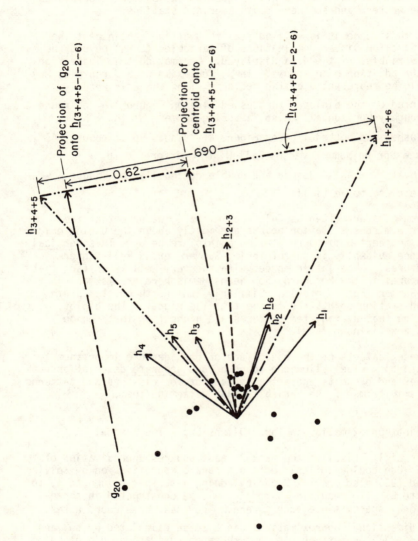

Figure 4. Biplot representation of linear combinations of variables.

Facilities exist on some computer installations that allow rotation of the higher dimensional approximation so that one gets successive two-dimensional views from different angles (Tsianco et al., 1981). These displays may be quite useful in revealing features of data that are not apparent from the original planar approximation.

As an illustration, consider again the $\underline{h}$ configuration in Figure 2. The annual cycle is represented by an upward movement from $\underline{h}_1$ through $\underline{h}_2$ and $\underline{h}_3$ to $\underline{h}_4$ and then a similar downward movement from $\underline{h}_4$ through $\underline{h}_5$ and $\underline{h}_6$ to $\underline{h}_1$. This movement suggests that there might be something like an elliptical orbit of the $\underline{h}$'s in three-space. Indeed, if one replots the $\underline{h}$ arrows along the second principal axis and an axis at 45° to the third and fourth principal axes (found by trial and error) one does find such an orbit - see Figure 5. True, the extra axis displayed in Figure 5 accounts for very little of the data's variability, but there is something satisfying in having a model which displays an annual cycle rather than a mere two season clustering. Indeed, this ellipse can be used to obtain a physically plausible model for the temperature data (see Section 6.3).

4. DATA ANALYSIS OF THE VARIABLES' CONFIGURATION

4.1 Purposes of Data Analysis

Data analysis aims at systematizing and summarizing data by noting regularities, tracing patterns, fitting models, etc. When the configuration of a set of variables is described by its variance matrix, a data analysis will attempt to elicit the salient features of variability and intercorrelation of the variables. Display of $\underline{h}$-vectors in a biplot, or in a higher dimensional bimodel, allows visual inspection of the configuration and may suggest grouping of highly correlated subsets of variables whose $\underline{h}$-vectors form tight sheaves. It may also indicate regular patterns, such as the elliptical orbit associated with the annual cycle of temperatures illustrated in Section 3.7. Such indications of regularity, whether suggested by visual inspection or otherwise, may lead to formulation of a "model" or systematic description of the set of variables.

4.2 Variables' Sheaves and Clustering Algorithms

The most common concept used for such descriptions is that of a "typical variable." When the variables group naturally into subsets such that there is high correlation between variables within subsets and much lower correlation from subset to subset, then one naturally thinks of a "typical" variable for each subset. Geometrically, when the $\underline{h}$-vectors separate into several tight sheaves, one may well

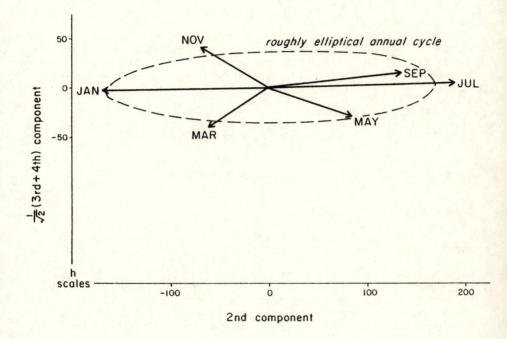

Figure 5. h arrow configuration of temperatures along special axes.

describe each sheaf by an arrow going through the center of the sheaf. Thus, in the temperature example in Section 3.2, the months' configuration seemed to cluster into a Fall-Winter sheaf and a Spring-Summer sheaf, and one could think of a typical variable for each.

Where there are too many variables for easy direct or graphical inspection, one may try to check for sheaves by some method of cluster analysis. If the variables do form separate tight sheaves, this situation will be revealed by any clustering algorithm. However, in many cases there are no tight and well separated sheaves and application of clustering techniques may not yield meaningful results. In such cases one would do better not to use the readily available algorithms to force "clustering." A good practical rule may be to use a number of alternative "clustering algorithms." Any "clusters" that are not revealed by most of the algorithms must be considered suspect; that is, they are likely to be artifacts.

4.3 Principal Components

One may define "typical" variables, or LCVs, in a sense akin to averaging. The "most typical" LCV is often taken to be that which has the highest average (squared) covariances with the observed variables. This correspondence is obviously an attractive descriptive property.

To obtain a second "typical" LCV one may consider the residuals from regressing each variable on the first "most typical" LCV. Then one may seek the LCV which has the highest average covariance with the residual parts of the variables - this combination will be the "second most typical" LCV. It will be found to be uncorrelated with the first. One may continue in this way, again taking residuals and obtaining a "third most typical LCV," etc.

The logic of looking at these successive residuals is not straightforward. Only the first of the "typical" LCVs is directly related to the original variables. For all the others it is not at all obvious if they can be considered "typical" of the original variables.

The coefficients giving the αth PC (PC_α for short) in terms of the original variables are well known to be the elements of the $\underline{g}_\alpha$ vectors of (18). Thus, the n "observations" or "scores" on PC_1 are given by

$$Y\underline{g}_1 = \lambda_1 \underline{p}_1, \tag{31}$$

which is the first column of G multiplied by λ_1. The scores of PC_2 are

$$Y\underline{q}_2 = \lambda_2\underline{p}_2, \tag{32}$$

the second column of G multiplied by λ_2. These two PCs are uncorrelated because

$$\lambda_1\underline{p}_1'\underline{p}_2\lambda_2 = \underline{q}_1'Y'Y\underline{q}_2 \tag{33}$$

and $\underline{p}_1$ and $\underline{p}_2$ are known to be orthogonal unless $\lambda_1 = \lambda_2$.

In general, PC_α is the LCV with observations vector $\lambda_\alpha\underline{p}_\alpha$ obtained by the solution of (18) with the αth largest root λ_α^2.

A well-known property of the PCs is that PC_1 has the largest variance of all normalized LCVs. That is,

$$\| Y\underline{q}_1 \|^2/n = \lambda_1\underline{p}_1'\underline{p}_1\lambda_1/n = \lambda_1^2/n. \tag{34}$$

Similarly, amongst all LCVs uncorrelated with PC_1, it is PC_2 which has the largest variance

$$\| Y\underline{q}_2 \|^2/n = \lambda_2\underline{p}_2'\underline{p}_2\lambda_2/n = \lambda_2^2/n, \tag{35}$$

and so on for other PCs.

PCs have simple least squares properties which were discovered by Householder and Young in 1938. In particular, the plane that best approximates the configuration is that going through the first two principal axes. In other words, the best fitting two-dimensional approximation of Y is

$$Y_{[2]} = \lambda_1\underline{p}_1\underline{q}_1' + \lambda_2\underline{p}_2\underline{q}_2', \tag{36}$$

which is a function of PC_1 and PC_2 only. It is because of this least squares property that these two principal axes were chosen to serve as horizontal and vertical axes of the biplot.

The generalization of these remarks to three-dimensional or higher order approximations is obvious. PC_3 provides the third axis of a bimodel.

The relation of the PCs to coordinates of the biplot is simply that the ith unit's observation on PC_α is

$$\underline{y}_i'\underline{q}_\alpha = \lambda_\alpha g_{i,\alpha}. \tag{37}$$

For the first two PCs this relationship follows from (24), (31), and (32). Table 9 gives the six principal coordinates for the twenty stations of the temperature data of Table 1. It is not obvious what interpretation one might want to put on these coordinates as such, though plotting the first few - scaled by λ_1, λ_2, etc. - has been found useful.

Table 9. Principal component scores.

Station	$\lambda_1 p_{1,i}$	$\lambda_2 p_{2,i}$	$\lambda_3 p_{3,i}$	$\lambda_4 p_{4,i}$	$\lambda_5 p_{5,i}$	$\lambda_6 p_{6,i}$
1	9.0	-58.5	-7.3	-7.5	-0.2	0.2
2	70.3	-16.7	11.0	-10.6	-19.0	-2.0
3	-7.3	-127.2	-23.8	41.5	-7.2	1.7
4	-130.1	-93.8	27.3	-4.4	-0.7	-1.0
5	-155.4	-69.4	2.2	2.6	8.5	1.4
6	68.8	-29.0	-8.3	-11.0	3.5	-2.0
7	99.8	-4.9	-1.7	-1.4	2.1	-7.3
8	70.2	-0.8	9.8	2.4	1.0	-0.9
9	50.0	-5.7	5.8	-0.6	3.7	-0.4
10	65.6	-7.1	1.4	-2.1	-10.4	7.3
11	-241.2	-36.3	13.1	-12.3	-0.8	0.9
12	-43.8	-53.1	-37.5	-14.0	5.0	-3.2
13	63.9	5.6	7.9	-1.7	2.0	-2.5
14	99.5	-0.1	0.9	-17.0	-1.7	5.7
15	91.6	11.3	-8.2	0.5	9.7	-1.9
16	83.1	39.3	5.9	7.6	11.4	3.5
17	12.9	76.0	-6.6	-1.4	1.5	7.7
18	22.1	63.9	36.4	21.4	2.3	-3.1
19	-76.7	127.1	-15.9	2.3	-12.8	-5.1
20	-152.3	179.1	-12.6	5.9	2.1	1.0

4.4 Principal Component Analysis

None of these mathematical properties make it clear why the PCs should be particularly interesting for an understanding of the variables' configuration. Clearly, PC_1 makes intuitive sense as an "average" or typical variable, or as the LCV with maximal variability. But what of PC_2, PC_3, etc.? They do not seem to have clear intuitive descriptive appeal. Their least squares property makes them useful for building approximations in the plane in terms of the biplot, in three dimensions in terms of the bimodel, etc., but that does not make them interesting individually. In the temperature example, the interesting "typical" LCVs seemed to be going at $45°$ and $-45°$ rather than at $0°$ and $90°$ to the horizontal. The fact that the PCs and principal axes are useful for plotting does not necessarily make them useful for interpretation. One usually does better by relating the g-points to the h's of the original variables rather than to the axes for the PCs.

A great many applications of PC analysis have been made through the years. This analysis is fine as a method of approximation in lower dimensional space, but as an interpretative device its popularity is surprising. What the method does is to express the original variables in terms of PCs in the form

$$\underline{y}_{(v)} = \sum_\alpha (\lambda_\alpha \underline{p}_\alpha)\, q_{\alpha,v}. \tag{38}$$

In view of (31) and (32), it is clear that the method also allows the PCs to be expressed in terms of the variables as

$$\lambda_\alpha \underline{p}_\alpha = \sum_v \underline{y}_{(v)} q_{\alpha,v}. \tag{39}$$

[Note that the weights in both linear combinations are the q's obtained by solving (18). They are therefore referred to as loadings.] But does this expression provide any insight? Does (38) "explain" the variables by the PCs or does (39) "explain" the PCs by the variables? Or do we expect, by circular reasoning, to have both "explanations"?

At best, consideration of the loadings q gives some insight into which pairs of variables are correlated (similar loadings on the first few PCs). What is puzzling are the attempts of many users of PCs to "reify" these mathematical constructs and ascribe "inherent," "underlying," or "explanatory" content to them. It is not evident how any such content follows from the mathematical definition of PCs and one may suspect that much of what has been published as PC analyses may have obscured rather than illuminated the configuration of the original variables which should have been studied.

The loadings of the monthly temperatures in the six PCs are shown in Table 10. The uniformly high PC_1 loadings for all months show PC_1 to be some average annual temperature factor which puts more emphasis on Fall-Winter. The latter is evident from the biplot in which all months' h-arrows point left, partly above and partly below the horizontal direction, but the Fall-Winter h's are closer to horizontal than the Spring-Summer h's. The PC_2 loadings are positive for Spring-Summer, negative for Fall-Winter, and thus indicate a seasonal component. Again, this conclusion was evident from the biplot configuration. Loadings on PC_3, PC_4, PC_5, and PC_6 are not so easily interpretable, although the joint consideration of PC_2, PC_3, and PC_4 could be interpreted and modeled in terms of an annual cycle. Nothing is revealed by consideration of these axes that was not seen by inspection of the h's themselves. The h-configuration is much more simply described by two sheaves, one for each half-year, than by two axes.

This example illustrates the shortcomings of PC analysis in considering each principal axis separately and the advantage of seeing the overall picture in a biplot or bimodel. It also illustrates the limitations of using orthogonal axes fitted by least squares - these axes do not necessarily provide the most readily interpretable references.

4.5 Some Further Comments on Principal Components

PCs depend on the scale of measurement of the original variables. This dependence is obvious from the definitions which depend on covariances, variances, and least squares fits. Much has been written on this dependence and how it limits the usefulness of PC analysis. All these writings seem beside the point. There is no real reason to consider PC analysis as a method for revealing the "underlying" structure or to regard PCs as "intrinsic" variables and hence it also does not matter that these "structures" and "intrinsic variables" are not scale independent.

Finally, PCs are often reified by reference to other variables extraneous to the original set. In the temperature illustration, it was not difficult to label PC_1 and PC_2, though no simple interpretation was evident for other PCs. Another illustration is a recent study of rainfall where PC_4 was noted to have a clear time trend associated with the spread of irrigation. It was therefore suggested that PC_4 was an "irrigation factor" and the possibility of using it as such a variable was considered. We suggest that direct use of the irrigation data - with which PC_4 has been found to be correlated - would have been simpler and more straightforward and would not have

Table 10. PC loadings $q_{i,\alpha}$ for temperature data.

$$PC_\alpha$$

Variable i	1	2	3	4	5	6
1(JAN)	0.452	-0.540	-0.405	-0.561	-0.158	-0.005
2(MAR)	0.436	-0.199	-0.224	0.700	0.028	-0.479
3(MAY)	0.370	0.272	-0.350	0.258	0.088	0.770
4(JUL)	0.293	0.601	-0.182	-0.349	0.505	-0.377
5(SEP)	0.374	0.432	0.247	-0.073	-0.772	-0.107
6(NOV)	0.492	-0.218	0.754	-0.034	0.339	0.156

required PC analysis at all. This example was quite typical of the "use" of PCs and the rationale of many other uses of PC analysis is equally puzzling.

To summarize, this author's view is that PCs are unlikely to have explanatory value in themselves. The most physically meaningful LCVs will not usually happen to lie along the principal axes of the configuration. This author sees the main usefulness of PCs as a tool to provide least squares approximations to data matrices and variables' configurations, and he would direct the scientific attention of investigators to what the approximation tells them about the original variables, and not to what it shows about the PCs. The investigator must have included his variables because he wants to know something about them, so let him discuss them instead of substituting mathematical artifacts.

4.6   The Rank or Dimension of a Configuration

When m variables are observed on more than m units, the configuration might be completely in a subspace of m-1, m-2, or fewer dimensions, but this situation is very unusual. It would indicate exact linear relations between the variables, even though these variables are affected by random variability and measurement error. In practice, when such things are observed, one usually finds that the original set of variables includes some repetitions of observations or sums, or averages, of other variables. It is rare and surprising to find such exact dependence otherwise. A set of m variables observed on n (> m) units almost invariably generates a configuration in m-space that cannot fit exactly into any lower dimensional subspace. It may well be approximated in a lower space, and perhaps even very closely approximated, but it is very unlikely to fit exactly.

The question of dimensionality rarely relates to the true configuration of variables, but usually makes sense only in the context of approximation. "Hypotheses" of reduced dimensionality are, in this author's opinion, rank nonsense. "Tests of significance" of PCs will therefore not be discussed here. If the hypothesis that a six-variate configuration is in a plane is physically implausible, it makes no sense to check it by testing the nullity of PC_3, PC_4, etc. Testing makes sense only if the hypotheses can be given credence.

This issue of dimensionality should correctly be addressed as one of approximation. The biplot plane fitted the six-variable temperature data to a goodness of fit of 0.967 and the three-dimensional bimodel had a 0.985 fit. That fit may well justify ignoring the remaining dimensions to all intents and purposes even if one is certain that there is some variability along those axes. There is no question of "testing" whether the data are in a plane or in three dimensions. The practical issue is simply that the fraction of real variation that lies outside the plane or three dimensions is negligible and need not be considered in interpreting the data, even though it is not assumed to be strictly null.

4.7 The Factor Analytic Model

Factor analysts postulate a model in which each variable can be written in the form

$$\underline{y}_{(v)} = \sum_{\alpha=1}^{r} \underline{f}_{(\alpha)} \ell_{\alpha,v} + \underline{e}_{(v)}, \tag{40}$$

as the sum of a linear combination of a few, say r, "factor" variables $\underline{f}_{(1)}, \underline{f}_{(2)}, \ldots$ with "loadings" $\ell_{\alpha,v}$ and errors $\underline{e}_{(v)}$ all of which are uncorrelated. Factor analysts usually postulate r to be quite small relative to m (so as to obtain parsimony in description), and then the model is most unlikely to fit exactly. They approximate the data by a model of type (40) with rank as low as will allow a reasonable fit. If we compare an approximation by model (40) with a PC approximation of the same rank, it is clear that the factor analytic model is very much less parsimonious in that it requires the $\underline{e}$ terms to be uncorrelated. The fit of its $\sum_{\alpha} \underline{f}_{\alpha} \ell_{\alpha,v}$ part is necessarily worse than that of the first r PCs.

One variant of factor analysis - MINRES - sets out directly to approximate the correlations. It is justified directly in terms of optimal approximation of the correlations - the off-diagonal elements of R.

As in the case of principal components, one has to ask whether the model as such makes physical sense so that the factors are "intrinsic" variables, or whether it serves as a mere vehicle of parsimonious approximation. Our answer to the first question should be similar to the one we have given for PC analysis. We see no a priori reason to think that the "factors" fitted in model (40) are any more "real" than the PCs.

The factor analytic model is no more plausible than the hypothesis of lower dimensionality which we discussed in connection with PC analysis. However, its saving grace is that it is so flexibly defined as to allow considerable manipulation which can on occasion be used to advantage. Axes can be rotated so as to yield factors correlated with extraneous variables or other information available to the investigator. It is difficult to see what "explanatory" function such a procedure possesses. The investigator had the "explanation" or extraneous variable anyway, and he could have correlated the original variables with it. Why bother to use factor analysis? Why not just take the multiple regression of the extraneous variable on the $y_{(v)}$'s as the "factor"?

Other methods of rotation, such as varimax, are built so as to make individual loadings as closely representative of sheaves of variables as possible. That procedure brings us back to the subject of applying clustering techniques to variables' configurations, an approach whose careful use may well yield important data analytic insight. However, it is difficult to see what role the factor analytic model plays in this process.

Our view of factor analysis differs sharply from that of its practitioners who talk about their model as though it had inherent reality. Even when they use a clearly approximative technique such as MINRES they try to reify the resulting factors.

5. ANALYZING THE SCATTER OF THE UNITS

5.1 A Batch of Units and Its Scatter

The units whose observations make up the rows of data matrix Z usually have identities which may be relevant to an analysis. Certain relations between units may be given a priori and it may be of interest to study if and how they are associated with statistical similarity of the corresponding rows of Z. Data analysis is often concerned as much with the units as with the variables. In the mean temperature example it is certainly as legitimate, and interesting, to study the scatter of stations as it is to study the variance configuration of months.

In modern statistics books this subject is hardly dealt with at all, and the idea of between units distance barely receives mention. This situation arises because these books deal exclusively with inference based on random samples, so that the units lose their individuality and become mere replications in a sampling process.

This sampling approach is undoubtedly appropriate for some experiments and in industrial quality control, where repeated observations are carried out regularly. But it is not appropriate to the study of batches of units with well-defined identities. Ignoring the information associated with these identities may stultify the analysis of such data.

In this section we consider, therefore, methods of analyzing the units' scatter and we choose to do so in terms of standardized distances $(d_{i,i'})^{1/2}$ between pairs of units (13) and $(u_{i,i})^{1/2}$

between units and the centroid (12). We will find it convenient to consider the scatter also in terms of the biplot approximations $\|g_i - g_{i'}\|$ and $\|g_i\|$ - (25) and (26) - of the above distances.

5.2 Use of Extraneous Information on the Units

When extraneous information is available (i.e., information on unit i other than the observation z_i), this information may be correlated with the observations. Thus, if the units fall into a number of categories, one may check whether these categories are associated with the statistical scatter of points. Do the categories form distinct groupings in m-space and/or on the biplot? Is there much or little overlap between categories?

A simple device is to mark the units of each category by a different mark, or color, on the biplot and see if the categories do separate. Figure 6 shows the g-points of the temperatures biplot (Figure 2) classified according to whether they are north or south of the equator. A clear separation is evident, showing that the temperature profiles of Northern Hemisphere stations differ from those in the Southern Hemisphere. The former are at the top of the biplot, whereas the latter are at the bottom. Recalling that the vertical direction on the biplot was a contrast between Spring-Summer and Fall-Winter (Section 3.5), one sees that the Northern versus Southern Hemisphere groupings reflect the difference in the season in which their maximum temperatures occur.

When the extraneous information is of a continuously variable character, one may record the extraneous measurements on the g-points of the biplot and see if they exhibit some regularity in the plane. Thus, in the temperature example we have marked the altitude on each g-point in Figure 7, and we see at once that the leftmost g-points are those of stations at high elevations; that is, there is a right-to-left trend in altitude. This trend is in a direction opposite to the general direction of the h-arrows for months' temperatures, confirming that temperatures are somewhat lower at higher elevations.

Some additional comments on this example will further illustrate uses of the biplot. The north-south differences of Figure 6 and the altitude differences of Figure 7 account for a great deal of the variability of the stations. However, a number of stations do not quite fit the pattern, especially stations 2 and 10 which are much farther to the right of the biplot than one would expect from their altitudes. Checking their locations on Figure 2 one sees that these stations are quite far inland on the South American continent. Evidently, in addition to altitude and to northern versus southern latitude, distance inland also plays a role in determining temperatures.

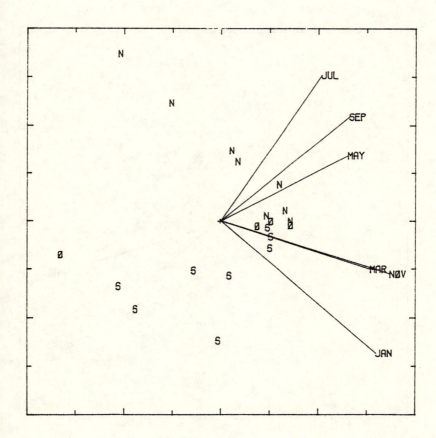

Figure 6.   Biplot of temperature data with indication of hemisphere of
            station (N = northern; 0 = within 2 degrees of equator; C =
            southern).

136

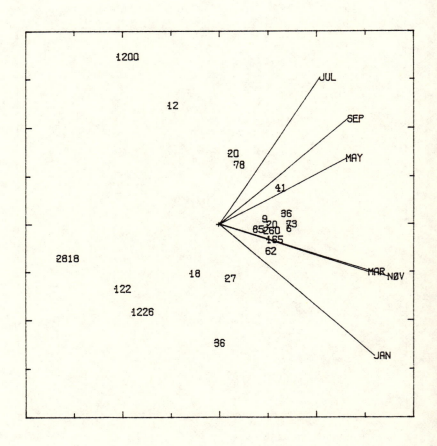

Figure 7. Biplot of temperature data with indication of altitude of
stations.

This illustration shows how the biplot can be used to check hunches about relationships to extraneous variables and how inspection of the biplot may suggest new things to look for. Of course, these are subjective impressions and their effective use depends very much on the ideas the investigator may be able to generate. The biplot may help him; it will not provide an objective substitute for his intuition.

5.3 Clustering of Units

Some groupings of units may be evident from the inter-unit distances themselves, rather than from extraneous information. Such data-dependent groups will be referred to as clusters, and methods of defining such groupings will be referred to as clustering algorithms. These methods differ from those used for locating sheaves of variables in that they relate to units rather than to variables and that the criterion for clustering is small inter-unit distances, whereas the criterion for forming sheaves was high intercorrelation of variables.

Many methods of clustering are available. A very simple one uses single linkage. To begin with, one clusters the nearest two points together. At the second stage one considers the next smallest distance. If it is between one of the first two points and a third point one clusters all three points together. On the other hand, if it is between two other points, one forms a second cluster of those two points. At each successive stage one considers the smallest of the distances between points which are not already in the same cluster. The points separated by this least distance are then linked together, and with them any other points clustered previously to either of them. Thus, if at a particular stage units 8 and 15 have least distance 0.28, and in previous stages unit 8 had been clustered with units 2, 6, 9, 10, and 13 whereas unit 15 had been clustered with units 7 and 14, then the new cluster consists of units 2, 6, 7, 8, 9, 10, 13, 14, and 15.

One may thus proceed step by step up to the largest distance between points, by which stage all units have become a single cluster. In practice one will presumably want to stop the clustering process before that, either when the number of clusters is small enough or when the remaining distances are too large.

The entire clustering process can be displayed by a dendogram, which is an inverted tree-like structure with a vertical scale corresponding to distance. It has a single stem on top at the height of the largest distance, when all units are clustered together. At the bottom, below the height of the least distance, it has n separate branches, one for each unit. In between, at the height of each distance, it has as many branches as there are clusters at that distance. Below that height the branch further branches and subbranches until the individual units' branches are reached.

The single linkage dendogram for the 20 stations of the mean monthly temperature example - corresponding to the standardized biplot distances -is given in Figure 8. For convenience, the order of the points has been rearranged to correspond as closely as possible to that of the biplot - in practical application this rearrangement is of course not possible since the "true" order is not known.

It will be seen that the dendogram of Figure 8 reproduces only some of the clusters evident from the biplot g-point scatter of Figure 2. Thus, if we divide the points into four clusters by eliminating the branches between heights 1.0 and 1.1, one cluster is of stations 4, 5, and 11, two are of the single stations 19 and 20, and one is of the remaining fifteen stations. This result is not very satisfactory because the last cluster is too large and spread out. The largest intra-cluster distance is 2.9 between stations 3 and 17. Such an elongated "cluster" is obtained because there is a "chain" of points at relatively small (below 1.0) distances from point 3 to point 17; that is, 3 to 1 (0.99), 1 to 6 (0.74), 6 to 15 (0.62), 15 to 16 (0.41), and 16 to 17 (0.89).

An alternative clustering criterion which would avoid such elongated clusters uses the complete linkage and clusters a set of units together, at distance d_0 and above, only if all units within the set are within d_0 of one another. The corresponding dendogram for the temperature data is shown in Figure 9. It is seen to differ from Figure 8 not only in that it shows less clustering for each given distance, but also in that it results in somewhat different clusters.

Thus, to obtain four distinct clusters, one would eliminate the branches at $d_0 = 2$ and the resulting clusters would be 5, 4, and 11 (as before); 19 and 20 (which had been separated before); 1, 3, and 12, and the remaining twelve stations (these last two clusters formed a single cluster by the previous method). The separation of that elongated cluster into two tighter clusters seems more satisfactory.

There are many other clustering criteria and computer programs to carry them out. Each criterion is supposedly "objective," but the choice of a particular one is subjective, and an investigator must make sure that he is using a method whose criterion is meaningful to him and appropriate to his purposes.

When the units cluster "naturally" into distinct tight groups, most clustering algorithms will reproduce that pattern. Often, however, the scatter does not reveal such obviously distinct clusters and the different algorithms will output different "clusters." Analysis into "clusters" may then become a game, especially if one tries out a variety of algorithms. The multiplicity of available

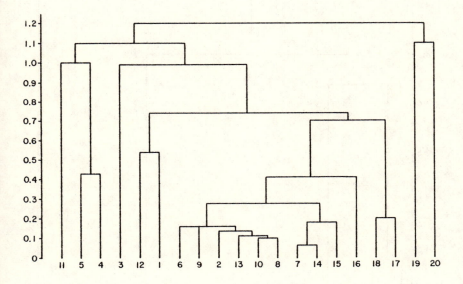

Figure 8. Temperature data: single linkage dendogram of biplot
distances.

140

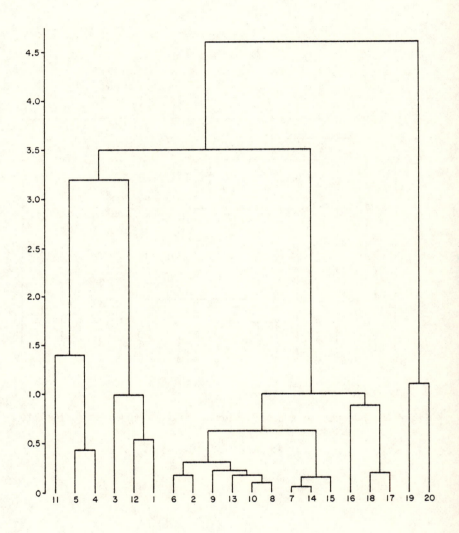

Figure 9. Temperature data: complete linkage dendogram of biplot
distances.

algorithms pretty much guarantees that any random scatter shall "cluster" nicely by some one criterion. Investigators should consider the clustering criterion carefully before committing data to an "objective" analysis into clusters.

The virtue of objectivity in data analysis is not obvious. A subjective approach which allows some capable researchers to obtain insights is certainly preferable to an objective method which usually fails to reveal anything worthwhile to any investigator. In the analysis of scatters of units (as well as that of configurations of variables), the capable investigator will usually approach his data with a great deal of prior knowledge, hunches, and hypotheses about patterns and relationships. He will do well to be guided by them and direct his analysis accordingly. If he wishes to cluster his data, he should not do so "objectively," merely on the basis of distance (or correlations) but should allow the interplay of observation with prior hypothesis. Specifically, if unit U_1 is about as distant from unit U_2 as from unit U_3, the investigator would do well to group it with the unit with which he has a priori reason to expect it to be more closely related.

5.4 Outliers

Distances $(d_{i,i'})^{1/2}$ are standardized by definition (Section 2.3). As a result, the scatter of points in m-space has the same variability in every direction, and its approximation on the biplot is essentially circular, unlike the well-known elliptic variability of raw variables.

In studying the form of the distribution, therefore, we should not look for differences in variability - these differences have been eliminated by standardization - but rather for other features such as clumping or clustering of points, special patterns, associations with external variables, outliers, etc.

If the g-scatter is roughly evenly distributed within radius unity, there is little to be said. If, however, one notes isolated points in one direction, with the remaining g's tightly bunched in the opposite direction, one should inspect the outlying points carefully for measurement or recording errors or perhaps for not belonging to the population under study. If that is found to be so, one might do well to omit such units from analysis and concentrate on the units that have a reasonable statistical scatter. Such a decision would, of course, mean recalculating the principal axes and g and h vectors after omission of the outliers.

A reasonable criterion for multivariate outliers is the distance $(u_{i,i})^{1/2}$ from the centroid. The distribution of these n

distances may indicate some clearly outlying units. Tests of significance are available for the multi-normal case (Gnanadesikan,
1977), but they should be used cautiously unless one is sure of normality.

It often happens that one finds one or more "outliers" but
checks do not reveal any reason why those observations should be
unusual. So one does not know whether the observations are extreme
values which do occur sometimes, though rarely, in the given observational situation, or whether they are erroneous records which do
not belong with the batch under study. One is in a quandary as to
whether to "reject" such outliers or not. Not to reject means
including observations that manifestly do not fit the statistical
distribution of the majority and vitiates the assumptions underlying
most statistical procedures. To reject exposes one to risks of
biasing the statistical analysis if the outliers were extremes from
the same distribution as the rest of the observations. An honest
rule would be always to report at least the number of rejected
outliers and preferably their entire observations, but not to include
them in the main statistical analysis.

5.5 The Distribution of Points in the g Scatter

Some idea of the distribution of the batch over the variables
may be obtained from considering the scatter of g-points on the
biplot. A more or less regular unimodal distribution should result
in a reasonably symmetrical biplot scatter with a concentration of
points about the centroid and gradual decreasing density towards the
edges.

Some other distributions obviously have different biplot scatters. A common case is that of multivariate J-shaped distributions
which have a mode near zero for all variables and a density which
decreases for higher values of each variable. Such distributions
will produce biplots of the type illustrated in Figure 10 - essentially a quadrant of points with a high concentration at the vertex
and along the edges. The h-arrows will be pointing in the direction
opposite to the vertex which represents the zero point of all
variables.

To check such distributions it is useful to project individual
numerical vectors, as in particular the zero vector, onto the biplot
scatter. For the zero vector, $\underline{z}_0 = \underline{0}$, the projection is

$$\underline{g}_0 = -\overline{\underline{z}}'F \tag{41}$$

as in (28). The rough location of the zero vector is indicated on
Figure 10 and confirms the supposition that these data are of a
multivariate J-shaped distribution.

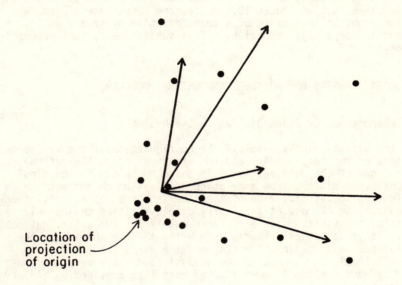

Location of
projection
of origin

Figure 10. Biplot of hypothetical data with 5 J-distributed
variables.

When the g-scatter does not show any special pattern, one may consider the distribution to be essentially random. It is not that regularities may not exist, but that they are not evident from the scatter. We know of no way of "testing" the biplot for normality, and would rather tend to use the normal model by default as a viable model in case nothing contrary emerges from biplot inspection.

Another intriguing issue is whether the biplot might be suggestive of a transformation to normality. All we can say at this time is that strongly skewed distributions should show up in a biplot looking like that of Figure 10. Hence, the appearance of such a scatter might be suggestive of a transformation by square roots, logarithms, or similar functions. This subject needs further examination.

6. JOINT ANALYSIS OF VARIABLES AND UNITS - MODELING

6.1. Importance of Joint Display in the Biplot

The biplot jointly displays the configuration of the variables (columns of the data matrix) and the standardized scatter of the units (rows of the data matrix). In doing these two things simultaneously, it differs from many other displays which concentrate on one feature to the exclusion of the other. For example, multidimensional scaling (Romney et al., 1972) displays either the correlation matrix of the variables or a distance matrix for the units, but not both. It is not usually feasible to bring the variables into the multidimensional scaling of units, or vice versa (see, however, Gabriel, 1978). As a result, the analysis and interpretation provided by such scaling is more limited than that provided by biplot representation.

Uses of the biplot for interpreting units' clusters in terms of variables have been discussed above. Analogously, correlations can sometimes be explained in terms of the scatter of units - specifically sometimes a single outlier in a particular direction can account for an increased correlation of the variables displayed in its direction. This situation is illustrated by the two parts of Figure 11. Figure 11a shows the biplot of a 5-observation 2-variate matrix, in which the low correlation ($r = 0.20$) is evident from the close to right angle ($78°$) between the two $\underline{h}$ vectors. The scatter of points, however, shows $\underline{g}_3$ to be fairly separate from $\underline{g}_1$, $\underline{g}_2$, $\underline{g}_4$, and $\underline{g}_5$ - evidently an outlier. Removal of the third point leads to a new configuration, biplotted in Figure 11b, and demonstrating a much higher correlation ($r = 0.94$) as evident from the $20°$ angle between $\underline{h}_1$ and $\underline{h}_2$.

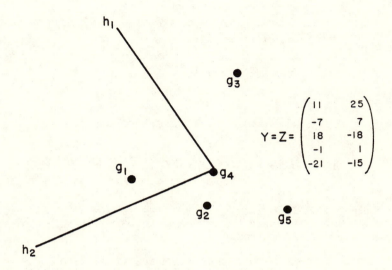

Figure 11a. A biplot with an outlier.

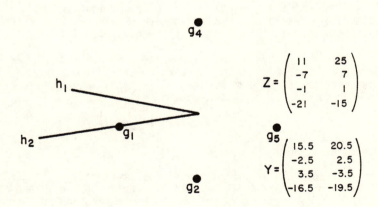

Figure 11b. The biplot after removal of the outlier.

6.2 Diagnosing Models by Means of the Biplot

Approximate functional fits of the data matrix may be identified by inspection of the biplot. Thus, if Z is approximately additive - that is, if

$$z_{i,v} = \bar{\bar{z}} + a_i + b_v + e_{i,v},\qquad(42)$$

for

$$\bar{\bar{z}} = \sum_i \sum_v z_{i,v}/nm\qquad(43)$$

and some $a_1, \ldots, a_n, b_1, \ldots, b_m$, and small e's, then the biplot of Z, or of $(z_{i,v} - \bar{\bar{z}})$, will have a simple form. In particular, the g-markers will be close to one straight line, the h-markers close to another such line, and these two lines will be at $90°$ to each other. Conversely, when the biplot markers display such a pattern, additivity can be inferred.

What is more, if some row markers are on one line and some column markers are on another line which is at $90°$ to the first, then one can infer that additivity holds for the sub-matrix of the corresponding rows and columns. For an illustration, consider the artificial air pollution data of Table 11 which are biplotted in Figure 12. It is immediately evident that the heads of the h arrows for the four years are very close to collinear and that the g-point for six of the stations are close to another line, pretty much at $90°$ to the h-arrowhead line - only the g-point for station F is far away from this line. One may therefore safely diagnose an additive model for the 6x4 table obtained by omitting station F. Inspection of the table will show that this conclusion is indeed appropriate.

This diagnostic method extends to some other models as well (Bradu and Gabriel, 1978). In particular, if the two above lines intersect at an angle other than $90°$, then a Tukey degree-of-freedom-for-non-additivity model holds; that is,

$$z_{i,v} = \bar{z} + a_i + b_v + \lambda a_i b_v + e_{i,v}\qquad(44)$$

for some λ.

This diagnostic use of the biplot is important since statisticians do not in general have adequate tools for such diagnosis. Textbooks give methods of estimating parameters and testing fit of given models, but do not usually provide techniques of choosing a model.

Biplot diagnosis of models rests on the matrix decomposition

$$Y \simeq GH'.\qquad(45)$$

Table 11. An air pollution index at seven localities (artificial data).

Station	Year			
	1960	1965	1970	1975
A	100	102	105	110
B	98	99	104	108
C	107	110	112	116
D	98	100	103	106
E	86	90	91	95
F	103	100	94	89
G	111	111	115	119

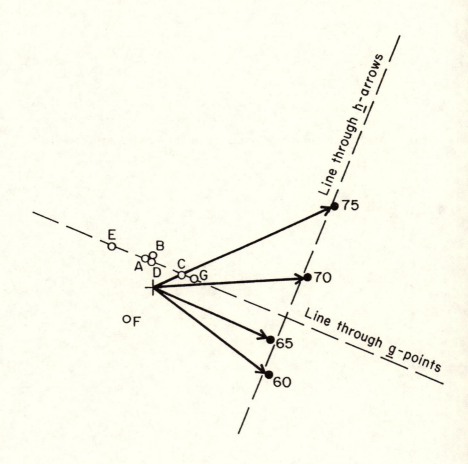

Figure 12. Biplot of fictitious air pollution data.

The rows of the latter two matrices are displayed in the biplot where visual inspection may lead to diagnoses of simple geometric descriptions. When these descriptions are formulated algebraically they can be entered into (45) and may be translated into a model for the data matrix itself.

6.3 An Example of Modeling by Means of the Biplot

As an example of the biplot's usefulness in modeling, consider the case where the vertices of the h-arrows are close to an ellipse. Writing $\underline{\mu}$ for the center of the ellipse, and $\underline{\alpha}$ and $\underline{\beta}$ for unit vectors along its principal axes, this situation means that there exists θ_v for each v, such that

$$\underline{h}_v = \underline{\mu} + \underline{\alpha} \cos \theta_v + \underline{\beta} \sin \theta_v. \tag{46}$$

Matrix H therefore becomes

$$H' = (\underline{\mu},\underline{\alpha},\underline{\beta}) \begin{pmatrix} 1 & \cdots & 1 & \cdots & 1 \\ \cos \theta_1 & \cdots & \cos \theta_v & \cdots & \cos \theta_m \\ \sin \theta_1 & \cdots & \sin \theta_v & \cdots & \sin \theta_m \end{pmatrix}, \tag{47}$$

and the data matrix is approximated by

$$Y \simeq G(\underline{\mu},\underline{\alpha},\underline{\beta}) \begin{pmatrix} \cdots & 1 & \cdots \\ \cdots & \cos \theta_v & \cdots \\ \cdots & \sin \theta_v & \cdots \end{pmatrix}. \tag{48}$$

Thus, the ith row is approximated by

$$\underline{y}_i' \simeq (\underline{g}_i'\underline{\mu},\ \underline{g}_i'\underline{\alpha},\ \underline{g}_i'\underline{\beta}) \begin{pmatrix} \cdots & 1 & \cdots \\ \cdots & \cos \theta_v & \cdots \\ \cdots & \sin \theta_v & \cdots \end{pmatrix}. \tag{49}$$

Writing

$$\sin \phi_i = \gamma_i/(\gamma_i^2 + \delta_i^2)^{1/2}, \tag{50}$$

$$\cos \phi_i = \delta_i/(\gamma_i^2 + \delta_i^2)^{1/2}, \tag{51}$$

and

$$\eta_i = \underline{g}'_i \underline{\mu}, \tag{52}$$

$$\gamma_i = \underline{g}'_i \underline{\alpha}, \tag{53}$$

$$\delta_i = \underline{g}'_i \underline{\beta}, \tag{54}$$

these approximations become

$$y_{i,v} \simeq \eta_i + \gamma_i \cos \theta_v + \delta_i \sin \theta_v. \tag{55}$$

Defining

$$\psi_i = (\gamma_i^2 + \delta_i^2)^{1/2} \tag{56}$$

and

$$\phi_i = \text{arc tan} (\gamma_i / \delta_i), \tag{57}$$

they become

$$y_{i,v} \simeq \eta_i + \psi_i \cos(\theta_v - \phi_i). \tag{58}$$

Thus, observation of the elliptical form of the h-configuration has led to diagnosing a harmonic model for the data matrix with constant and amplitude depending on the rows and phase depending on the columns. An example where such a model was appropriate was given in Section 3.6, where the elliptical configuration of the months' arrows was found. Those annual temperature data could therefore be fitted by a harmonic model with constants and amplitudes depending on the station and the phase depending on the month (Tsianco and Gabriel, 1982).

Let it be stressed that the function of the biplot, or bimodel, is merely to suggest a suitable model, not to provide estimates of its parameters. Once a model such as (42), (44), or (58) is suggested, standard estimation techniques should be reverted to, such as least squares or its robust counterparts.

7. OTHER LITERATURE

For a more general review of multivariate statistical techniques the reader is referred to Mardia et al. (1979), to the more elementary text by Morrison (1976), or to Essenwanger's (1976) volume which is directed to atmospheric scientists. For non-normal situations Gnanadesikan's (1977) work is useful, as are Everitt's (1978) and Gordon's (1981) books on graphical methods and Barnett's (1981) collection of application oriented papers.

ACKNOWLEDGMENTS

This work was supported in part by ONR Contract N00014-80-C-037 on Biplot Multivariate Graphics (K. R. Gabriel, Principal Investigator). Computations were carried out by program BIPLOT and package BGRAPH (Tsianco et al., 1981), both available from the author. The help of M. Tsianco, D. Gheva, and S. Plumb with the examples is gratefully acknowledged.

REFERENCES

Barnett, V., 1981: Interpreting Multivariate Data. London, Wiley.

Bradu, D., and K. R. Gabriel, 1978: The biplot as a diagnostic tool for models of two-way tables. Technometrics, 20, 47-68.

Essenwanger, O., 1976: Applied Statistics in Atmospheric Science. Amsterdam, Elsevier.

Everitt, B. S., 1978: Graphical Techniques for Multivariate Analysis. London, Heinemann.

Gabriel, K. R., 1971: The biplot - graphic display of matrices with application to principal component analysis. Biometrika, 58, 453-467.

Gabriel, K. R., 1978: The complex correlational biplot. In Theory Construction and Data Analysis in the Behavioral Sciences (S. Shye, ed.). San Francisco, Jossey-Bass, pp. 350-370.

Gnanadesikan, R., 1977: Methods for Statistical Data Analysis of Multivariate Observations. New York, Wiley.

Gordon, A. D., 1981: Classification. London, Chapman and Hall.

Mardia, K. V., J. T. Kent, and J. M. Bibby, 1979: Multivariate Analysis. London, Academic Press.

Morrison, D. F., 1976: Multivariate Statistical Methods, Second Edition. New York, McGraw-Hill.

Romney, A. K., R. N. Shepard, and S. B. Nerlove, 1972: Multidimensional Scaling, Vols. I and II. New York, Seminar Press.

Tsianco, M. C., and K. R. Gabriel, 1982: Modeling temperature data: an illustration of the use of biplots and bimodels in non-linear modeling. Submitted to Journal of Applied Meteorology.

Tsianco, M. C., K. R. Gabriel, C. L. Odoroff, and S. Plumb, 1981: BGRAPH: A program for multivariate graphics. Proceedings of the Interface Symposium (Pittsburgh).

Tsianco, M. C., and K. R. Gabriel, 1982: Modeling temperature data: an illustration of the use of biplots and bimodels in non-linear modeling. Submitted to <u>Journal of Applied Meteorology</u>.

4
Multivariate Comparisons of Data from Several Batches

K. Ruben Gabriel

1. COMPARING SEVERAL BATCHES OF OBSERVATIONS

1.1 Joint Inspection of the Two Batches' Scatters

Observations coming from several different sources, or populations, need different methodology and analysis than single batches of multivariate data. For each single batch, one may be concerned with description and analysis of the configuration of variables and of the scatter of units and with consideration of distributions, outliers, models, and other summarizations. The new aspect that appears when several batches of data are available is that of <u>comparing</u> batches. Problems arise with the search for, and identification of, characteristics on which the batches differ, with the measurement of "distances" between batches, with the appraisal of the significance or possible randomness of observed differences, and with the classification of additional units as being similar to one or another of the batches.

One may begin with the most straightforward case of comparing two batches. Generally the individuality of the units is ignored and they are considered as mere members of one batch or the other. Each batch is regarded as a <u>sample</u> and the units lose their identities and become mere replicate observations. We use the term "batch" for any assemblage of units, however obtained, and reserve the term "sample" for data obtained by random sampling. Clearly, significance tests cannot be applied validly to batches that are not samples, but descriptive techniques can still be used.

In comparing two samples, and in using them for testing population differences, one considers the within sample, inter-unit differences, mainly as providing estimates of <u>random</u> variation, or "noise," against which to judge inter-sample differences (averaged over units). Thus, in a comparison of 1977 winter storms with 1978 winter storms, the individual storms of each year are averaged for the main comparison, and the variability from storm to storm within each year serves as a yardstick against which one may measure the averages' comparison. A study of the special features of individual storms of either season would be part of each batch's analysis, not part of the batch-to-batch comparison.

As an example, consider the data in Table 1 relating to 26 storms occurring in the summer of 1973. Pielke and Biondini (1977) treated these storms as two samples, 13 with geostrophic wind speeds above 3m and 13 with slower geostrophic wind speeds. In comparing these two samples, the individual storms are averaged for comparison, and the storm-to-storm variation within each batch provides estimates of random variability. A study of the special features of each sample's individual storms is not a main part of the batch-to-batch comparison. Table 2 gives the five-variate means of each sample and the variance-covariance estimates.

Note that the means and variances in Table 2 do not relate directly to the variables in the form computed by Pielke and Biondini (see Table 1), but to various transformations of these variables. A preliminary check by means of probability plots (Gnanadesikan, 1977) showed three of the variables to have very skewed distributions, especially the first one. Transformation by a fractional power was therefore indicated. After some trial and error, transformations were chosen which produced reasonably symmetric distributions. Such preliminary inspection and transformation of variables is quite important. Without it one might apply least squares methods to variables which are highly skewed and for which these methods would be quite unsuitable.

One way of representing two batches of multivariate observations is by regarding them as distinct scatters of units in the same space of variables. An approximating display - GH' biplot - may be constructed for the matrix of both batches' multivariate observations and the g-points of the two batches may be distinguished on the biplot by some special marks or colors. The summer 1973 storms are biplotted accordingly in Figure 1 - again using the data transformed as noted in Table 2.

It is immediately evident from this biplot that the scatters of the two types of storms differ. The most obvious difference is that the g-points for "fast" storms are mostly higher up on the biplot than the g-points for the "slow" storms. The vertical direction is that of the two variables: rainfall (R) and wind direction (D). Evidently, slow storms have more rainfall and larger angle of wind direction than fast storms. The two superimposed batch scatters are next examined for differences in distributions. If the two scatters are completely disjoint, one may be sure of a clear between-batch difference. If there is some overlap, the distinction is less obvious and may need testing for significance if the data are from samples - more about that topic later. At this stage, one does well

Table 1. Summer 1973 storms.

Date	$R^a(m^3)$	Wind speed $(m\ sec^{-1})$	D^b $(°)$	T^c $(°C\ cm^{-1})$	S^d $(°C)$	$P^e(mb)$
July 3	49.61	4	90	0.019	5.99	432
July 4	172.83	2	70	0.032	5.98	453
July 5	20.72	2	225	0.054	5.77	371
July 6	59.93	4	270	0.032	10.77	515
July 7	26.12	3	135	0.048	11.58	494
July 15	75.48	3	100	0.082	8.41	336
July 16	51.71	5	170	0.068	14.14	267
July 17	56.33	6	100	0.042	10.57	477
July 18	23.66	5	100	0.081	13.36	326
July 20	62.95	3	135	0.076	12.70	357
July 23	31.13	6	120	0.048	9.35	539
July 24	17.09	10	90	0.096	10.44	404
July 25	14.61	6	85	0.081	3.64	356
July 26	37.29	2.5	110	0.074	11.57	304
July 27	84.06	1	180	0.047	11.04	316
July 28	77.22	1	170	0.039	9.22	329
July 29	108.71	0	180	0.053	12.97	295
July 31	93.88	6	180	0.120	16.59	280
Aug 1	38.66	1	180	0.116	15.53	252
Aug 2	75.61	3	190	0.113	12.54	242
Aug 6	79.98	4	100	0.070	15.53	317
Aug 10	127.04	3	140	0.028	4.46	564
Aug 11	24.85	4	135	0.058	9.59	401
Aug 12	17.66	8	90	0.081	15.85	427
Aug 13	33.15	7	100	0.043	9.19	338
Aug 14	97.53	3	135	0.050	7.80	532

[a] $20 \times \ln(rainfall)$.

[b] Surface level geostrophic wind direction.

[c] Gradient of equivalent potential temperature.

[d] Difference between saturation equivalent potential temperature and equivalent potential temperature.

[e] Depth of convective instability.

Table 2. Means and variances-covariances of storm data (transformed).[a]

Fast geostrophic wind: speed > 3m/sec - 13 storms

		R	D	T	S	P
Means		215.10	125.38	248.03	111.55	294.82
Standard deviations		35.17	51.12	54.96	37.03	31.43
Variances-covariances (correlations below diagonal)	R	<u>1237.06</u>	910.08	-405.95	540.62	-194.38
	D	0.506	<u>2613.31</u>	-327.15	470.87	139.69
	T	-0.210	-0.116	<u>3020.83</u>	1032.91	-999.24
	S	0.415	0.249	0.408	<u>1370.91</u>	-459.90
	P	-0.176	0.087	-0.579	-0.395	<u>987.61</u>

Slow geostrophic wind: speed $\leq$ 3m/sec - 13 storms

		R	D	T	S	P
Means		251.09	150.00	244.35	99.67	287.04
Standard deviations		36.05	40.76	52.47	31.85	38.27
Variances-covariances (correlations below diagonal)	R	<u>1299.37</u>	-603.74	-719.94	-330.53	296.38
	D	-0.411	<u>1661.54</u>	499.50	321.12	-649.04
	T	-0.381	0.234	<u>2752.82</u>	1183.67	-1472.29
	S	-0.288	0.247	0.708	<u>1014.57</u>	-855.09
	P	0.215	-0.416	-0.733	-0.701	<u>1646.74</u>

[a]Transformations: $R \leftarrow 60\ln(\text{rainfall})$; $D \leftarrow D$; $T \leftarrow 1000(T)^{1/2}$; $S \leftarrow 10S$; $P \leftarrow 15(P)^{1/2}$

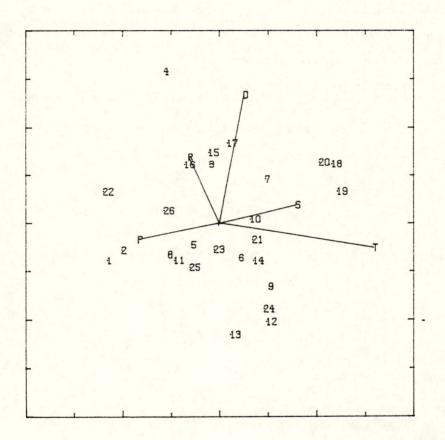

Figure 1.   Biplot of storm data (Table 1).

to inspect the shape of the two scatters as well as their approximate location. (a) If the centroids differ, a difference in mean level on some variables is indicated; the particular variables can be identified by considering the vector from one of the centroids to the other and projecting it onto the h-arrows to look for long intercepts. (b) If the extent or shape of the two scatters differs, different variability is indicated; the particular variables on which the variability differs are indicated by identifying the h-arrows in the direction of differing scatter.

An aid in inspecting and comparing the scatter of samples of points is the construction of concentration ellipses. For a batch of m units whose biplot g-points coordinates are $(g_{i,1}, g_{i,2})$, i = 1, ..., m, the concentration ellipse is defined as the locus of

$$\bar{g} + \beta_1 \cos \theta + \beta_2 \sin \theta, \quad (0 \le \theta \le 2) \tag{1}$$

with center at point

$$\bar{g}' = (1/m) \sum_{i=1}^{m} (g_{i,1}, g_{i,2}) \tag{2}$$

and β_1 and β_2 being obtained in the following manner. Calculate the matrix

$$V = \frac{1}{m} \begin{pmatrix} \sum_i g_{i,1}^2 & \sum_i g_{i,1} g_{i,2} \\ \sum_i g_{i,1} g_{i,2} & \sum_i g_{i,2}^2 \end{pmatrix} - \begin{pmatrix} \bar{g}_1^2 & \bar{g}_1 \bar{g}_2 \\ \bar{g}_1 \bar{g}_2 & \bar{g}_2^2 \end{pmatrix}, \tag{3}$$

and solve

$$V \underline{g}_\alpha = \lambda_\alpha^2 \underline{g}_\alpha \quad (\alpha = 1,2) \tag{4}$$

for the maximum and minimum eigenvalues λ_1^2 and λ_2^2, respectively. Then set

$$\underline{\beta}_\alpha = \lambda_\alpha \underline{g}_\alpha \quad (\alpha = 1,2). \tag{5}$$

The center of the ellipse - and the centroid of the batch of m $\underline{g}_i$ points - is at $\bar{g}$ and the maximum and minimum diameters are, respectively, of lengths $2\lambda_1$ and $2\lambda_2$ and in directions $\underline{g}_1$ and $\underline{g}_2$ from the centroid.

The concentration ellipses for storms of each type are drawn onto the biplot in Figure 2. They clearly show the vertical displacement of the two samples, confirming the impression gained from

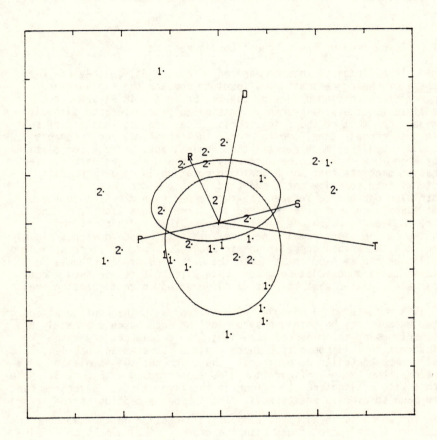

Figure 2. Biplot of storms with identification of fast (1) and slow
(2) storms and concentration ellipses.

inspection of the g-points themselves. They also indicate no horizontal displacement, confirming that the two types of storms do not
differ appreciably on variables T, S, and P. The correctness of
these graphical impressions can be verified from the means in Table
2.

1.2   Comparison of Two Batches' Configurations

In addition to this comparison of centroids, one may use the
shape of the concentration ellipses to compare the variance and
correlation configurations of the two batches. In Figure 2 the
ellipse for the slow storms is considerably squatter and slightly
wider than that of the fast storms. Recalling the relation of scatters to correlations, one may infer that the T, S, and P correlations
would not differ much between the samples, but that the correlations
of R and D with each other and with T, S, and P might differ. The
biplot suggests that the R and D correlation is higher for fast
storms than for slow ones, and that is indeed the most
striking difference between the two correlation matrices in Table 2.
It also suggests that both R and D's correlations with T, P, and S
are smaller in magnitude for fast storms than for slow storms,
although the signs remain the same. This inference does not clearly
reflect the actual correlations in Table 2. Evidently, comparison of
ellipses can be used to indicate the existence of differences in
variances and correlations, but it is difficult to use such a comparison to infer what the actual differences in configuration are.

A more sensitive display of differences in the configurations of
two batches may be obtained by biplotting each batch separately and
superimposing their h-configurations (the g scatters are of no
interest in this context). These h-plots (Corsten and Gabriel, 1976)
allow more detailed comparisons. Thus, for the two samples of 1973
storms, the two h-configurations are superimposed in Figure 3. Note
that with a slight rotation four of the five pairs of h-vectors can
be made to overlap pretty well. The obvious exception is the h_R

vector which is in almost the opposite direction in the two configurations - its correlations with the other variables must be of virtually opposite signs in the two samples. This conclusion agrees
pretty well with Table 2.

We caution that before comparing batches of multivariate data,
one should check the batch scatters to see whether they are reasonably elliptic in character. If there seem to be few outliers, long
tails, and/or strong concentration at one edge or corner (likely to
be the zero point of the variables if the measurements are all
non-negative), then the data should be readjusted in a manner similar
to single batch data with such properties. If the variability seems
to be systematically longer for the batch with larger means, transformations may be called for. A comparison of ln(standard

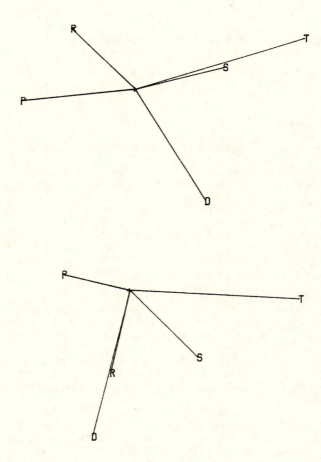

Figure 3. H-plots of fast storms (above) and slow storms (below).

deviations) against ln(mean) may show a non-horizontal slope for some of the variables - these variables' observations are likely to be more regularly scattered (i.e., have more equal variabilities) if they are re-expressed as

$$(variable)^{1-slope} = (re\text{-}expressed\ variable).$$

Such re-expression is a rough and ready method and the exponent should in general be rounded to the nearest 1/2. Note that for a slope around unity $(\cdot)^{1-slope}$ is to be read as $\ln(\cdot)$ (Tukey, 1977, Chapters 3 and 4).

1.3 Comparison of Three or More Batches

When multivariate observations occur in several batches, or have been classified a priori into several categories, one may wish to compare these several sets of observations. Such comparisons of three or more batches are analogous to the comparison of two batches described above. There are essentially two approaches to such comparisons: (a) comparing the several batches' configurations of variables without reference to the location of the scatters, or (b) comparing the location of several batches' unit-scatters against the background of one configuration of variables. Note that these two approaches correspond to univariate comparisons of scale and location. A detailed illustration is given in Section 1.4.

For comparisons of configurations one would need separate variances and covariances to be displayed for each batch. The h-configuration of each batch would be obtained from its GH'-biplot. These several configurations might then be displayed alongside one another and compared visually. Section 1.2 illustrated a comparison of two batches' 4-variate configurations. The comparison of, say, six batches' 10-variate configurations would be much more difficult. Unfortunately, we cannot suggest a simpler way of making such comparisons.

For location comparisons one might begin by pooling all batches to obtain an overall estimate of the variables' configuration; that is, the h-configuration in the GH'-biplot of all the data. One would then compare the batches by classifying the g-points according to the batches whose units they represent. And again, as in Section 1.1, one might summarize the scatter of each batch by a concentration ellipse, which permits convenient comparison of locations and shapes.

An alternative approach to location comparisons concentrates on the means of the several batches on each of the variables. A suitable metric for such comparisons is that of the "within batches" sums of squares of products (its use assumes that the variance-covariance configuration of the different batches are much the same).

Thus, a biplot of the means would show which batches differ from which other batches and on what variables these differences are evident.

Such an approach is analogous to MANOVA (multivariate analysis of variance). An application to meteorology has been studied in the context of the Israeli rainfall stimulation experiment (Gabriel, 1972).

1.4 An Example - Different Techniques for Comparing Batches

To evaluate the three different ways of comparing several batches of multivariate data, we will study an example in some detail. We use historical data of annual precipitation in Illinois to simulate a weather modification situation. Suppose a "cloud seeding" operation had taken place in the years 1955-60 in the southern Illinois area, and that another such operation had been carried out during 1970-78 in the northwestern part of Illinois. Also, suppose that no cloud seeding was carried out in Illinois at any other time or place. Precipitation in central Illinois could, therefore, serve as con-comitant observations to indicate "natural" precipitation; it would not have been "seeded" in either period. (The quotes are used since the data relate to simulated "operations," not to real ones).

To evaluate the effect of both "operations" it may be proposed to use 50 years' data, 1929-78, for the following five stations: Dubuque (DUB) and Moline (MOL) to represent northwestern Illinois, "seeded" in Period 4 - 1970-78; St. Louis (STL) to represent southern Illinois, "seeded" in Period 2 - 1955-60; Peoria (PEO) and Springfield (SPR) to represent central Illinois - never "seeded." These 50 years also pro-vide two "unseeded" periods for comparison; that is, 1 - 1929-54 and 3 - 1961-69, as set out in Table 3. The corresponding data for annual precipitation are shown in Table 4. Note that the data are actual precipitation observations except that in the "operational" years each "target" station's precipitation was augmented to simulate the effects of "seeding."

We are using simulated data for illustration because that allows us to anticipate the findings and then see how, and to what extent, the analyses recover the simulated "effects." Thus, we should expect that during "operations" the precipitation at target stations would be higher and more variable.

We begin by examining the entire data set, irrespective of batches (i.e., operational or other years). Means, standard deviations, covariances, and correlations are shown in Table 5, and the coordinates for the GH'-biplot are given in Table 6. The biplot has been fitted to residuals from the five-variate centroid. This biplot is displayed in Figure 4.

Table 3. Areas and periods of "operations" and comparisons.

Period	Number of years	Southern Illinois "target" (St. Louis)	Northwestern Illinois "target" (Dubuque, Moline)	Central Illinois control (Peoria, Springfield)
1. 1929-54	26	Unseeded	Unseeded	Unseeded
2. 1955-60	6	"Seeded"	Unseeded	Unseeded
3. 1961-69	9	Unseeded	Unseeded	Unseeded
4. 1970-78	9	Unseeded	"Seeded"	Unseeded
Total	50			

Table 4. Annual precipitation at five Illinois stations, 1929-78.[a]

Station

Year	DUB	MOL	PEO	SPR	STL
1929	24.26	34.71	39.66	33.68	46.30
1930	28.35	30.01	24.03	24.32	23.23
1931	29.52	31.39	37.75	36.21	37.30
1932	25.97	34.49	33.68	32.05	38.01
1933	28.70	28.31	34.07	36.47	34.77
1934	34.50	36.85	30.43	35.68	29.19
1935	32.55	35.58	40.15	41.22	39.36
1936	26.77	30.08	30.91	28.92	26.14
1937	31.77	30.96	29.89	34.63	35.87
1938	47.63	43.75	42.62	36.98	41.22
1939	29.53	28.50	38.27	33.05	40.15
1940	33.90	25.20	24.16	22.88	25.00
1941	32.50	36.94	42.39	44.72	32.12
1942	35.57	32.88	37.86	43.36	45.14
1943	31.92	32.16	32.81	32.36	33.60
1944	42.50	38.93	35.93	33.28	33.51
1945	38.78	33.84	36.13	43.40	49.82
1946	32.51	38.32	38.89	39.91	57.12
1947	42.28	35.63	39.17	36.48	35.78
1948	33.35	34.35	30.13	30.86	42.26
1949	31.51	34.56	33.33	37.52	45.76
1950	32.33	32.88	37.30	32.05	37.63
1951	45.01	48.60	37.23	39.51	36.37
1952	27.26	28.64	35.43	30.39	25.67
1953	34.95	26.47	28.83	23.98	20.59
1954	38.21	38.86	41.96	26.67	27.61
1955	28.07	26.09	29.99	34.15	40.73
1956	24.08	20.20	25.62	31.21	44.76
1957	38.82	32.92	36.99	41.97	61.31
1958	26.07	24.45	31.45	30.56	48.59
1959	54.36	42.10	30.63	35.98	36.80
1960	43.36	39.45	37.63	38.91	41.31
1961	63.39	45.90	39.45	37.91	41.20
1962	42.77	33.85	24.82	30.62	34.61
1963	35.44	30.78	25.66	28.89	28.62
1964	36.14	35.67	28.95	31.02	32.16
1965	61.42	49.59	48.26	39.08	28.26
1966	39.23	37.68	33.14	30.70	32.34
1967	52.97	42.36	35.95	36.31	41.30
1968	39.96	31.85	33.89	31.67	32.49
1969	33.70	41.79	33.70	34.68	43.72
1970	47.80	67.24	44.72	38.25	36.20
1971	48.22	49.97	26.38	27.62	33.73
1972	51.71	60.64	36.23	32.03	33.74
1973	51.43	73.27	50.22	44.29	39.82
1974	50.15	60.88	42.51	40.82	36.83
1975	42.26	37.64	41.22	37.66	40.21
1976	30.65	32.46	31.23	25.70	23.46
1977	50.74	54.55	38.41	42.71	43.41
1978	40.30	40.65	32.09	31.83	37.71

[a]These data are actual precipitation observations as obtained from
the Illinois State Water Survey, except for the 1955-60 figures for
St. Louis and the 1970-78 figures for Dubuque and Moline which are
equal to 130% of the recorded natural precipitation.

Table 5. Measures of location and dispersion of the entire data set. (Covariances above and correlations below diagonal in matrix).

	Station				
	Dubuque	Moline	Peoria	Springfield	St. Louis
Means	38.111	37.897	35.043	34.503	37.507
Standard deviations	9.626	10.805	6.013	5.474	8.203
	Covariances				
	-	77.643	24.937	18.916	1.022
	0.7465	-	39.215	26.563	5.620
Correlations	0.4308	0.6036	-	22.878	13.405
	0.3590	0.4491	0.6951	-	27.287
	0.0129	0.0634	0.2717	0.6077	-

Table 6. GH'-biplot coordinates for entire set.

i	g_{i1}	g_{i2}	i	g_{i1}	g_{i2}	i	g_{i1}	g_{i2}
1929	-0.083	-0.193	1946	0.014	-0.338	1963	-0.114	0.154
1930	-0.182	0.232	1947	0.023	0.003	1964	-0.058	0.095
1931	-0.085	-0.068	1948	-0.070	-0.058	1965	0.259	0.161
1932	-0.104	-0.046	1949	-0.051	-0.163	1966	-0.014	0.097
1933	-0.124	-0.034	1950	-0.068	-0.032	1967	0.130	-0.007
1934	-0.048	0.099	1951	0.133	0.020	1968	-0.046	0.077
1935	-0.017	-0.114	1952	-0.149	0.118	1969	0.005	-0.095
1936	-0.158	0.137	1953	-0.158	0.271	1970	0.300	0.057
1937	-0.102	-0.007	1954	0.000	0.156	1971	0.101	0.178
1938	0.129	-0.048	1955	-0.154	-0.094	1972	0.235	0.156
1939	-0.106	-0.094	1956	-0.234	-0.145	1973	0.397	-0.020
1940	-0.185	0.226	1957	0.017	-0.389	1974	0.269	0.035
1941	-0.001	-0.042	1958	-0.173	-0.195	1975	0.051	-0.067
1942	-0.014	-0.193	1959	0.115	0.078	1976	-0.127	0.211
1943	-0.092	0.032	1960	0.064	-0.070	1977	0.227	-0.064
1944	0.029	0.071	1961	0.230	0.020	1978	0.019	0.031
1945	0.012	-0.237	1962	-0.042	0.098			

j	h_{j1}	h_{j2}
DUB	59.235	14.807
MOL	71.587	8.645
PEO	29.079	-12.736
SPR	22.569	-24.452
STL	10.577	-54.246

$\lambda_1 = 100.50$

$\lambda_2 = 63.22$

$\sum_{\alpha=1}^{5} \lambda_\alpha^2 = 16798.25$

Goodness of fit: 0.8392

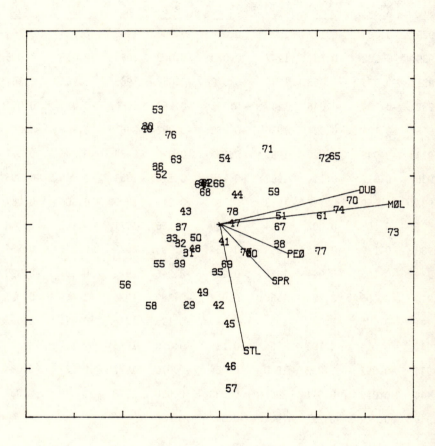

Figure 4. GH'-biplot of 50 years' Illinois rainfall (g-points
identified by years; h-arrows by station).

Mean precipitation in Table 5 is quite uniform over the five Illinois stations, although it is perhaps a little lower in the center. Variability changes more strikingly, the standard deviations being appreciably smaller in central Illinois. Correlations reflect the geographical location, the highest correlations being found for adjacent stations; that is, Dubuque with Moline, Moline to a lesser extent with Peoria, Peoria with Springfield, and St. Louis with Springfield. Generally, correlation tapers off with distance between stations - thus the St. Louis correlations with Dubuque and Moline are very low.

The biplot in Figure 4 reflects this configuration of variation and covariation (since this GH'-biplot is mean-centered it conveys no information on the overall means). The h-arrows for the central Illinois stations are shorter (less variability) than those for the stations in northern and southern Illinois. The order of the arrowheads reflects the geographical location of the five stations and so the angles subtended at the centroid are smaller for nearby stations and larger for far away stations. Thus, the cosines decrease with distance, reflecting the decrease of correlations with distance. This biplot is therefore seen to provide a simple display of both the pattern of variation and the configuration of the correlations.

We next turn to the scatter of g-points in Figure 4, which displays the distribution of the 50 years about their five-variate mean. The scatter is pretty evenly spread - no obvious outliers are evident, except perhaps 1957 at the bottom of the biplot. This display shows unusually high precipitation in 1957 at St. Louis. (Note that St. Louis was a "seeding target" in this year!) Indeed, on closer examination, we note that four out of the six years "seeded" at St. Louis have g-points far out in the direction of the h-arrow for that station. Also, we see that six out of the nine years of northwestern Illinois "seeding" have g-points far out in the biplot direction of the Dubuque and Moline h-arrows. This configuration is suggestive of "seeding effects."

The distinction between the four periods may be accentuated by suppressing the dates on the biplot and substituting the number of the period (i.e., 1, 2, 3, or 4, at each g-point). The results of this process are displayed in Figure 5. That display emphasizes the predominance of g-points of Periods 2 and 4 in the directions of the h-arrows for, respectively, St. Louis and Dubuque/Moline.

Figure 6 is another version of this same GH'-biplot in which the individual years' g-points have been replaced by concentration ellipses for each of the periods. Now the comparisons are much easier to grasp. The average level of Period 2 precipitation is seen to be highest on the St. Louis target and that of Period 4 on the Dubuque and Moline targets. The two unseeded periods, 1 and 3, have fairly similar ellipses which are not particularly high at any one of the stations.

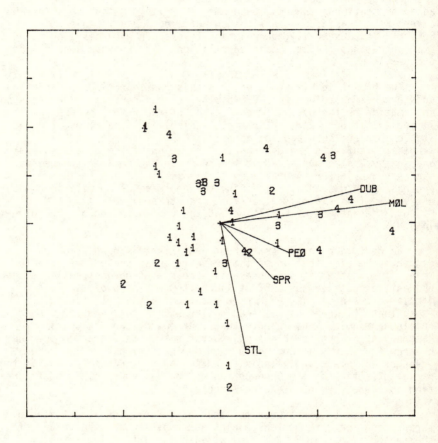

Figure 5. GH'-biplot of Illinois rainfall (g-points identified by period).

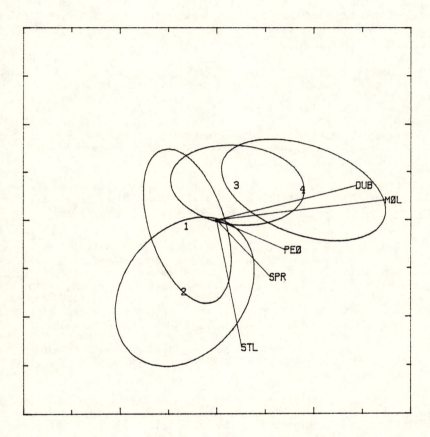

Figure 6. GH'-biplot of Illinois rainfall with concentration
ellipses for periods.

Also note the different shapes of the ellipses, indicating differences in variability. The elongation of the Period 4 ellipse along h_{DUB} and h_{MOL} suggests that the variance in northwestern Illinois and the correlation between the stations must have been higher in the period when it was the "seeding target." Similarly, the ellipse for Period 2 is somewhat elongated along the direction of h_{STL}. The latter indicates that when St. Louis was being "seeded" its variability was rather high.

Inspection of the GH' biplot of the entire data set has revealed differences in location as well as in variability and correlations. In most analyses this biplot is likely to be the single most useful display. However, we will also illustrate the other two displays; namely, the set of batch h-plots which is designed specifically for comparisons of variability and correlation, and the MANOVA biplot which displays comparisons of means standardized for within batch variability.

For the comparison of periods, Table 7 gives the means, standard deviations, covariances, and correlations and Table 8 indicates the coordinates for the h-plots of all periods. The four periods' h-plots are shown together in Figure 7.

The h-plots for the four periods look rather different at first glance, as do the standard deviations and correlations in Table 8. This impression is mostly due to the random variability between such small batches of data. It is well known that correlations based on samples of as few as 6 and 9 observations fluctuate wildly. Indeed, the comparison of Periods 1 and 3, which were both "unseeded," shows how large random variability really is. Note that the Dubuque-Springfield correlation is 0.250 in Period 1 versus 0.842 in Period 3! This illustration should serve as a warning against drawing far reaching conclusions about variability and correlation on the basis of small data sets.

Despite the smallness of the samples, there is some consistency in the four h-plots of Figure 7. The geographical gradient from northwestern through central to southern Illinois is shown consistently in all periods except Period 3 in which there is one inversion in the geographical order - between Moline and Peoria. The orientation of the geographical gradient changes from period to period, but the gradient persists, illustrating that some general patterns may be revealed even from small samples of data.

It is difficult to find the expected "effects of seeding" in these displays. "Target" variability was expected to increase during "seeding" - the Moline h-arrow is unusually long in Period 4. But the Dubuque h-arrow is rather short in Period 4 and the St. Louis h-arrow is not particularly long in Period 2. Nor does the angle between h_{DUB} and h_{MOL} seem unusually low in Period 4 - as it should

Table 7. Measures of location and dispersion of four periods of
years.

Means

Stations

Period	DUB	MOL	PEO	SPR.	STL
1. 1929-54	33.558	33.957	35.116	34.253	36.135
2. 1955-60	35.793	30.868	32.052	35.463	45.585
3. 1961-69	45.002	38.830	33.758	33.431	34.967
4. 1970-78	45.919	53.033	38.112	35.657	36.123

Standard deviations (diagonal), covariances (above diagonal), correlations (below diagonal).

Period 1	5.886	19.998	9.298	8.738	4.234
	0.679	5.000	15.158	15.019	16.554
	0.309	0.593	5.115	19.073	21.562
	0.250	0.506	0.628	5.934	35.309
	0.083	0.380	0.484	0.683	8.708
Period 2	11.854	100.791	25.574	32.601	-25.972
	0.973	8.739	24.873	26.371	-18.066
	0.474	0.626	4.550	16.728	15.964
	0.620	0.680	0.829	4.435	15.637
	-0.253	-0.238	0.405	0.407	8.671
Period 3	11.361	57.407	62.972	34.954	8.756
	0.779	6.487	41.067	22.489	12.777
	0.768	0.877	7.215	23.575	2.621
	0.842	0.949	0.895	3.652	8.416
	0.135	0.344	0.063	0.403	5.721
Period 4	6.999	81.964	23.608	29.373	26.095
	0.840	13.946	70.461	60.135	32.450
	0.453	0.678	7.451	43.173	21.676
	0.632	0.649	0.872	6.645	30.404
	0.656	0.410	0.512	0.805	5.682

Table 8. The $\underline{h}$-plot coordinates for the four periods.

Station

Period 1	DUB	MOL	PEO	SPR	STL	
h'_{j1}	10.9	17.0	18.5	25.4	39.2	$\lambda_1 = 54.1$
h'_{j2}	24.9	15.9	5.6	-0.5	-16.1	$\lambda_2 = 34.1$
					Goodness of fit = 0.8236	

Period 2	DUB	MOL	PEO	SPR	STL	
h'_{j1}	26.3	19.4	5.5	6.4	-5.3	$\lambda_1 = 34.2$
h'_{j2}	-0.1	-0.9	-6.6	-6.5	-18.4	$\lambda_2 = 20.7$
					Goodness of fit = 0.9596	

Period 3	DUB	MOL	PEO	SPR	STL	
h'_{j1}	30.9	16.7	18.3	9.8	3.5	$\lambda_1 = 40.9$
h'_{j2}	3.8	-4.1	1.6	-2.4	-15.5	$\lambda_2 = 16.7$
					Goodness of fit = 0.9072	

Period 4	DUB	MOL	PEO	SPR	STL	
h'_{j1}	17.2	38.2	16.6	15.2	9.6	$\lambda_1 = 48.5$
h'_{j2}	2.7	9.5	-9.4	-10.6	-9.4	$\lambda_2 = 19.6$
					Goodness of fit = 0.9120	

175

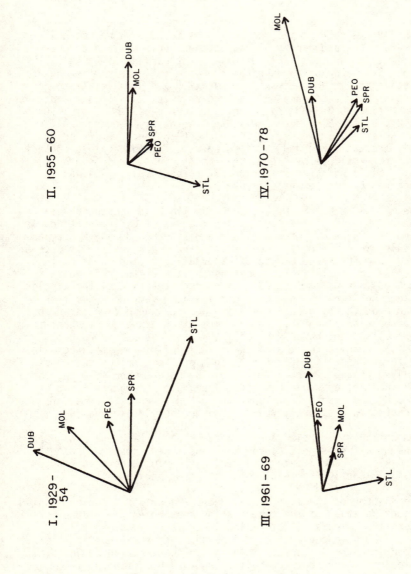

Figure 7. Illinois rainfall: h-plots for each of the four periods.

have been if "seeding" had increased correlation between "target" stations. Indeed, if we check back to Table 7, we note that these "expected effects" did not occur. It is not the h-plot display that obscured them, but the magnitude of random fluctuations in small samples.

Finally, we turn to the comparison of means - shown above in Table 7 - as standardized by within batches variances and covariances. This is the multivariate analysis of variance (MANOVA) approach. Standardization is effected by weighting with the inverse of an estimated variance-covariance matrix. The usual estimate is the "within" matrix of variances and covariances. In this example it would be estimated by pooling the bottom four panels of Table 7 with weights 25, 5, and 8, respectively, to a total of 46 degrees of freedom for error. However, in the present instance we prefer to pool only the two "unseeded" periods so as to avoid possible contamination of the estimate by "seeding" effect. Thus, we pooled panels 1 and 3 of Table 7 with weights 25 and 8, respectively, yielding 33 error degrees of freedom - see Table 9.

The MANOVA calculations are shown in Table 10 and the corresponding JK'-biplot of the four period means at the five stations is displayed in Figure 8. Each period mean is surrounded by a "comparison circle" which gives an idea of the random variability of each of those period means. The method of calculation of the radii of these circles is shown in Table 10; for a discussion of the rationale of these methods see Gabriel (1972). The interpretation of the circles is simple. Any two periods whose circles intersect do not differ more than expected by chance. Any two periods whose circles are disjoint differ significantly; that is, more than expected by chance. In this application chance variability is read as 95% of random variability overall; thus, a 5% chance - level of significance - is allowed for finding significance on some pair of periods that does not really differ. Other levels could be chosen; for example, for a 1% level the circles would be larger - because a larger θ would be read from Heck's charts (see Table 10) - and fewer differences found significant. This strategy would be safer, but less revealing.

The comparison circle significance tests on Figure 8 show Period 4's means to differ significantly from the other three periods' means. Periods 1 and 3 barely differ, and Period 2 does not quite differ significantly from either of these two periods.

The scatter of the four period means - j-points - can be related to the configuration of the five station measurements - h-arrows. It is evident from Figure 8 that Period 4 had large means in northwestern Illinois, especially in Moline and less so at Dubuque. This result confirms the "effect of seeding" in that area in Period 4, though the difference between Moline and Dubuque is unexpected. The small, and non-significant, difference between Period 2 and the

Table 9. Estimate of "error" variance based on two periods without operations.[a]

	Stations				
Stations	DUB	MOL	PEO	SPR	STL
DUB	7.585	29.067	22.310	15.094	5.330
MOL	0.692	5.541	21.439	16.830	15.638
PEO	0.516	0.679	5.696	20.165	16.970
SPR	0.364	0.555	0.647	5.469	28.789
STL	0.087	0.349	0.368	0.651	8.086

[a]Covariances above diagonal, standard deviations on diagonal, correlations below diagonal.

Table 10. Calculations for MANOVA and JK'-biplot of means.

X: Batch means

Period i	Sample size n_i	Station DUB	MOL	PEO	SPR	STL
1	26	33.558	33.957	35.116	34.253	36.135
2	6	35.793	30.868	32.052	35.463	45.585
3	9	45.002	38.830	33.758	33.431	34.967
4	9	45.919	53.033	38.112	35.657	36.123

S^{-1}: Inverse of error variances

Station	DUB	MOL	Station PEO	SPR	STL
DUB	0.036 037	-0.032 422	-0.004 803	-0.004 409	0.008 005
MOL	-0.032 422	0.092 402	-0.029 702	-0.007 462	-0.008 463
PEO	-0.004 803	-0.029 702	0.072 456	-0.032 675	0.003 077
SPR	-0.004 409	-0.007 462	-0.032 675	0.089 663	-0.028 855
STL	0.008 005	-0.008 463	0.003 077	-0.028 855	0.028 573

$X'NXS^{-1}$

Station	DUB	MOL	Station PEO	SPR	STL
DUB	-1.475 78	102.070 99	-46.893 78	-16.250 12	-8.097 50
MOL	-35.447 94	188.119 76	-57.863 27	-21.508 86	-23.885 08
PEO	-13.071 80	39.692 80	-6.663 76	-2.790 28	-7.861 94
SPR	-2.942 10	9.749 88	-2.943 06	-1.247 80	0.224 44
STL	10.422 75	-31.365 39	1.239 63	0.365 76	13.901 73

Table 10.   Continued

First two eigenvectors[a]

		Station		
	DUB	MOL	PEO	SPR	STL
$\underline{w}_1'$	-2.42	-4.16	-0.84	-0.19	0.72
$\underline{w}_2'$	-4.62	-0.81	1.26	0.06	-1.71
Column markers:	$\underline{k}_{DUB}$	$\underline{k}_{MOL}$	$\underline{k}_{PEO}$	$\underline{k}_{SPR}$	$\underline{k}_{STL}$

[a]These eigenvectors are standardized so that $\underline{w}_1'S^{-1}\underline{w}_1 = \underline{w}_2'S^{-1}\underline{w}_2 = 1$, $\underline{w}_1'S^{-1}\underline{w}_2 = 0$.

$J = XS^{-1}(\underline{w}_1, \underline{w}_2)$:   Row markers

   1: $\underline{j}_1' = (0.822, 0.597)$

   2: $\underline{j}_2' = (1.990, -1.002)$

   3: $\underline{j}_3' = (-0.087, -1.108)$

   4: $\underline{j}_4' = (-3.615, 0.051)$

θ: Critical value $\theta = 0.420$ $33\theta/(1-\theta) = 23.892$

   This is the upper 5% point of the maximum characteristic root distribution for 5 variables, 4 samples, and 33 degrees of freedom for error (Heck, 1960).

Comparison circle:

		Period		
i	1	2	3	4
Radius of $[33\theta/(1-\theta)2n_i]^{1/2}$	0.679	1.411	1.152	1.152

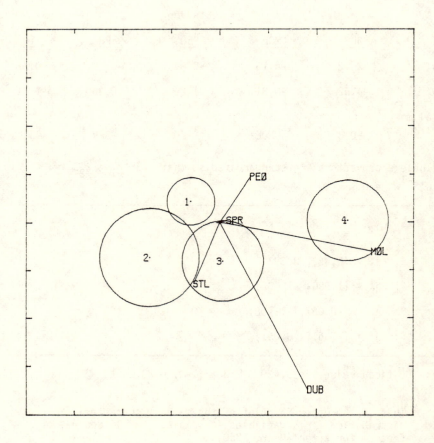

Figure 8. JK'-biplot of period means of Illinois rainfall (with 5%
comparison circles).

"unseeded" Periods 1 and 3 is mostly in the direction of $\underline{k}_{STL}$, indicating higher precipitation at St. Louis in that period - which is as it should be since that was where "seeding" was carried out. The other small, though significant, difference is between the two unseeded periods, 1 and 3; it is not quite clear what this difference is due to and it may well be a "Type I error." That is, it may be a falsely significant finding when no true difference exists.

It is evident that much the same general picture was obtained from the comparison of means on the MANOVA JK'-biplot of Figure 8 and from the comparison of scatters (ellipses of concentration) in the GH'-biplot of Figure 6. Indeed, both these biplots are projections of the data matrix, with the four batches of points and five columns, onto different two-dimensional planes. The GH'-biplot describes the entire variability of the data, whereas the JK'-biplot shows only the scatter of means. The latter therefore emphasizes the <u>differences</u> between the periods rather than what they have in common. To the extent that the latter configuration is different from the former, it is because of this different emphasis.

An idea of how the GH'-biplot of units differs from the JK'-biplot of means may be obtained by considering the three-dimensional GH'-bimodel with the concentration ellipsoids. Figure 9 shows three orthogonal aspects of this bimodel. (The XY-aspect is a reflection about the X-axis of the biplot of Figure 6.) It is easy to visualize the three-dimensional bimodel with its four ellipsoids. Thus, as compared to the XY-aspect, the bimodel has ellipses 1 and 4 to the front, 2 and 3 behind, $\underline{h}_{PEO}$ and $\underline{h}_{MOL}$ in front and $\underline{h}_{STL}$ and $\underline{h}_{DUB}$ behind. It is readily apparent that the JK'-biplot - Figure 8 - essentially corresponds to the XZ-aspect of the GH' bimodel. Thus, the MANOVA display has stressed different aspects from those stressed by the overall display. Apparently, the second axis Y is one of general variability rather than one differentiating the four periods - hence it is ignored in the planar MANOVA representation of Figure 8. It is also interesting to note that the principal axis of both displays is very much the same - although the one in Figure 8 differentiates Period 2 from Period 1 whereas that in Figure 9 does not - showing the JK'-biplot's greater emphasis of batch differentials.

It must be remarked that the approximate significance tests illustrated in Figure 8 are valid only to the extent that the required assumptions are satisfied. These assumptions are (a) multinormal distribution of precipitation at the five stations, (b) equal variances-covariances, and (c) independence of observations. For annual precipitation data (a) may be a reasonable approximation and (b) would probably hold pretty well unless seeding effects were large. Whether successive annual precipitation amounts are independent is more doubtful - although a recent study (Gabriel and Petrondas, 1982) suggests that assumption (c) is not entirely

182

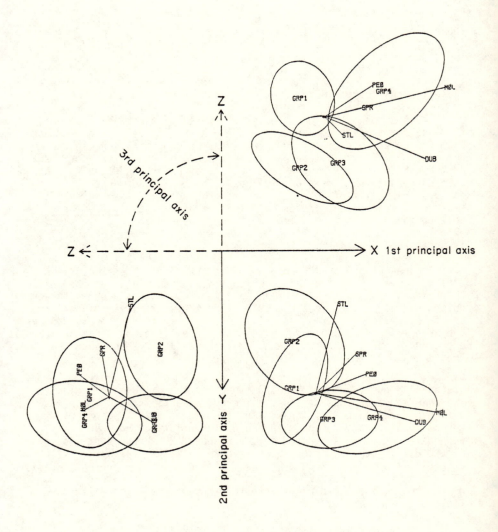

Figure 9. GH'-bimodel of Illinois rainfall with concentration
          ellipsoids for periods. (For 3-D visualization fold up the
          top and left panels along 1st and 2nd principal axes so 3rd
          axis points up; both dashed arrows will then coincide.)

appropriate. It is not quite right to regard the four periods as
random samples and the significance tests may be quite radical (i.e.,
result in too many significant tests). That point is crucial in many
meteorological applications; it is often doubtful whether successive
observations can be considered independent and thus the application
of significance tests is suspect. The emphasis in Chapters 3 and 4
was therefore on exploratory data analysis rather than on significance testing of hypotheses - the former seems to be of more use in
meteorological research.

1.5 Classification of Data into Categories - Discriminant Analysis

A common situation requires the classification of a new unit
into one of several populations from which it might have originated.
Thus, storms may be of a number of synoptic types and radar observations may be available for batches of earlier storms of each type.
A new storm now occurs and one is asked to use its radar observations
in order to allocate it to a synoptic type. Statistically, one would
want to classify the new storm into the type whose batch's radar
observations match the new storm's observations most closely.
Statisticians generally refer to this problem as "discrimination" and
the techniques they use go under the name of discriminant analysis.
The subject is too large to explore here. Instead we refer the
reader to Miller's (1964) monograph, written for meteorologists, to
Lachenbruch's (1975) volume on discriminant analysis, and to Gabriel
and Pun's (1978) description of and program for two-category discrimination by logistic techniques.

A more difficult problem is that of classifying a collection of
multivariate observations. The difficulty usually stems from the
vagueness of the objective of such a procedure. For example, synoptic maps are highly multivariate since each map contains readings at
a large number of localities. Meteorologists often want an "objective" statistical classification of such maps. They would like a
statistical method of reducing the evident complexity of these data
into few and meaningful clusters. They hope and expect a statistical
data analysis to reveal systematic order and pertinent classes.
Unfortunately, methods of cluster analysis do not reveal clear synoptic "types" but leave much overlap and uncertainty. Since "natural"
clusters are not evident, and since meteorologists are often at a
loss to state precisely what they want the classification for, the
problem may be ill-defined.

2. ON TESTS OF SIGNIFICANCE

2.1 The Logic of Significance Testing

A test of significance provides a decision on whether to regard
an observed phenomenon as "random" or real. In other words, could

the phenomenon have arisen in a manner analogous to the outcome of a game of chance, or does it reflect a real pattern? The issue of randomness versus real effects is often of great importance. Are there real periodicities in precipitation, trends in temperature, etc., or could the claimed effect of cloud seeding programs be merely due to chance? The use of significance tests to resolve these questions is not as straightforward as might be thought. A few words on this topic are in order.

Significance tests are designed to disentangle real from random effects. They do so by checking whether the observations seem "nonrandom" in the direction in which real effects are thought a priori likely to occur. Thus, when a cloud seeding experiment is designed, the hypothesis of no effect is to be tested against that of augmented precipitation subsequent to seeding. When this expected effect is precisely defined in terms of location of precipitation, time, method of measurement, etc., a significance test can properly be applied.

Significance tests are of more doubtful validity when they are applied to "effects" which were first observed during the experiment itself. For example, the Swiss Grossversuch III was designed to reduce hail, but observation of increased rainfall led to significance testing of augmented precipitation. It is common practice that apparent effects are first observed in a particular area or at a particular time - for example, after some change in seeding protocol - and then these particular "effects" are tested for significance. The validity of such testing is in doubt because it does not take into account the fact that the most striking phenomenon observed on the data was singled out for testing. A multiplicity of other phenomena were not tested because they did not happen to occur in such an extreme form on those particular data. Significance tests are not usually designed to accommodate selection of effects for testing. When such selection does occur, the multiplicity of possible choices dilutes the significance and leads to spuriously "significant" results.

When non-experimental data are tested for significance, one should have even greater concern for the validity of inferences. Why was a particular phenomenon chosen for testing? Surely because it was observed to be remarkable. If so, the results of significance tests are strongly biased in favor of deciding on non-randomness. Tests would be valid only if carried out on new data sets, independent of those which suggested the phenomenon.

A convenient terminology is that distinguishing confirmatory from exploratory analyses (Tukey, 1977). The latter are essentially inductive, and involve sifting through data for leads, patterns, suggestions, and ideas. The former are of a more deductive and rigidly defined character - they follow a protocol laid out in advance for the confirmation or refutation of a particular issue - as in the prior hypothesis on precipitation to be confirmed or rejected by a cloud seeding experiment (Gabriel, 1981).

No doubt there is much more exploration than confirmation in scientific work, especially in non-laboratory situations. And these situations are common in meteorology. Application of significance tests in exploratory analyses cannot be regarded as a rigorous, well-defined procedure. At best it serves to give vague indications of the relative roles of randomness and real effects.

2.2 The Exploratory Nature of Multivariate Analysis

Multivariate analysis, by definition, deals with a multiplicity of measures, none of which has been identified as the unique or principal bearer of the information sought. If a problem were closely defined and circumscribed, a single variable or function of variables would have been likely to emerge as the measure most relevant to the problem at hand. The analysis then would have lost its multivariate character. The simultaneous study of several variables implies that the subject is not narrowly focused and a definite hypothesis about the phenomena under study has not yet emerged. Hence multivariate analyses are unlikely to be confirmatory. Conversely, a confirmatory study is most likely to be univariate; the topic to be tested has been formulated precisely and allows confirmation. Exploratory studies are often multivariate, and they allow the investigator to search for effects among a multiplicity of variables.

2.3 Significance Tests in Multivariate Analysis

We have argued that multivariate analysis is mostly exploratory, and that exploratory studies do not in general depend much on significance tests. Hence the role of significance testing in multivariate analysis is likely to be minimal. This chapter has therefore not stressed topics of significance testing. Readers who still wish to apply tests of significance to multivariate data are referred to Morrison's (1976) excellent elementary test, to Essenwanger's (1976) more advanced volume, or to the recent book by Mardia et al. (1979). They will find tests for the types of comparisons discussed here as well as for other types of multivariate analyses of data from normal distributions. For a description of methods which are more robust against non-normality, readers are referred to Gnanadesikan (1977), Everitt (1978), Gordon (1981), and Barnett (1981). The present author hopes that the convenience of summarizing complex data by a single significance level will not deter readers from exploring their data carefully. He trusts that Chapters 3 and 4 may help the readers to look at their data and discover what they have to tell.

ACKNOWLEDGMENTS

This work was supported in part by ONR Contract N00014-80-C-037 on Biplot Multivariate Graphics (K. R. Gabriel, Principal Investigator). Computations were carried out by program BIPLOT and plotting package BGRAPH (Tsianco et al., 1981) both available from the author. The help of Mike Tsianco, David Gheva, and Sandra Plumb with the examples and Michael Greenacre's critical reading is gratefully acknowledged.

REFERENCES

Barnett, V., 1981: Interpreting Multivariate Data. London, Wiley.

Corsten, L. C. A., and K. R. Gabriel, 1976: Graphical exploration in comparing variance matrices. Biometrics, 32, 851-863.

Essenwanger, O., 1976: Applied Statistics in Atmospheric Science. Amsterdam, Elsevier.

Everitt, B. S., 1978: Graphical Techniques for Multivariate Data. London, Heinemann.

Gabriel, K. R., 1972: Analysis of meteorological data by means of canonical decomposition and biplots. Journal of Applied Meteorology, 11, 1071-1077.

Gabriel, K. R., 1981: On the role of physicists and statisticians in weather modification experimentation. Bulletin of the American Meteorological Society, 62, 62-69.

Gabriel, K. R., and Petrondas, D., 1982: On using historical comparisons in evaluating cloud seeding operations. Submitted to Journal of Applied Meteorology.

Gabriel, K. R., and C. F. Pun, 1978: Binary prediction of weather events with several predictors. Technical Report submitted to National Weather Service.

Gnanadesikan, R., 1977: Methods for Statistical Data Analysis of Multivariate Observations. New York, Wiley.

Gordon, A. D., 1981: Classification. London, Chapman and Hall.

Heck, D. L., 1960: Charts of some upper percentage points of the distribution of the largest characteristic root. Annals of Mathematical Statistics, 31, 625-642.

Lachenbruch, P. A., 1975: Discriminant Analysis. New York, Hafner.

Mardia, K. V., J. T. Kent, and J. M. Bibby, 1979: Multivariate Analysis. London, Academic Press.

Miller, R. G., 1964: "Regression estimation of event probabilities." Technical Report No. 1. Contract CWB-10704, The Travelers Research Center, Inc., Hartford, Conn., 153 pp.

Morrison, D. F., 1976: _Multivariate Statistical Methods_, Second
    Edition. New York, McGraw-Hill.

Pielke, R. A., and R. Biondini, 1977: Rainfall in the EML target
    area as a function of synoptic parameter. Unpublished report.

Tsianco, M. C., K. R. Gabriel, C. L. Odoroff, and S. Plumb, 1981:
    BGRAPH: A program for multivariate graphics. _Proceedings of
    the Interface Symposium_ (Pittsburgh).

Tukey, J. W., 1977: _Exploratory Data Analysis_. Reading, Mass.,
    Addison-Wesley.

5
Time Series Analysis—
Frequency Domain

Richard H. Jones

1. INTRODUCTION

Meteorological data are collected in time, and time series analysis can be defined as the analysis of data collected in time. The standard statistical tests, such as a t-test for testing whether two means are significantly different, assume that the observations are statistically independent. When data are collected close together in time, such as daily observations, the data are correlated and the t-test is invalid. However, this serial correlation has advantages; it permits forecasting into the future. In meteorology, the investigator is usually interested in inferring physical principles or forecasting.

Frequency domain and time domain approaches to time series analysis are complementary and it is essential to have a feeling for both. Basic concepts are introduced in this chapter, but there is strong interaction between this chapter and the next chapter, so they should be taken as a pair and not read in isolation. The approach of these chapters is deliberately elementary, with basic concepts being stated without proof since more details can be found in many papers and books.

2. TERMINOLOGY

A stochastic process is a set of random variables $x(t)$ indexed by time (or possibly space). A realization of this stochastic process, which is a set of observations, is called a time series. The expectation operator "E" will be used to denote expected value or ensemble averages, averages over independent realizations of the process. Using this notation, the mean-value function of the process is

$$\mu(t) = E[x(t)]. \qquad (1)$$

The correlation structure of the data can be expressed by defining the covariance between any two time points s and t as

$$C(s,t) = E\{[x(s) - \mu(s)][x(t) - \mu(t)]\}. \tag{2}$$

If the two time points are the same, this function is the <u>variance</u>

$$Var[x(t)] = C(t,t), \tag{3}$$

and its square root is the standard deviation of the process

$$\sigma(t) = [C(t,t)]^{1/2}. \tag{4}$$

The correlation between the values of the process at times s and t is

$$\rho(s,t) = C(s,t)/[\sigma(s)\sigma(t)]. \tag{5}$$

A second-order <u>stationary</u> process has a mean-value function which does not depend on time and can be expressed as

$$\mu(t) = \mu, \tag{6}$$

and a covariance function which depends only on time difference. In this case, the covariance function is written with only one argument representing the time difference as

$$C(t) = E\{[x(s+t) - \mu][x(s) - \mu]\}, \tag{7}$$

and the process variance is constant with

$$Var[x(t)] = C(0). \tag{8}$$

In meteorology, processes are rarely stationary. The mean and variance, as well as the covariances and correlations, depend on time of day and time of year. There is also the possibility of long-term trends, but this issue will not be considered here. The diurnal and annual variations can be handled in part by properly subtracting out the mean-value function or "normal values" and dealing with the residuals or anomalies. The time-varying variances, covariances, and correlations are usually handled by methods such as using daily observations for part of a year, replicated over several years of data.

The <u>spectral density</u> of a stationary process is the Fourier transform of the covariance function; that is,

$$S(f) = \int_{-\infty}^{\infty} C(t)\cos(2\pi ft)dt, \tag{9}$$

where f denotes frequency in cycles per unit time. The spectral density is a decomposition of the variance of the process as a function of frequency. From the inverse transform

$$C(t) = \int_{-\infty}^{\infty} S(f)\cos(2\pi ft)df, \tag{10}$$

it can be seen that the area under the spectral density is the variance of the process $C(0)$.

If the process is sampled at equally spaced time points with a spacing of h, the "aliased" spectral density is defined only in the interval $|f| \leq 1/2h$ as

$$S(f) = h \sum_{k=-\infty}^{\infty} C(kh)\cos(2\pi khf), \tag{11}$$

and the inverse transform is

$$C(kh) = \int_{-1/2h}^{1/2h} S(f)\cos(2\pi khf)df. \tag{12}$$

These concepts will be explained more fully in the sections on estimation.

3. PERIODIC MEAN FUNCTIONS

Diurnal and annual variations of mean-value functions are two of the most characteristic properties of meteorological time series. For annual variation, these means are often referred to as "normal" values for a given date. Ignoring the possibility of long-term climatic trends, these means would be the values obtained by averaging over a large number of years for a given date and perhaps applying a little day-to-day smoothing. Handling leap-year by assuming that February 29 has the same mean as March 1, this mean-value function is expressed by the 365 numbers

$$\mu(t), \quad t = 1, 2, \ldots, 365. \tag{13}$$

The actual observation on any given day in year i would be

$$y_i(t) = \mu(t) + x_i(t), \tag{14}$$

where $x_i(t)$ denotes the anomaly or deviation from the mean or "normal." Skill in forecasting is measured by how well this anomaly $x_i(t)$ can be forecast. Predicting that it will be hot in the summer and cold in the winter is stating what is known about the mean-value function and does not represent skill. Therefore, care must be exercised to ensure that the periodic mean-value function is not creeping into the forecasting equations and generating what falsely appears to be skill.

A periodic mean-value function has discrete Fourier representation

$$\mu(t) = A_0 + \sum_{\nu=1}^{182} [A_\nu \cos(2\pi t\nu/365) + B_\nu \sin(2\pi t\nu/365)], \qquad (15)$$

where A_0 is the mean of $\mu(t)$ and the index ν denotes the harmonic; that is, $\nu = 1$ is the first harmonic with frequency one cycle per year, $\nu = 2$ is the second harmonic, two cycles per year, etc. This is an exact representation of the 365 numbers $\mu(t)$ in terms of 365 other numbers, 183 A_ν's and 182 B_ν's. Since the A_ν's and B_ν's are coefficients of cosines and sines of various frequencies, they are referred to as the frequency domain representation of the time sequence $\mu(t)$. The A_ν's and B_ν's can be calculated from $\mu(t)$ by

$$A_0 = \frac{1}{365} \sum_{t=1}^{365} \mu(t),$$

$$A_\nu = \frac{2}{365} \sum_{t=1}^{365} \mu(t)\cos(2\pi t\nu/365), \qquad (16)$$

$$B_\nu = \frac{2}{365} \sum_{t=1}^{365} \mu(t)\sin(2\pi t\nu/365),$$

$\nu = 1, 2, \ldots, 182.$

An alternative form of the Fourier representation is

$$\mu(t) = A_0 + \sum_{\nu=1}^{182} C_\nu \cos(2\pi t\nu/365 - \phi_\nu), \qquad (17)$$

where C_ν is the amplitude of harmonic ν and ϕ_ν is the phase angle. From the formula for the cosine of the sum of two angles,

$$A_\nu = C_\nu \cos(\phi_\nu),$$

$$B_\nu = C_\nu \sin(\phi_\nu),$$

$$C_\nu = (A_\nu^2 + B_\nu^2)^{1/2}, \qquad (18)$$

$$\phi_\nu = \text{arc tan } (B_\nu/A_\nu).$$

The discrete Fourier transform is usually expressed using complex numbers. If $\{x_k: k = 0, 1, \ldots, n-1\}$ are n numbers, the discrete Fourier transform (DFT) of these numbers is

$$z_\nu = \sum_{k=0}^{n-1} x_k e^{2\pi i k \nu/n}, \quad \nu = 0, 1, \ldots, n-1, \tag{19}$$

where $i = (-1)^{1/2}$. The inverse transform is

$$x_k = \frac{1}{n} \sum_{\nu=0}^{n-1} z_\nu e^{-2\pi i k \nu/n}. \tag{20}$$

That this is an exact relationship can be verified by changing the summation index on the second equation to μ, substituting x_k into the first equation and using the sum of a finite geometric series

$$\sum_{k=0}^{n-1} \rho^k = \begin{cases} \dfrac{1 - \rho^n}{1 - \rho} & \text{if } \rho \neq 1, \\ n & \text{if } \rho = 1, \end{cases} \tag{21}$$

where

$$\rho = e^{2\pi i(\nu-\mu)/n},$$

so $\rho^n = 1$ if $\nu = \mu$.

Algorithms which take advantage of the factorability of n to reduce the computing time for calculating the z_ν from the x_k are known as fast Fourier transforms (FFT). The computing time is proportional to n times (sum of the factors of n), rather than proportional to n^2.

The relationship between the real and complex forms of the discrete Fourier transform is

$$A_0 = z_0/n$$

since z_0 is real, and

$$A_\nu = \frac{2}{n}\text{Re}(z_\nu), \quad B_\nu = \frac{2}{n}\text{Im}(z_\nu),$$

$$z_\nu = \frac{n}{2}(A_\nu + iB_\nu), \quad \nu = 1, 2, \ldots, [n/2], \tag{22}$$

where "Re" denotes the real part, "Im" denotes the imaginary part, and the brackets denote "the integer part." If n is even, the highest harmonic n/2 has no imaginary part and

$$A_{n/2} = \frac{1}{n}z_{n/2}.$$ (23)

There are various conventions for the factors of two and n in these definitions, but the important thing is to be consistent so that when numbers are transformed, then inverse transformed, the original numbers are recovered. Consistency requires dividing by n at some stage of the calculation. In the complex form of the discrete Fourier transform, there is a symmetry about $\nu = [n/2]$; if the original data are real,

$$z_\nu = \overline{z_{n-\nu}},$$ (24)

where the bar denotes complex conjugate. All the information is in $\{z_0, z_1, \ldots, z_{[n/2]}\}$. The factor of two in the real version of the transform is also in the complex version, since the range of summation is doubled.

As an example of removing the effect of annual variation, consider daily observations over a portion of a year repeated for several years. This could be 90 daily observations per year beginning December 1 each year for 10 years. Let $y_j(t)$ denote the observation on day t, t = 0, 1, ..., n-1, in year j, j = 1, 2, ..., m. Let y(t) be the mean of the m observations on day t,

$$\hat{y}(t) = \frac{1}{m} \sum_{j=1}^{m} y_j(t).$$ (25)

Let $\overset{\vee}{y}_j(t)$ be the data with the means subtracted,

$$\overset{\vee}{y}_j(t) = y_j(t) - \hat{y}(t).$$ (26)

The information about the mean-value function is now contained in the n numbers $\hat{y}(t)$ and the residuals $\overset{\vee}{y}_j(t)$ are free of the influence of the periodic annual mean (Jones, 1964). This is true even if $\hat{y}(t)$ is a poor estimate of the true mean or "normal" for the date because of a small number of years of data.

If the residuals $\overset{\vee}{y}_j(t)$ are expanded in Fourier series, each year separately,

$$\overset{\vee}{z}_\nu(j) = \sum_{t=0}^{n-1} \overset{\vee}{y}_j(t)e^{2\pi i t\nu/n},$$ (27)

so that

$$\tilde{y}_j(t) = \frac{1}{n} \sum_{\nu=0}^{n-1} \tilde{z}_\nu(j) e^{-2\pi it\nu/n}, \tag{28}$$

different coefficients $\tilde{z}_\nu(j)$ are obtained for each year j. These are Fourier expansions of data that have the periodic mean-value function removed; that is, they represent the random variation about the mean. If the sample mean function is expanded in a Fourier series,

$$\hat{z}_\nu = \sum_{t=0}^{n-1} \hat{y}(t) e^{2\pi it\nu/n},$$

$$\hat{y}(t) = \frac{1}{n} \sum_{\nu=0}^{n-1} \hat{z}_\nu e^{-2\pi it\nu/n}, \tag{29}$$

exactly the same coefficients $\hat{z}_\nu$ would be obtained as if the original data were transformed before subtracting the day-by-day means giving coefficients $z_\nu(j)$, which are then averaged

$$\hat{z}_\nu = \frac{1}{m} \sum_{j=1}^{m} z_\nu(j),$$

and $\tag{30}$

$$\tilde{z}_\nu(j) = z_\nu(j) - \hat{z}_\nu.$$

The spectral density of the time series is the variance of the residual $\tilde{z}_\nu(j)$'s, properly scaled. It is the variance of the time series with the mean-value function removed, decomposed as a function of frequency. The scaling is such that the area under the spectral density curve is equal to the variance of the original data after removal of the mean-value function. As when estimating the variance of a random variable, the spectral density can be estimated using the sum of squares of deviations from the mean. Since the Fourier transform is complex, two degrees of freedom for each harmonic are available from each year of data, and two degrees of freedom are lost by estimating the mean.

4. ESTIMATING THE SPECTRAL DENSITY

In a Fourier representation, each index ν corresponds to a frequency f, in cycles per unit time, a sine wave being written as $\sin(2\pi ft)$. If the sampling interval, or time between observations, is $\Delta t = h$, the harmonic ν of a Fourier series corresponds to frequency

$$f = \frac{\nu}{nh}.$$

In the above example, if the time unit is taken as days, $h = 1$ and the index ν corresponds to a frequency of ν/n cycles per day. If the time unit is taken as years, $h = 1/365$ and ν corresponds to a frequency of $365 \, \nu/n$ cycles per year.

When $\nu = n/2$, the highest value of the Fourier representation, the frequency is $f = 1/2h$, or one cycle in two time intervals. This is the highest frequency that can be resolved with the given sampling interval and is known as the Nyquist frequency. Any higher frequencies in the data appear exactly the same as frequencies in the interval $[0, 1/2h]$. This confounding of frequencies or aliasing is described in many books, for example, Blackman and Tukey (1958).

The spectral density can be estimated from the $z_\nu^{(j)}$'s obtained by Fourier transforming the data after subtraction of the mean. Using the example of the previous section where m years of data of length n are available,

$$\hat{S}(\frac{\nu}{nh}) = \frac{1}{m-1} \sum_{j=1}^{n} I_\nu^{(j)},$$

where   (31)

$$I_\nu^{(j)} = \frac{h}{n}|z_\nu^{(j)}|^2$$

is the periodogram calculated from year j (Jones, 1965). The factor h/n is the proper scale factor for estimating the spectrum using the definition of the Fourier transform given here. The factor $1/(m-1)$ is necessary for obtaining an unbiased estimate if the day-by-day mean has been subtracted from the data, or equivalently, if the $z_\nu^{(j)}$'s are calculated from each year of raw data and their mean subtracted. This case is different from the usual situation in time series analysis where it is assumed that the time series has a constant mean which affects only z_0, rather than a periodic mean which can affect all harmonics.

The vertical bars denote absolute value of the complex number $z_\nu^{(j)}$, so for $0 < \nu < n/2$,

$$I_\nu^{(j)} = \frac{nh}{4}\{[A_\nu^{(j)}]^2 + [B_\nu^{(j)}]^2\}.$$

Also,

$$I_0(j) = nh[A_0(j)]^2 \tag{32}$$

and, if n is even,

$$I_{n/2}(j) = nh[A_{n/2}(j)]^2.$$

Again, in the usual situation, where a single realization of the time series is observed, and a constant mean is estimated and subtracted, I_0 would be equal to zero. If more than one realization is observed and the time point by time point average is subtracted because of a periodic mean-value function, I_0 can be used to esti- the spectral density at zero frequency. However, as will be seen later, this method costs one realization worth of data in degrees of freedom.

Spectrum estimates are characterized by <u>bandwidth</u> and <u>variability</u>. Bandwidth is a measure of how well frequencies which are close together can be resolved. The bandwidth is the frequency interval over which the spectral density is averaged to obtain an estimate. It is impossible to estimate the spectral density at a point, just as it is impossible to estimate a probability density function at a point (hence the histogram, which averages over an interval). The averaging function for spectrum estimation is called the <u>spectral window</u>, and this function for the periodogram is shown in Figure 1. The bandwidth of the periodogram spectral window is inversely proportional to the length of the span of data.

Variability of a spectrum estimate is usually expressed in terms of equivalent degrees of freedom (edf). A spectrum estimate is an estimate of variance, the variance of the time series associated with frequencies inside the spectral window. In statistics, when estimating a variance from independent observations of a Gaussian distribution, the distribution of the estimated variance is proportional to chi-squared with n-1 degrees of freedom. The distribution of a spectrum estimate is not exactly proportional to chi-squared. The approximation is better as the spectral window becomes more flat topped as in a histogram. The larger the equivalent degrees of freedom, the less the variability of the spectrum estimate. The best way to express this variability is to plot a confidence interval when plotting an estimated spectrum. This helps avoid declaring peaks in the spectrum as real when they are due to statistical variability.

Confidence limits of a variance estimate are obtained from percentage points of the chi-squared distribution. From tables of chi-squared percentage points, let dof be the degrees of freedom obtained by truncating the equivalent degrees of freedom to an integer. To obtain 95% confidence interval, let a and b be the lower and upper values obtained from the table such that

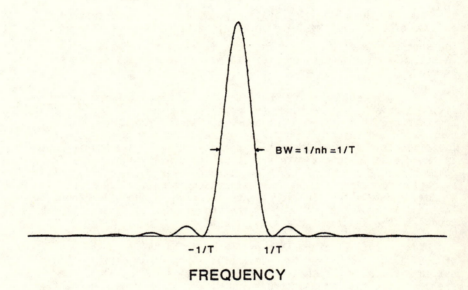

Figure 1. Spectral window of the periodogram based on a sample of size n and a sampling interval of h giving a total data span of T = nh.

$$Pr\{\chi^2_{dof} \leq a\} = 0.025,$$

$$Pr\{\chi^2_{dof} \geq b\} = 0.025. \tag{33}$$

For example, for 30 degrees of freedom, a = 16.79 and b = 46.98. Some tables supply values of chi-squared divided by degrees of freedom. In this example,

$$a/30 = 0.550,$$

$$b/30 = 1.566. \tag{34}$$

If $\hat{S}(f)$ is a spectrum estimate with equivalent degrees of freedom between 30 and 31, the approximate 95% confidence interval is

$$Pr\{\frac{\hat{S}(f)}{1.566} < S(f) < \frac{\hat{S}(f)}{0.550}\} = 0.95. \tag{35}$$

The proper interpretation of the confidence interval at some frequency f is that if the spectrum estimate were calculated many times from independent data, the confidence interval would be different each time, and 95% of the time it would include the true value of the spectral density.

This confidence interval is for a single frequency chosen in advance of the data analysis. If a peak in the estimated spectral density is noticed after estimating the spectrum, wider confidence intervals are appropriate since they apply to all frequencies. In this case, if the number of observations is n, and the estimates have edf equivalent degrees of freedom, there are approximately n/edf independent spectrum estimates, so an overall 95% confidence band would be obtained using chi-squared values such that

$$Pr\{\chi^2_{dof} \leq a\} = 0.025edf/n,$$

$$Pr\{\chi^2_{dof} \geq b\} = 0.025edf/n. \tag{36}$$

Spectrum estimates should be plotted on a log scale, since it is a variance stabilizing transformation and the same confidence interval can be applied to every frequency. The most common log scale in engineering is the db scale,

$$db(f) = 10 \log_{10}[\hat{S}(f)], \tag{37}$$

scaled relative to some point such as the maximum value of $\hat{S}(f)$,

$$db(f) = 10 \log_{10}[\hat{S}(f)/\hat{S}_{max}]. \tag{38}$$

Now the width of the confidence interval does not depend on the value of $\hat{S}(f)$ (see Figure 2).

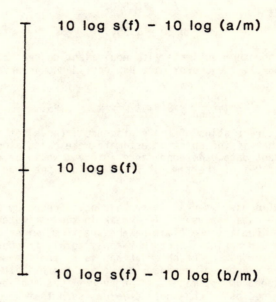

10 log s(f) − 10 log (a/m)

10 log s(f)

10 log s(f) − 10 log (b/m)

Figure 2. Confidence interval of a spectrum estimate on a db scale,
where a and b are the lower and upper values of a chi-
squared distribution with degrees of freedom equal to the
equivalent degrees of freedom of the spectrum estimate
(note that the interval is not symmetric with the upper
leg being longer).

If the periodogram or periodogram averaged over realizations is too variable, the variability can be decreased at the cost of increasing the bandwidth by averaging the periodogram over adjoining frequencies. Unless there are very strong peaks in the vicinity of the frequencies being averaged, the adjoining periodogram estimates are nearly uncorrelated so degrees of freedom add. Averaging over five frequencies produces estimates with ten degrees of freedom. If this estimate is averaged over three realizations, the result is an estimate with thirty degrees of freedom and a bandwidth five times as wide as the periodogram, 5/nh.

If there are strong peaks in the spectrum, the side lobes of the periodogram spectral window can cause problems. A strong peak in a side lobe appears as additional power at the center frequency of the spectral window. Data tapers can be used to reduce this problem when the periodogram method is used to estimate the spectrum. A data taper is a set of weights, which multiply the data near the ends of the data span in order to taper the data towards zero (after the mean is subtracted) to reduce the "ringing" caused by truncation. One popular data taper is the cosine taper used over about 5-10% of the data at each end. Choose $a \leq n/2$, and let

$$
b_k = \begin{cases}
\frac{1}{2}[1-\cos(\pi k/a)] & \text{if } k < a, \\
1 & \text{if } a \leq k \leq n-a, \\
\frac{1}{2}\{1-\cos[\pi(n-k)/a]\} & \text{if } a > n-k.
\end{cases} \tag{39}
$$

The data are multiplied by these weights after subtracting the mean but before applying the FFT. After applying a data taper the periodogram still has two degrees of freedom but adjoining frequencies are correlated; that is, the average of the periodogram from k adjoining frequencies has less than 2k degrees of freedom. The cost of lowering the side lobes is a slight increase in variability. The actual spectral window can be calculated by forming an array of the b_k's, extending it with zeros to form a series of length several times the length of the original series and calculating the periodogram of the array.

The effects of a data taper can be determined by calculating the total degrees of freedom (TDF),

$$
TDF = (\sum_{k=1}^{n} b_k^2)^2 / \sum_{k=1}^{n} b_k^4. \tag{40}
$$

If the b_k's are all equal, TDF = n. But if the b_k's are not equal, TDF < n. For the cosine taper,

$$
TDF = n - 34a/35. \tag{41}
$$

If a = n/2, so that the taper is over the whole data span, TDF = 18n/35. If a data taper is used, and the periodogram averaged over k adjoining frequencies, the adjustments to the equivalent degrees of freedom and bandwidth are

$$edf = 2[1 + (k-1)TDF/n],$$

$$BW = \frac{1}{h}[\frac{1}{TDF} + \frac{(k-1)}{n}].$$

(42)

An interesting special case of this result is when m zeros are annexed to the data before transforming in order to make the supplemented sample size n+m more factorable, or a power of two, or to give the periodogram a closer frequency spacing so that pure sine waves cannot slip between the peaks. When n-1 zeros have been annexed, all the information is available and nothing is gained by annexing more zeros. In the case of annexing m zeros, (42) becomes

$$edf = 2[1 + (k-1)TDF/(n+m)],$$

$$BW = \frac{1}{h}[\frac{1}{TDF} + \frac{(k-1)}{n+m}],$$

(43)

and if no data taper is used on the portion of data available of length n, TDF = n.

Another possibility for smoothing the periodogram is to use unequal weights

$$\hat{S}(\frac{v}{nh}) = \sum_{\mu} W_{\mu} I_{v-\mu},$$

(44)

where $W_{\mu} \geq 0$ and

$$\sum_{\mu} W_{\mu} = 1.$$

(45)

Usually the weights would be symmetrical about zero and taper to zero for increasing $|\mu|$. In this case,

$$edf = 2/\sum_{\mu} W_{\mu}^2,$$

(46)

which would be reduced by a factor of TDF/n if a data taper were used. The bandwidth would be

$$BW = edf/2nh.$$

(47)

5. ESTIMATION VIA THE COVARIANCE FUNCTION

Assuming that the mean-value function has been properly subtracted from the data, the covariance function can be estimated as

$$\hat{C}(\tau) = \frac{1}{n} \sum_{t=1}^{n-\tau} x(t+\tau)x(t). \tag{48}$$

It is possible to estimate the covariance function for values of the lag τ from 0 to n-1. However, when τ approaches n, the estimate becomes less stable since it is based on fewer observations. Dividing by n, rather than the actual number of terms in the sum n-τ, gives estimates which appear less variable for large lags and produces an estimated covariance function which has all the properties of an actual covariance function. An important property of a covariance function is that it is non-negative definite so that its transform (11) produces a non-negative spectral density. This would be the case if $C(\tau)$ is estimated for all possible lags, 0 to n-1 and substituted into (11) with the summation index going from -(n-1) to (n-1). In fact, for values of f equal to the periodogram frequencies of

$$f = \frac{\nu}{nh}, \quad \nu = 0, 1, \ldots, [n/2], \tag{49}$$

this estimate is algebraically identical to the periodogram and therefore has the same characteristics of high variability and narrow bandwidth.

If the estimated covariance function is truncated at some lag m < n-1 before substitution into (11), a new problem arises. Truncation in the time domain produces "ringing" in the frequency domain; that is, a spectral window with large negative and positive side lobes. This property can be seen by writing the truncated series as

$$S_T(f) = h \sum_{k=-m}^{m} C(kh)\cos(2\pi khf)$$

$$= h \sum_{k=-\infty}^{\infty} w_k C(kh)\cos(2\pi khf), \tag{50}$$

where the weights w_k are

$$w_k = \begin{cases} 1 & \text{if } |k| \leq m, \\ 0 & \text{if } |k| > m. \end{cases} \tag{51}$$

This multiplication of the covariance function by a weight function in the time domain results in a convolution in the frequency domain,

$$S_T(f) = \int_{-1/2h}^{1/2h} S(f-u)W(u)du, \tag{52}$$

where

$$W(u) = h \sum_{k=-\infty}^{\infty} w_k \cos(2\pi khu). \tag{53}$$

For the weights of (51), (53) can be expressed in closed form as

$$W(u) = h \sum_{k=-m}^{m} \cos(2\pi khu) = h \frac{\sin[\pi(2m+1)hu]}{\sin(\pi hu)}. \tag{54}$$

Note that where the denominator is zero, the numerator is also zero and the limit of the ratio as these points are approached is $h(2m+1)$; that is,

$$W(0) = h(2m+1). \tag{55}$$

This function is cyclic with period $1/h$. A slight modification which has side lobes that decay faster involves replacing the end weights by 1/2, giving

$$w_k = \begin{array}{ll} 1 & \text{if } |k| < m, \\ \frac{1}{2} & \text{if } |k| = m, \\ 0 & \text{if } |k| > m, \end{array} \tag{56}$$

and

$$W(u) = h \frac{\sin(2\pi mhu)}{\tan(\pi hu)}. \tag{57}$$

A plot of the first few side lobes is shown in Figure 3.

To overcome this side lobe problem, Blackman and Tukey (1958) advised truncating at some lag m, calculating the transform at a proper frequency spacing and smoothing in such a way that the side lobes nearly canceled. The choice of m necessitated a compromise between small bandwidth (large m) and low variability (small m). They also pointed out that this is equivalent to multiplying the estimated covariance function by weights (lag window) that start at one for lag zero and taper to zero by some truncation lag m. Then came a period of "window carpentry" when various lag windows were suggested in the literature. The main feature of a good weighting function is that its transform, the "spectral window," has small side lobes. Another desirable feature is that side lobes which exist be positive to avoid the possibility of negative spectrum estimates, which sometimes occurred with Blackman and Tukey's estimates.

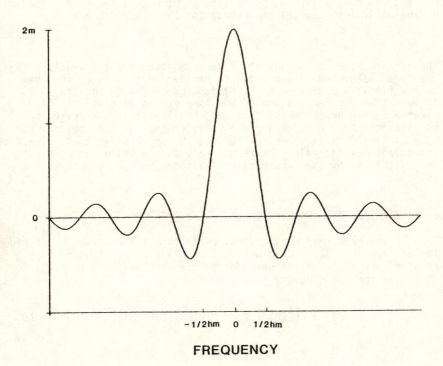

FREQUENCY

Figure 3. Spectral window when truncating the covariance function at lag m and replacing lag m by half its value.

Parzen's window is an example of a weighting function whose spectral window has small positive side lobes. The lag window is

$$
w_k = \begin{cases}
1 - 6(k/m)^2 + 6(k/m)^3 & \text{if } k < m/2, \\
2(1 - k/m)^3 & \text{if } m/2 \le k < m, \\
0 & \text{if } k \ge m.
\end{cases} \tag{58}
$$

For a given sample size n, truncation point m < n, and sample spacing h, the equivalent degrees of freedom and the bandwidth are

$$
\text{edf} \simeq 3.7n/m, \quad BW = 1.85/hm. \tag{59}
$$

The edf are reduced to half at frequencies zero and the Nyquist frequency of $1/2h$, and an extra degree of freedom is lost at zero by subtracting the sample mean. (59) is used to design the analysis. If more than one realization of the time series is available, the covariance can be estimated from each realization and averaged over realizations before weighting and transforming, or the spectrum can be estimated from each realization and the spectra averaged. The two procedures are equivalent, with the only change being that the edf are multiplied by the number of realizations, giving a more stable estimate.

6. ESTIMATING THE COVARIANCE FUNCTION VIA TWO PASSES OF THE FFT

It was mentioned earlier that estimating the covariance function for all possible lags, 0 to n-1, using (48) and substituting the results into (11) for the periodogram frequencies is another way to calculate the periodogram, resulting in

$$
I_\nu = h[\hat{C}(0) + 2 \sum_{k=1}^{n-1} \hat{C}(k)\cos(2\pi k\nu/n)], \tag{60}
$$

$$
\nu = 0, 1, \ldots, [n/2].
$$

If the inverse transform is calculated from

$$
\hat{C}_c(k) = \frac{1}{hn}[I_0 + 2 \sum_{\nu=1}^{[(n-1)/2]} I_\nu \cos(2\pi k\nu/n) + I_{n/2}\cos(\pi k)], \tag{61}
$$

where the brackets denote the integer part and the last term is present only when n is even, then the result is the so-called "circular" estimate of the covariance function. As if the data were wrapped around a circle,

$$\hat{C}_c(1) = \frac{1}{n}[x_1 x_2 + x_2 x_3 + \cdots + x_{n-1} x_n + x_n x_1],$$

$$\hat{C}_c(2) = \frac{1}{n}[x_1 x_3 + x_2 x_4 + \cdots + x_{n-2} x_n + x_{n-1} x_1 + x_n x_2]. \tag{62}$$

This "wrap around" problem can be solved by annexing zeros to the data before transforming. If the covariance function is to be estimated for lags up to m, at least m zeros must be annexed to the data to avoid this wrap around problem.

It may seem surprising that estimating a covariance function may be faster using two passes of a FFT than by directly accumulating the lag product sums. This is often true because of the number of multiplications required by the direct method. The algorithm is as follows:

(a) Remove the sample mean or any cyclic mean from the data.

(b) Annex m zeros to the data, forming a sequence of length n+m

$$x_1, \; x_2, \; \ldots, \; x_n, \; 0, \; \ldots, \; 0$$

(c) Fourier transform this sequence obtaining the z_ν's, and form the periodogram from

$$I_\nu = \frac{h}{n} |z_\nu|^2. \tag{63}$$

The frequency corresponding to the index ν is now $\nu/(n+m)h$ because of the artificial lengthening of the series.

(d) The estimated covariance function is then calculated from (61), except that each n is replaced by n+m and the results are discarded for lags greater than m because of wrap around.

There are various tricks for carrying out these calculations. For example, if the data after subtracting the mean and augmenting by zeros are stored in the real part of the complex array of the FFT, and the FFT called, the periodogram can be calculated in this array by replacing the real part by the sum of squares of the real and imaginary part and replacing the imaginary part by zero. The FFT can then be called again producing the covariance function, except for a constant multiple, in the real part of the array with the elements not affected by wrap around (lags 0 to m) in locations 1 to m+1. The spectral density can be estimated by multiplying the covariance function by the lag window w_k. Remembering the symmetry condition, the values in locations 2 and n+m would be multiplied by w_1, locations 3 and n+m-1 by w_2, ..., locations m and n+2 by w_{m-1}, and the rest of the middle part of the array gets replaced by zero. The imaginary part of the array should also be zeroed since it contains "round off zeros." A third call to the FFT gives the estimate of the spectral density in the real part of locations 1 to $[(m+n)/2]$ + 1, corresponding to frequencies 0, $1/h(n+m)$, $2/h(n+m)$, ..., $1/2h$. Since the spectral density is often plotted on a log scale normalized relative to the highest value, it is not necessary to worry about constants during these calculations.

7. THE EFFECTS OF FILTERING

Good references are available on digital filtering, such as the book by Hamming (1977). What needs to be stressed is that the researcher must be aware of the effects of any preprocessing of the data on the final conclusions. Subtracting a mean leaves a hole in the spectrum around zero frequency. Subtracting an equally weighted moving average induces peaks in the spectrum.

A linear digital filter is a set of weights c_j used in the operation

$$y_k = \sum_j c_j x_{k-j}. \tag{64}$$

In practice the limits of the summation are finite. If $c_j = 0$ for $j < 0$, the filter involves the current and past values of x_k and is called realizable since the filtering can be carried out in real time as the data become available. However, for non-real time applications in a digital computer, future values can be used and c_j is often chosen symmetric

$$c_j = c_{-j} \tag{65}$$

and the span finite

$$y_k = \sum_{j=-m}^{m} c_j x_{k-j}. \tag{66}$$

The frequency response function of the filter is

$$H(f) = \sum_j c_j e^{2\pi i j h f}, \tag{67}$$

which is real if the c_j's are symmetric, but in general is complex. The absolute value of this function is called the gain of the filter and, if the c_j's are not symmetric, the phase shift is

$$\theta(f) = \tan^{-1}\{[\sum_j c_j \sin(2\pi jhf)]/[\sum_j c_j \cos(2\pi jhf)]\}. \qquad (68)$$

Both the gain and phase shift are a function of frequency. If the input to the filter is a pure sine wave of frequency f, the output will be a pure sine wave of frequency f (since this is a linear filter), and the gain is the factor by which the amplitude is changed. The frequency response function is of interest when purchasing stereo amplifiers, and should be of equal interest whenever linear operations are applied to data.

The effect of filtering on the spectrum is the square of the frequency response function. This <u>variance transfer function</u> multiplies the original spectrum to give the spectrum of the process after filtering

$$S_0(f) = |H(f)|^2 S(f). \qquad (69)$$

A common filtering method used to remove slowly varying components in the data is an equally weighted moving average. This is probably the worst possible method and is <u>not</u> recommended. The effects of this method of trend removal are shown here to stress the bad effects. An equally weighted moving average with smoothing span of 2m+1 is

$$y_k = \frac{1}{2m+1} \sum_{j=-m}^{m} x_{k-j}, \qquad (70)$$

with frequency response function

$$H(f) = \frac{\sin[(2m+1)\pi hf]}{(2m+1)\sin(\pi hf)}, \qquad (71)$$

which is very similar to the curve shown in Figure 3. The variance transfer function has a shape similar to Figure 1, except that the first zero appears at f = 1/h(2m+1) rather than at f = 1/T.

When an equally weighted moving average is subtracted from the data for the purpose of trend removal, the frequency response function is 1 - H(f), and the variance transfer function is

$$|1 - H(f)|^2, \qquad (72)$$

which is shown in Figure 4.

The first peak induced by this procedure is approximately at the frequency 1.44/(2m+1)h. Since 2m+1 is the smoothing span and h the sampling interval, (2m+1)h is the time span of the moving average. In order to smooth out the ripple induced by subtracting a moving

210

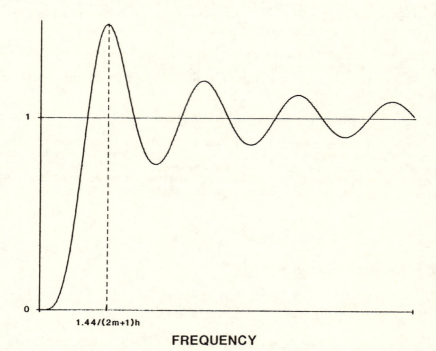

FREQUENCY

Figure 4. Variance transfer function when an equally weighted moving average with span 2m+1 is subtracted from the data.

average, the bandwidth of the spectrum estimate must be at least $2/(2m+1)h$.

An interesting special case of the effect shown in Figure 4 is the result of subtracting the sample mean from the data. This is the frequency response function if $2m+1$ is replaced by the length of the series n. The variance transfer function passes through 1.0 at the periodogram frequencies of ν/nh. The bandwidth of the periodogram is $1/nh$. The periodogram estimate of the spectrum is not greatly biased by the ripple, since the spectral window includes both a low area and a high area. However, if zeros are annexed to the data after subtracting the mean, the periodogram is at a closer spacing. If the spectrum is estimated by averaging the periodogram over realizations, the ripples in the variance transfer function of subtracting the mean may show up in the estimated spectrum.

Another method of inducing an artificial peak in the spectrum is to subtract a properly weighted moving average from data with a "red noise" spectrum. This would be data with positive serial correlation so that the spectral density is higher at lower frequencies. Figure 5 shows a "red" spectral density, the variance transfer function of a reasonable high pass filter, and the resulting spectrum after filtering. To avoid this problem, estimates should only be interpreted above the cutoff frequency of the filter, and should probably not even be plotted below this frequency.

It is important to realize that apparent peaks in estimated spectra can be due to either digital preprocessing of the data or to statistical variability of the estimate itself. Plots of both the variance transfer function of any data processing and confidence intervals for the estimates will help avoid these problems.

8. TESTING HYPOTHESES

A frequency domain test for stationarity can be obtained by dividing the data into segments such as years and estimating the spectral density separately for each section. If the spectrum estimates are calculated at a frequency spacing greater than or equal to the bandwidth of the estimates, and the estimates at zero frequency and the Nyquist frequency are omitted because of the decrease in degrees of freedom, an analysis of variance test can be used on the logarithms of the spectrum estimates (Priestly and Rao, 1969). Taking logarithms of the estimates is a variance stabilizing transformation, and the transformed estimates are closer to a Gaussian distribution. Also the variance of the transformed spectrum estimate is known and depends only on edf (Jones et al., 1972). The variance of the log periodogram (edf = 2) is $\pi^2/6 \simeq 1.645$. For larger degrees of freedom, a good approximation for the variance of the logarithm of a spectrum estimator is

212

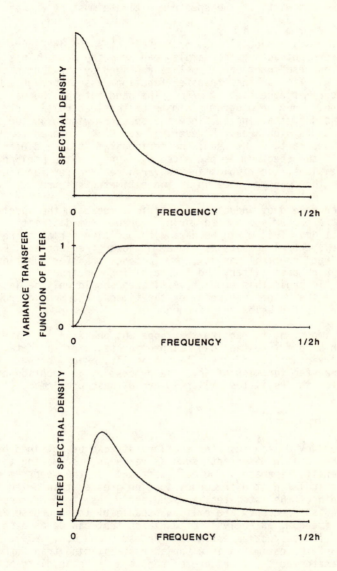

Figure 5. The effect of removing slowly varying component of a time series with a "red noise" spectral density using a high pass digital filter which induces a peak.

$$Var(\ell) \approx \frac{2}{edf}(1 + \frac{1}{edf}).\qquad(73)$$

The analysis of variance test is carried out by arranging the logarithm of the spectrum estimates in a two-way table with frequency in one direction and time segment in the other. In analysis of variance terms, the "frequency effect" would be a test for constant spectral densities at all frequencies; that is, "white noise." The "time effect" tests the hypothesis that the spectral density at each frequency is constant over time; that is, the process is stationary. Since the error variance is known, the usual F-tests can be replaced by chi-squared tests.

A common situation is when the data are known to be non-white, and it is of interest to test only for stationarity. This hypothesis can be tested quite simply by testing whether the estimated variance of the log of the spectrum estimates is significantly greater than the theoretical variance for a stationary process. In Table 1, ℓ_{ij} indicates the natural logarithm of the spectral density from time segment i and frequency j, and $\overline{\ell}_{.j}$ denotes the average of column j. Calculating the sum of squares of deviations about these means,

$$SS = \sum_{i=1}^{n_1} \sum_{j=1}^{n_2} (\ell_{ij} - \overline{\ell}_{.j})^2.\qquad(74)$$

Then the following statistic can be used as a test for stationarity:

$$SS/Var(\ell) \sim \chi^2_{n_2(n_1-1)}.\qquad(75)$$

Note that one degree of freedom is lost from each column, because of subtracting the sample mean of the column.

A test for white noise based on a single realization of a time series is the cumulated periodogram (Durbin, 1967; Jones and Kearns, 1976). The periodogram is calculated and cumulated starting at the first frequency. It is then normalized so that the sum at the Nyquist frequency is one. If the spectrum is flat, the cumulated periodogram should not wander too far from the diagonal. Two lines can be drawn on each side of the diagonal at the proper distance, which depends on the sample size and significance level of the test. The distance can be obtained from tables of the Kolmogorov-Smirnov test.

It is possible to test whether a harmonic of a periodic mean function is zero. We assume that the true value of harmonic ν of the periodic mean-value function is z_ν and is estimated from m realizations by $\hat{z}_\nu$ as in (30). The variance of the estimate of the mean

Table 1. Data array of logarithms of spectrum estimates to be used
as a test for stationarity. The frequency spacing should
be at least as wide as the bandwidth of the estimates and
should exclude zero and the Nyquist frequencies. Time
denotes estimates based on different time segments of the
time series.

Frequency

$$\ell_{11} \quad \ell_{12} \quad \cdots \quad \ell_{1n_2}$$

$$\ell_{21} \quad \ell_{22} \quad \cdots \quad \ell_{2n_2}$$

Time

$$\ell_{n_1 1} \quad \ell_{n_1 2} \quad \cdots \quad \ell_{n_1 n_2}$$

Column means $\quad \bar{\ell}_{.1} \quad \bar{\ell}_{.2} \quad \cdots \quad \bar{\ell}_{.n_2}$

$$\text{Var}(\hat{z}_\nu) = E(|\hat{z}_\nu - z_\nu|^2) \tag{76}$$

can be estimated as

$$\hat{V} = \frac{1}{m(m-1)} \sum_{j=1}^{m} |z_\nu^{(j)} - \hat{z}_\nu|^2. \tag{77}$$

Except for a constant, this is the periodogram and has $2(m-1)$ degrees of freedom. The test for harmonic ν being zero is

$$|\hat{z}_\nu|^2/\hat{V} \sim F_{2,2(m-1)}. \tag{78}$$

An observed value of the test statistic larger than a value obtained from tables of the F distribution leads to rejection of the hypothesis that this harmonic is zero. It is possible to carry out this test from a single realization which extends over several cycles by estimating the variance from neighboring frequencies, making sure that harmonics of the frequency being tested are not included.

9. MULTIVARIATE TIME SERIES

If more than a single variable is observed at each time, the data can be arranged as a column vector

$$X(t) = \begin{bmatrix} x_1(t) \\ x_2(t) \\ \vdots \\ x_d(t) \end{bmatrix}. \tag{79}$$

The mean of the process is now a vector

$$E[X(t)] = \begin{bmatrix} \mu_1(t) \\ \mu_2(t) \\ \vdots \\ \mu_d(t) \end{bmatrix}. \tag{80}$$

For a stationary process, the means of the components must be constant, but often in meteorology there will be diurnal and annual components which must be handled for each variable. Assuming that the means have been subtracted, the covariance function is a sequence of d by d matrices, defined at lag t as

$$C(t) = E\{X(s+t)[X(s)]'\}. \tag{81}$$

The prime in (81) denotes the transposed vector, so this is the "outer product" of two vectors consisting of all possible products of the elements. For a two-dimensional process,

$$C(t) = \begin{bmatrix} E[x_1(s+t)x_1(s)] & E[x_1(s+t)x_2(s)] \\ \\ E[x_2(s+t)x_1(s)] & E[x_2(s+t)x_2(s)] \end{bmatrix} = \begin{bmatrix} C_{11}(t) & C_{12}(t) \\ \\ C_{21}(t) & C_{22}(t) \end{bmatrix}. \tag{82}$$

The diagonal elements of this matrix are the covariance functions of the two components of the process. The off-diagonal elements are the cross covariance functions and satisfy the relation

$$C_{21}(t) = C_{12}(-t), \tag{83}$$

so the covariance function matrices satisfy

$$C(-t) = [C(t)]'. \tag{84}$$

The cross covariance function is useful for determining leading or lagging relationships between two series.

For a two-dimensional process, the spectral density matrix

$$S(f) = \begin{bmatrix} S_{11}(f) & S_{12}(f) \\ \\ S_{21}(f) & S_{22}(f) \end{bmatrix} \tag{85}$$

has elements which are Fourier transforms of the corresponding covariance

$$S_{j\ell}(f) = h \sum_{k=-\infty}^{\infty} C_{j\ell}(kh)e^{-2\pi ikhf}. \tag{86}$$

The diagonal elements, $j = k$, are real because of the symmetry of the covariance function,

$$C_{jj}(kh) = C_{jj}(-kh), \tag{87}$$

and are the spectral densities of the component processes. The off-diagonal elements are complex, and are called the cross spectral densities with the following symmetry condition:

$$S_{j\ell}(f) = \overline{S_{\ell j}(f)}, \tag{88}$$

where the bar denotes complex conjugate (the spectral density matrix is Hermitian). The real part of the cross spectral density is the cospectral density

$$co_{j\ell}(f) = h \sum_{k=-\infty}^{\infty} C_{j\ell}(kh)\cos(2\pi khf), \qquad (89)$$

and the imaginary part is the <u>quadrature spectral density</u>

$$q_{j\ell}(f) = h \sum_{k=-\infty}^{\infty} C_{j\ell}(kh)\sin(2\pi khf). \qquad (90)$$

The cospectral density and quadrature spectral density are usually hard to interpret, because they have dimensions equal to the product of the dimensions of the two series. Therefore, they are usually reexpressed as <u>coherence</u>

$$Ch_{j\ell}(f) = \frac{co_{j\ell}^2(f) + q_{j\ell}^2(f)}{S_{jj}(f)S_{\ell\ell}(f)}, \qquad (91)$$

and <u>phase</u>

$$\theta_{j\ell}(f) = \arctan\,[q_{j\ell}(f)/co_{j\ell}(f)]. \qquad (92)$$

The coherence is in the range

$$0 \leq Ch(f) \leq 1, \qquad (93)$$

and measures the relationship between the two series as a function of frequency, similar to a correlation squared. The phase measures the lead or lag as a function of frequency.

As in the univariate case, the multivariate spectral density can be estimated either by transforming the weighted estimated covariance functions and cross covariance functions or by direct Fourier transforming of each component of the multivariate time series. If the transform of each component is written in vector form as

$$Z_\nu = \sum_{k=0}^{n-1} X(kh)e^{2\pi ik\nu/n}, \qquad (94)$$

the multivariate periodogram is a d by d matrix obtained by forming the outer product of this vector with its complex conjugate (the complex conjugate transposed vector will be denoted by *)

$$I_\nu = \frac{h}{n} Z_\nu Z_\nu^*. \qquad (95)$$

As in the univariate case, this estimate with two degrees of freedom has very high variability. In fact, it cannot be used to estimate the coherence, since this estimate would always be one. It is necessary to increase the degrees of freedom by averaging I_ν over realizations or adjoining frequencies.

To test the hypotheses that the true coherence at some frequency is equal to zero, the α significance level (e.g., $\alpha = 0.05$), can be calculated from

$$Ch_\alpha = 1 - \alpha^{2/(edf-2)}. \tag{96}$$

If the estimated coherence is greater than this value, the hypothesis of zero coherence is rejected. Again, it must be remembered that this test is for a single frequency chosen in advance of collecting the data. If a test for all frequencies is desired at the 5% level, α should be replaced by

$$1 - (1-\alpha)^{n/edf} = 0.05,$$
$$\alpha \simeq 0.05edf/n, \tag{97}$$

since there are approximately n/edf independent tests. Approximate confidence intervals on the phase can be obtained by the method of Groves and Hannan (1968) as

$$\hat{\theta}(f) \pm \arc\sin \{t_\alpha[\frac{1}{edf-2} (\frac{1}{Ch(f)} - 1)]\}^{1/2}, \tag{98}$$

where t_α is from two-tailed t-tables with edf minus two degrees of freedom. If the quantity in brackets is greater than one, the confidence interval includes the entire circle. This is most likely to be the case in frequency bands of low coherence where the phase is poorly determined.

Another important topic treated by Hannan and Thomson (1973) is the estimation of the time lag (or group delay) between two time series. This is the time lag for which the cross correlation is greatest. Often, simply calculating the cross correlation and looking for the peak gives very poor resolution of this time lag since the peak can be broad and flat. The cross spectral density is weighted by a function of the coherence, then transformed to give a modified cross covariance function which is searched for its maximum

$$\tilde{C}_{12}(kh) = \int_{-1/2h}^{1/2h} S_{12}(f) \frac{Ch(f)}{|S_{12}(f)|[1-Ch(f)]} e^{2\pi ikhf}df. \tag{99}$$

This procedure weights the cross spectrum in such a way as to maximize the information for determining time lag.

If $x(t)$ and $y(t)$ are the input and output of a real linear system with additive noise, the system can be expressed as

$$y(t) = \sum_{k=0}^{\infty} w(k)x(t-k) + n(t), \qquad (100)$$

where the $w(k)$'s are the linear weights relating the input to the output known as the _impulse response function_, and $n(t)$ is a stationary noise process which is uncorrelated with the $x(t)$ process. Multiplying (100) by $x(t-\tau)$ and taking expected values gives

$$C_{yx}(\tau) = \sum_{k=0}^{\infty} w(k)C_{xx}(\tau-k), \qquad (101)$$

where C_{yx} is the cross covariance function between the input and output processes and C_{xx} is the covariance function of the input process. The noise term vanishes because of the assumption of zero correlation with the input. Taking Fourier transforms gives

$$S_{yx}(f) = H(f)S_{xx}(f), \qquad (102)$$

where

$$H(f) = \sum_{k=0}^{\infty} w(k)e^{-2\pi ikhf} \qquad (103)$$

is called the _frequency response function_ or _transfer function_ of the linear system and is a complex function with amplitude (called the _gain_) and phase.

Estimation of the frequency response function in the frequency domain has been considered by Goodman et al. (1961) and Akaike and Yamanonchi (1962), and can be treated as regression analysis with complex variables over frequency bands that are not so large that the frequency response function changes significantly. Let $\hat{S}_{xx}(f)$, $\hat{S}_{yy}(f)$, and $\hat{S}_{yx}(f)$ be estimates of the input, output, and cross spectral densities with edf greater than two degrees of freedom. By a complete analogy to regression analysis, the estimate of the frequency response function is

$$\hat{H}(f) = \hat{S}_{yx}(f)/\hat{S}_{xx}(f), \qquad (104)$$

and the noise spectral density can be estimated by the analog of the residual sum of squares as

$$\hat{S}_{nn}(f) = \frac{edf}{edf-2}[\hat{S}_{yy}(f) - |\hat{S}_{yx}(f)|^2/\hat{S}_{xx}(f)] \qquad (105)$$

and has edf minus two degrees of freedom. The F-test for the hypothesis that $H(f) = 0$ is

$$\frac{edf|\hat{S}_{yx}(f)|^2}{2\hat{S}_{xx}(f)\hat{S}_{nn}(f)} \sim F_{2,edf-2}, \tag{106}$$

is equivalent to the test that the coherence is zero (96).

REFERENCES

Akaike, H., and Y. Yamanonchi, 1962: On the statistical estimation of frequency response function. Annals of the Institute of Statistical Mathematics, 14, 23-56.

Blackman, R. B., and J. W. Tukey, 1958: The Measurement of Power Spectra from the Point of View of Communications Engineering. New York, Dover Publications.

Durbin, J., 1967: Tests of serial independence based on the cumulated periodogram. Bulletin of the International Statistical Institute, 42, 1041-1048.

Goodman, N. R., S. Katz, B. H. Kramer, and M. T. Kuo, 1961: Frequency response from stationary noise. Two case histories. Technometrics, 3, 245-268.

Groves, G. W., and E. J. Hannan, 1968: Time series regression of sea level on weather. Reviews of Geophysics, 6, 129-174.

Hamming, R. W., 1977: Digital Filters. Englewood Cliffs, N.J., Prentice-Hall.

Hannan, E. J., and P. J. Thomson, 1973: Estimating group delay. Biometrika, 60, 241-253.

Jones, R. H., 1964: Spectral analysis and linear prediction of meteorological time series. Journal of Applied Meteorology, 3, 45-52.

Jones, R. H., 1965: A reappraisal of the periodogram in spectral analysis. Technometrics, 7, 531-542.

Jones, R. H., D. H. Crowell, J. K. Nakagawa, and L. Kapuniai, 1972: Statistical comparisons of EEG spectra before and during simulation in human neonates. Computers in Biomedicine, Supplement to Proceedings of Fifth Hawaii International Conference on System Sciences. North Hollywood, Calif., Western Periodicals, pp. 18-21.

Jones, R. H., and J. P. Kearns, 1976: Fortaleza, Ceara, Brazil rainfall. _Journal of Applied Meteorology_, _15_, 307-308.

Priestly, M. D., and T. S. Rao, 1969: A test for non-stationarity of time series. _Journal of the Royal Statistical Society, Series B_, _31_, 140-149.

6
Time Series Analysis— Time Domain

Richard H. Jones

1. INTRODUCTION

Time domain methods of time series analysis use parametric models to represent the stochastic structure of the process, for the purposes of prediction, spectrum estimation, hypothesis testing, and helping to understand relationships between series. The book by Box and Jenkins (1976) is an excellent reference for these methods. However, it must be remembered that Box and Jenkins developed these techniques mainly from business, economic, and chemical engineering applications. Because many of these processes do not have a tendency to return to some mean level, differencing this type of data in order to try to obtain a stationary process may be quite appropriate. On the other hand, meteorological processes do have a tendency to return to a mean level, even though diurnal and annual variations may be present. Differencing this type of data is not appropriate, because information about how far an observation is from the mean is thrown away.

The oldest parametric methods and the easiest to understand and calculate are the autoregressive methods; that is, methods that involve regressing on the past history of the time series. There are now several computing algorithms for applying autoregressive methods. These algorithms can be best explained by considering a first-order autoregression.

2. FIRST-ORDER AUTOREGRESSION

A first-order autoregression for a zero-mean process can be expressed as

$$x(t) = \alpha x(t-1) + \varepsilon(t), \tag{1}$$

where $\varepsilon(t)$ is a random input at time t; that is, having zero mean, being uncorrelated at different times, and having variance σ^2. For the process to be stationary, it is necessary that $|\alpha| < 1$. It is most common for α to be in the range $0 < \alpha < 1$. In this case, the process is sometimes referred to as "red noise," since the spectral density

$$S(f) = h\sigma^2[1 + \alpha^2 - 2\alpha\cos(2\pi hf)]^{-1}, \quad 0 \le f \le 1/2h, \quad (2)$$

is higher at lower frequencies (longer wavelengths) similar to red light. Here h is the time interval between observations.

There are various possibilities for estimating the two parameters α and σ^2. One possibility is to use linear regression. Let $x_1, x_2, \ldots, x_n$ be the observations after subtracting the mean or removing a periodic mean function. The data can be arranged for regression as follows:

$$x_2 = \alpha x_1 + \varepsilon_2, \; x_3 = \alpha x_2 + \varepsilon_3, \; \ldots, \; x_n = \alpha x_{n-1} + \varepsilon_n. \quad (3)$$

The least squares approach is to choose the estimate, a_{LS} say, of α, so that the sum of squares of residuals

$$RSS = \sum_{t=2}^{n} (x_t - a_{LS}x_{t-1})^2 \quad (4)$$

is minimized, giving

$$a_{LS} = (\sum_{t=2}^{n} x_{t-1}x_t)/(\sum_{t=2}^{n} x_{t-1}^2). \quad (5)$$

The estimate of σ^2 is then

$$s^2 = RSS/(n-3). \quad (6)$$

The three in the denominator represents the n-1 terms in the sum minus the degrees of freedom lost by subtracting the sample mean and estimating α. This linear regression or least squares approach has the advantage of giving an unbiased estimate of α, but the disadvantage that the estimate may not be in the range $-1 < a_{LS} < 1$, so the estimate may not produce a stationary process.

A second method of estimating α is based on the estimated variance and lag-one covariance of the process. We note that in (5) the numerator is a sum of products of lag one, a quantity that is used to estimate the lag-one covariance. The denominator is the sum of squares used to estimate the variance, except that the term x_n^2 is missing. The covariance function of a first-order autoregression can be calculated by multiplying (1) by x(t-k) for k > 0 and taking expected values, obtaining

$$C_k = \alpha C_{k-1}. \tag{7}$$

(7) is called the Yule-Walker equation for a first-order autoregression and shows the exponential decay with increasing lag, a property which is used by Box and Jenkins for the identification of this process. In particular,

$$C_1 = \alpha C_0. \tag{8}$$

Solving for α and substituting in the estimated covariances

$$\hat{C}_k = \frac{1}{n} \sum_{t=1}^{n-k} x_{t+k} x_t, \quad k = 0, 1, \tag{9}$$

gives the estimate for α of

$$a_{YW} = (\sum_{t=2}^{n} x_{t-1} x_t) / (\sum_{t=1}^{n} x_t^2), \tag{10}$$

which is the same as (5) except for an extra term in the denominator. This estimate (10), based on the Yule-Walker equations and positive definitive estimate of the covariance function (9), always produces a stationary process (i.e., $-1 < a_{YW} < 1$), but has the disadvantage that the estimate is biased towards zero.

A third method for estimating α was used by Burg (1967) in his maximum entropy algorithm. It has the advantages of the two above methods without the disadvantages. It produces an unbiased estimate a_{ME} in the interval $-1 \leq a_{ME} \leq 1$. It should be pointed out that all these methods are asymptotically equivalent as n tends to infinity.

Burg noted that if time were to go backwards, the autoregression model would be the same,

$$x_t = \alpha x_{t+1} + \eta_t, \tag{11}$$

where the random input η_t is not the same as ε_t, but has the same mean and variance. The least squares estimate based on a backwards autoregression is

$$a_B = (\sum_{t=1}^{n-1} x_{t+1} x_t) / (\sum_{t=1}^{n-1} x_{t+1}^2). \tag{12}$$

The numerator of this estimate is the same as the numerator of the forward regression estimate (5), but the denominators differ as to which term is dropped from the sum. Burg's maximum entropy estimate a_{ME} is obtained by averaging the two denominators, yielding

$$a_{ME} = (\sum_{t=1}^{n-1} x_{t+1}x_t)/(\frac{1}{2}x_1^2 + \sum_{t=2}^{n-1} x_t^2 + \frac{1}{2}x_n^2). \tag{13}$$

One way to see the differences among the three estimates, (5), (10), and (13), is to look at the special case of two observations. For $n = 2$,

$$a_{LS} = \frac{x_1 x_2}{x_1^2} = \frac{x_2}{x_1},$$

which is in the range $-\infty < a_{LS} < \infty$;

$$a_{YW} = \frac{x_1 x_2}{x_1^2 + x_2^2},$$

which is in the range $-1/2 \le a_{YW} \le 1/2$; and

$$a_{ME} = \frac{x_1 x_2}{\frac{1}{2}(x_1^2 + x_2^2)},$$

which is in the range $-1 \le a_{ME} \le 1$. Although this last estimate could fall on the nonstationary boundary of $a_{ME} = \pm 1$, this event would occur with probability zero.

Another estimate is the exact maximum likelihood estimate under the assumption that the errors are Gaussian. The likelihood function is

$$L(\alpha, \sigma^2) = \frac{1}{(2\pi)^{n/2}|C|^{1/2}} e^{-\frac{1}{2}x'C^{-1}x}, \tag{14}$$

where x is a column vector of the n observations, x' denotes its transpose, C is the n by n covariance matrix of the observations, and $|C|$ its determinant. Here C is a Toeplitz matrix, constant along every diagonal; that is,

$$C = \begin{bmatrix} c_0 & c_1 & c_2 & \cdots & c_{n-1} \\ c_1 & c_0 & c_1 & \cdots & c_{n-2} \\ c_2 & c_1 & c_0 & \cdots & c_{n-3} \\ \cdot & \cdot & \cdot & & \cdot \\ \cdot & \cdot & \cdot & & \cdot \\ c_{n-1} & c_{n-2} & c_{n-3} & \cdots & c_0 \end{bmatrix} . \tag{15}$$

The relationship between c_0 and the parameters α and σ^2 can be obtained by multiplying (1) by $x(t)$ and taking the expected value. Noting that $E[x(t)\epsilon(t)] = \sigma^2$,

$$c_0 = \frac{\sigma^2}{1 - \alpha^2}. \tag{16}$$

The matrix C can be written as

$$C = \frac{\sigma^2}{1-\alpha^2} \begin{bmatrix} 1 & \alpha & \alpha^2 & \cdots & \alpha^{n-1} \\ \alpha & 1 & \alpha & \cdots & \alpha^{n-2} \\ \alpha^2 & \alpha & 1 & \cdots & \alpha^{n-3} \\ \cdot & \cdot & \cdot & & \cdot \\ \cdot & \cdot & \cdot & & \cdot \\ \cdot & \cdot & \cdot & & \cdot \\ \alpha^{n-1} & \alpha^{n-2} & \alpha^{n-3} & \cdots & 1 \end{bmatrix} . \tag{17}$$

The inverse of this matrix is tridiagonal (Siddiqui, 1958) and constant along the diagonals except for the two corners; specifically,

$$C^{-1} = \frac{1}{\sigma^2} \begin{bmatrix} 1 & -\alpha & & & 0 \\ -\alpha & 1+\alpha^2 & -\alpha & & \\ & -\alpha & \ddots & \ddots & \\ & & \ddots & 1+\alpha^2 & -\alpha \\ 0 & & & -\alpha & 1 \end{bmatrix} . \tag{18}$$

It can be factored into the product of an upper triangular matrix, with only the diagonal and super diagonal non-zero, and its transpose (Jones, 1977); that is,

$$C^{-1} = UU', \tag{19}$$

where

$$
U = \frac{1}{\sigma}
\begin{bmatrix}
(1-\alpha^2)^{1/2} & -\alpha & & & & \\
0 & 1 & \ddots & & & \\
 & \ddots & \ddots & \ddots & & 0 \\
 & & \ddots & \ddots & \ddots & \\
0 & & \ddots & \ddots & -\alpha & \\
 & & & \ddots & & \\
 & & & 0 & & 1
\end{bmatrix} .
\tag{20}
$$

Now

$$
x'U = \frac{1}{\sigma}[(1-\alpha^2)^{1/2}x_1, \; x_2 - \alpha x_1, \; \ldots, \; x_n - \alpha x_{n-1}],
\tag{21}
$$

which is a vector of one-step prediction errors divided by the one-step prediction standard deviation σ, except that the first observation (which cannot be predicted) is divided by the standard deviation of the process. If a slightly modified residual sum of squares is defined as

$$
RSS = (1-\alpha^2)x_1^2 + \sum_{t=2}^{n}(x_t - \alpha x_{t-1})^2,
\tag{22}
$$

then the quadratic form in (14) can be expressed as

$$
x'C^{-1}x = (x'U)(x'U)' = RSS/\sigma^2.
\tag{23}
$$

The determinant also can be easily calculated from the factored matrix. Since it is a triangular matrix, the determinant of U is the product of the diagonal elements, giving

$$
|U| = [(1-\alpha^2)^{1/2}]/\sigma^n,
$$
$$
|C| = [(\sigma^2)^n]/(1-\alpha^2).
\tag{24}
$$

The maximum likelihood estimates of α and σ^2 can be obtained by maximizing $L(\alpha,\sigma^2)$ with respect to α and σ^2, replacing the vector x by the data (which are assumed to have zero mean). It is equivalent (and more convenient) to minimize -2 ln(likelihood) dropping constants that do not depend on the parameters. In this case,

$$
\ell(\alpha,\sigma^2) \equiv -2 \ln[L(\alpha,\sigma^2)] = \ln|C| + x'C^{-1}x
$$
$$
= n \ln(\sigma^2) - \ln(1-\alpha^2) + RSS/\sigma^2.
\tag{25}
$$

Differentiating with respect to σ^2 and setting the result equal to zero gives

$$\sigma^2 = RSS/n. \tag{26}$$

Substituting (26) back into (25) gives the function to be minimized with respect to α. Dropping terms which do not depend on α,

$$\ell(\alpha) = n \ln(RSS) - \ln(1-\alpha^2). \tag{27}$$

Differentiating (27) with respect to α and equating the result to zero gives a cubic to be solved for α. A root will be close to the estimates obtained by the other methods and typically (27) is solved by numerical optimization programs using one of the other estimates as an initial guess. For any reasonable sample size, this approach should only give a slight refinement of the estimate since all four methods are asymptotically equivalent.

A recurring problem in meteorology is that the parameters of a model may be different in the winter than in the summer, or different during the night than in the day. This periodic structure is discussed by Jones and Brelsford (1967). It is a different phenomenon than periodic means or normal values, as it applies to the correlation structure of the data. For example, the day-to-day correlation may be higher during seasons when the weather is more predictable. The usual approach to this problem is to divide the data into seasons and analyze each season separately. An alternative for parametric models is to allow the parameters to be periodic. In this case,

$$x(t) = \alpha(t)x(t-1) + \varepsilon(t), \tag{28}$$

where $\alpha(t)$ is periodic with period n [i.e., $\alpha(t+n) = \alpha(t)$] and has a Fourier representation

$$\alpha(t) = \alpha_0 + \sum_{k=1}^{m} [\alpha_k \cos(2\pi kt/n) + \beta_k \sin(2\pi kt/n)]. \tag{29}$$

Substituting (29) into (28) gives a model which is linear in the parameters α_k and β_k so that they can be simply estimated.

3. HIGHER-ORDER AUTOREGRESSIONS

The model for an autoregression of order p for a zero-mean process is

$$x(t) = \sum_{k=1}^{p} \alpha_k x(t-k) + \varepsilon(t), \tag{30}$$

which is a linear regression on the past. The errors $\varepsilon(t)$ have zero mean, are uncorrelated, and have variance σ^2. The methods of the previous section generalize to higher-order autoregressions. Discussion of the maximum likelihood method will be postponed until Section 8.

Given n observations x_1, x_2, ..., x_n, lagged values of the data can be used in a usual linear regression equation

$$y = X\beta + \varepsilon, \tag{31}$$

where

$$y = \begin{bmatrix} x_{p+1} \\ x_{p+2} \\ \vdots \\ x_n \end{bmatrix}, \quad X = \begin{bmatrix} x_p & x_{p-1} & \cdots & x_1 \\ x_{p+1} & x_p & \cdots & x_2 \\ \vdots & \vdots & & \vdots \\ x_{n-1} & x_{n-2} & \cdots & x_{n-p} \end{bmatrix},$$

$$\tag{32}$$

$$\beta = \begin{bmatrix} \alpha_1 \\ \alpha_2 \\ \vdots \\ \alpha_p \end{bmatrix}, \quad \varepsilon = \begin{bmatrix} \varepsilon_{p+1} \\ \varepsilon_{p+2} \\ \vdots \\ \varepsilon_n \end{bmatrix}.$$

The least squares estimates of the regression coefficients are the solutions of the so-called "normal equations"

$$(X'X)b = X'y \tag{33}$$

or

$$b = \begin{bmatrix} a_1 \\ a_2 \\ \vdots \\ a_p \end{bmatrix} = (X'X)^{-1}X'y. \tag{34}$$

The covariance matrix of the estimated autoregression coefficients is

$$\text{Cov}(b) = \sigma^2(X'X)^{-1}, \tag{35}$$

and σ^2 can be estimated from the residual sum of squares

$$\hat{\sigma}^2 = (y - Xb)'(y - Xb)/(n - 2p - 1)$$

$$\tag{36}$$

$$= y'(y - Xb)/(n - 2p - 1).$$

Since p data points are lost at the beginning of the series in order to start the predictions, the degrees of freedom in the denominator are the n-p observations minus p degrees of freedom for the estimated coefficients and one for the subtraction of the sample mean.

Any available regression program may be used for this analysis including stepwise and subset selection routines (Furnival and Wilson, 1974). The latter is referred to as subset autoregression (McClave, 1975). In applying these procedures, the X'X matrix and X'y vector would be calculated using the maximum value of p to be considered.

The elements of the p by p X'X matrix and p by one X'y vector are

$$(X'X)_{jk} = \sum_{t=p+1}^{n} x_{t-j} x_{t-k},$$

$$(X'y)_j = \sum_{t=p+1}^{n} x_t x_{t-j}. \tag{37}$$

These differ only by the elimination of a few terms from the ends of the summation of the lag product sums

$$\Gamma_k = \sum_{t=1}^{n-k} x_{t+k} x_t, \quad k \geq 0. \tag{38}$$

The positive definite estimate of the covariance function [see Eq. (48) of Chapter 5] is

$$\hat{C}(k) = \Gamma_k/n. \tag{39}$$

The Yule-Walker equations for an autoregression of order p are

$$C_k = \alpha_1 C_{k-1} + \alpha_2 C_{k-2} + \cdots + \alpha_p C_{k-p}, \quad k > 0. \tag{40}$$

These equations, for k = 1, 2, ..., p, give a system of equations for the α_k's in terms of the covariances as follows:

$$
\begin{bmatrix}
C_0 & C_1 & C_2 & \cdots & C_{p-1} \\
C_1 & C_0 & C_1 & \cdots & C_{p-2} \\
C_2 & C_1 & C_0 & \cdots & C_{p-3} \\
\vdots & \vdots & \vdots & & \vdots \\
C_{p-1} & C_{p-2} & C_{p-3} & \cdots & C_0
\end{bmatrix}
\begin{bmatrix}
\alpha_1 \\
\alpha_2 \\
\alpha_3 \\
\vdots \\
\alpha_p
\end{bmatrix}
=
\begin{bmatrix}
C_1 \\
C_2 \\
C_3 \\
\vdots \\
C_p
\end{bmatrix}. \tag{41}
$$

Replacing the covariances by their positive definite estimates [i.e., replacing the C_k's in (40) by the lag product sums r_k], gives

$$
\begin{bmatrix}
r_0 & r_1 & r_2 & \cdots & r_{p-1} \\
r_1 & r_0 & r_1 & \cdots & r_{p-2} \\
r_2 & r_1 & r_0 & \cdots & r_{p-3} \\
\cdot & \cdot & \cdot & & \cdot \\
\cdot & \cdot & \cdot & & \cdot \\
\cdot & \cdot & \cdot & & \cdot \\
r_{p-1} & r_{p-2} & r_{p-3} & \cdots & r_0
\end{bmatrix}
\begin{bmatrix}
a_1 \\
a_2 \\
a_3 \\
\cdot \\
\cdot \\
\cdot \\
a_p
\end{bmatrix}
=
\begin{bmatrix}
r_1 \\
r_2 \\
r_3 \\
\cdot \\
\cdot \\
\cdot \\
r_p
\end{bmatrix}
\quad (42)
$$

These equations differ from the normal equations of linear regression only by end effects, which become small as the sample size increases. However, there is an important difference between the two estimation techniques. The estimated autoregression coefficients based on (42) always produce a stationary process, whereas this is not necessarily true for the estimates based on the linear regression procedure. In this regard, we note that a process is stationary if the roots of the characteristic equation

$$
1 - \sum_{k=1}^{p} a_k z^k = 0 \quad (43)
$$

are outside the unit circle. Another advantage of the estimate based on the Yule-Walker equations and positive definite estimate of the covariance function is that the system of equations to be solved contains only $p+1$ distinct numbers $r_0, r_1, \ldots, r_p$, which reduces the computations. Also, the Toeplitz form of the matrix, constant along every diagonal, allows a simplified stepwise solution. As in the first-order autoregression, this estimate has the disadvantage that for small sample sizes it is biased.

The stepwise solution due to Levinson (1947) and Durbin (1960) fits autoregressions of increasing order. Let $a_k^{(p)}$ be a_k calculated when fitting an autoregression of order p, and S_p be the analog of the residual sum of squares in linear regression. The recursion starts with

$$
a_1^{(1)} = r_1/r_0, \quad S_1 = r_0 - a_1^{(1)} r_1. \quad (44)
$$

The general step starts by first calculating the largest order coefficient, known as the partial autocorrelation coefficient or reflection coefficient, using the recursive relationship

$$a_p^{(p)} = [\Gamma_p - \sum_{k=1}^{p-1} a_k^{(p-1)} \Gamma_{p-k}]/S_{p-1}. \tag{45}$$

The other coefficients are then updated by

$$a_k^{(p)} = a_k^{(p-1)} - a_p^{(p)} a_{p-k}^{(p-1)}, \quad k = 1, 2, \ldots, p-1, \tag{46}$$

and

$$S_p = \{1 - [a_p^{(p)}]^2\} S_{p-1}, \tag{47}$$

completing the recursion.

Burg's maximum entropy algorithm makes use of the above recursion, but is based on residuals from forward and backward autoregressions. It produces stationary estimates without the bias of the Yule-Walker method when the sample size is small. Suppose the maximum entropy estimate for a first-order autoregression from (13) is used to generate residuals

$$e_t^{(1)} = x_t - a_1^{(1)} x_{t-1}, \quad t = 2, 3, \ldots, n, \tag{48}$$

from forward predictions and residuals

$$n_t^{(1)} = x_t - a_1^{(1)} x_{t+1}, \quad t = 1, 2, \ldots, n-1, \tag{49}$$

from backward predictions. Here the superscript denotes that the autoregression is first order. The general step in the maximum entropy recursion can now be written, assuming that the coefficients and forward and backwards residuals are available for an autoregression of order p-1, as

$$a_p^{(p)} = [2 \sum_{t=p+1}^{n} e_t^{(p-1)} n_{t-p}^{(p-1)}]/\{\sum_{t=1}^{n-p}[n_t^{(p-1)}]^2 + \sum_{t=p+1}^{n}[e_t^{(p-1)}]^2\}. \tag{50}$$

The residuals can be updated without actually calculating all of the new autoregression coefficients by

$$e_t^{(p)} = e_t^{(p-1)} - a_p^{(p)} n_{t-p}^{(p-1)}, \quad t = p+1, p+2, \ldots, n,$$

$$n_t^{(p)} = n_t^{(p-1)} - a_p^{(p)} e_{t+p}^{(p-1)}, \quad t = 1, 2, \ldots, n-p. \tag{51}$$

The estimate (50) is the value of $a_p^{(p)}$ which minimizes the mean of the forward and backward residual sum of squares

$$S_p = \frac{1}{2} \sum_{t=p+1}^{n} [e_t^{(p)}]^2 + \frac{1}{2} \sum_{t=1}^{n-p} [n_t^{(p)}]^2. \tag{52}$$

This procedure can be started at order zero by letting

$$\varepsilon_t^{(0)} = n_t^{(0)} = x_t, \quad S_0 = \Gamma_0. \tag{53}$$

Now the residual sum of squares can be updated at each step using (47). It is possible to calculate all the autoregression coefficients for each order from (46), but this is not necessary for the recursion and can be postponed until the order of the autoregression to be used has been selected. The recursion gives the partial autoregression coefficients, the only information necessary to calculate the auto-regression coefficients of any order.

One of the reasons for fitting autoregressions to data is to make predictions. If the autoregression coefficients were known exactly, a one-step prediction would be

$$x^{(1)}(t+1) = \sum_{k=1}^{p} \alpha_k x(t+1-k) \tag{54}$$

and would have variance

$$E\{[x(t) - x^{(1)}(t)]^2\} = \sigma^2. \tag{55}$$

When the autoregression coefficients are estimated, this prediction variance increases slightly, the increase depending on the sample size; the larger the sample size, the less the increase. This problem will be discussed in the next section under order selection, and the σ^2 used here should be replaced by the Final Prediction Error (FPE) introduced in the next section.

Predictions of any number of steps into the future can be made, but the prediction variance increases, and in the limit approaches the variance of the process (indicating no skill). A two-step prediction is

$$x^{(2)}(t+2) = \alpha_1 x^{(1)}(t+1) + \sum_{k=2}^{p} \alpha_k x(t+2-k). \tag{56}$$

Let the variance of a k-step prediction be

$$V_k = E\{[x(t+k) - x^{(k)}(t+k)]^2\}, \tag{57}$$

noting that $V_1 = \sigma^2$. The variance of a two-step prediction can be calculated by substituting

$$x(t+2) = \sum_{k=1}^{p} \alpha_k x(t+2-k) + \varepsilon(t+2), \tag{58}$$

yielding

$$V_2 = E\{[\epsilon(t+2) + \alpha_1\epsilon(t+1)]^2\}$$

$$= \sigma^2(1+\alpha_1^2), \tag{59}$$

since the innovations $\epsilon(t)$ are uncorrelated. An autoregression can be inverted into a one-sided moving average

$$x(t) = \epsilon(t) + \sum_{k=1}^{\infty} \psi_k\epsilon(t-k) \tag{60}$$

by substitution. The following recursion gives the ψ_k's in terms of the α_k's:

$$\psi_1 = \alpha_1,$$

$$\psi_k = \alpha_k + \sum_{j=1}^{k-1} \psi_j\alpha_{k-j}, \quad k > 1. \tag{61}$$

The k-step prediction variance can be calculated from the recursion

$$V_k = V_{k-1} + \psi_{k-1}^2\sigma^2, \quad k > 1, \tag{62}$$

with

$$\lim_{k\to\infty} V_k = C_0, \tag{63}$$

the variance of the process.

Another use of autoregression is called "prewhitening" (sometimes referred to as "predictive deconvolution"). When the autoregressive coefficients have been estimated, the errors (or "innovations") $\epsilon(t)$ are estimated by residuals

$$e(t) = x(t) - \sum_{k=1}^{p} a_k x(t-k). \tag{64}$$

The $e(t)$ series is the prewhitened series and, if the model is correct, should be very close to white noise. In fact, the fit of the model to the data can be evaluated by testing these residuals for whiteness using the cumulative periodogram (see Chapter 5). Prewhitening is often useful to remove major features in the spectrum before spectrum estimation (see Chapter 5) to reduce sidelobe leakage problems.

The frequency response function of the prewhitening operation is

$$A(f) = 1 - \sum_{k=1}^{p} a_k e^{2\pi ikhf}. \tag{65}$$

If the model is correct and the spectral density of the prewhitened series e(t) is truly white, it will have a spectral density which is a constant; namely,

$$S_e(f) = h\sigma^2, \quad |f| \leq 1/2h. \tag{66}$$

Recalling the results on the variance transfer function of a filter [see Eq. (69) of Chapter 5],

$$S_e(f) = |A(f)|^2 S_x(f), \tag{67}$$

so the spectral density of the x(t) series is

$$S(f) = h\sigma^2 |A(f)|^{-2}. \tag{68}$$

This spectral density is a function of the autoregressive coefficients in A(f); the innovation variance σ^2, which is estimated from the sum of squares of residuals; and the sampling interval h. Calculating S(f) from estimated parameters is known as "autoregressive spectrum estimation" or, if the maximum entropy algorithm is used, "maximum entropy spectrum estimation."

If prewhitening is used before estimating the spectrum with a windowed method (see Chapter 5), it need only be remembered that a linear operation with variance transfer function $|A(f)|^2$ was applied to the data before estimation so the resulting spectrum estimate should be divided by this function to obtain an estimate of the spectrum of the original data. This raises the question of why estimate the spectrum by windowed methods after prewhitening, when the result should be white noise with a flat spectrum and the spectrum already has been estimated by autoregressive methods? One reason is that the statistical variability of autoregressive spectrum estimates is very complicated. The estimates at widely separated frequencies can be highly correlated, as opposed to windowed estimates where the estimates are essentially uncorrelated if the spectral windows do not overlap. The confidence intervals for windowed estimates are more accurate if the spectral density is relatively flat with no strong peaks or valleys. Prewhitening by using autoregressions is an easy way to improve these estimates. Experience has shown (Jones, 1974) that estimating the spectrum by both autoregressive and windowed methods (with or without prewhitening) and plotting both estimates superimposed, with a confidence interval applying only to the windowed estimate, gives a good visual feeling for the true spectrum.

Predictive deconvolution is used to detect changes or errors in the time series. By removing the predictable part of each observation, any change is easier to detect. Normalizing the prediction residual through dividing by its standard deviation gives a dimensionless quantity $e(t)/\sigma$, which can be plotted. A common rule used to detect suspicious observations is to flag an observation if the absolute value of this normalized residual is greater than three. Predictive deconvolution has become an advanced art in seismology where it is used to map underground strata (Robinson, 1957).

4. ORDER SELECTION

When fitting autoregressions to data, the selection of an appropriate order of the autoregression is very important. This does not mean the identification of the "correct" order, because in the real world the correct order is often infinite with coefficients after a certain order being very small. Selecting a finite order is a compromise which approximates the real situation. Reasonable methods of order selection are based on how well the estimates will hold up on independent data, taking into account the variability introduced by estimating parameters. The critical point is to avoid overfitting, which fits the dependent data very well but not independent data.

Akaike's Final Prediction Error (FPE) estimates the one-step prediction variance when the estimated autoregression coefficients are used to make predictions on new (or independent) data (Akaike, 1969). An unbiased estimate of the one-step prediction variance (or mean squared error), $\hat{\sigma}^2$ say, is multiplied by a factor which represents the increase in prediction variance caused by estimating p parameters; namely,

$$FPE_p = \hat{\sigma}^2 (1 + \frac{p}{n}). \tag{69}$$

The order of the autoregression is chosen for which FPE_p is a minimum. The proper method of obtaining $\hat{\sigma}^2$ depends on the method used to estimate the autoregression coefficients. For the Yule-Walker method using the Levinson-Durbin stepwise procedure, the usual method is to take

$$\hat{\sigma}^2 = \frac{1}{n-p} S_p, \tag{70}$$

so that

$$FPE_p = \frac{1}{n} S_p \frac{n+p}{n-p}. \tag{71}$$

If the sample mean has been subtracted from the data, there are actually p+1 coefficients to be estimated, so

$$FPE_p = \frac{1}{n} S_p \frac{n + p + 1}{n - p - 1}. \tag{72}$$

Here the subscript p denotes the order of the autoregression. If several harmonics of a periodic mean were estimated from the data by regression and subtracted, the "one" in (72) would be replaced by the number of estimated coefficients. In the case where m realizations of length n are used and a periodic mean is removed by subtracting the mean at each time as in Eq. (26) of Chapter 5, the mean-value function estimated in this manner would probably not be used to make predictions. The procedure would be used to determine the order of the autoregression with the effect of the periodic mean removed. In this case, one realization worth of degrees of freedom has been donated to remove this periodic mean completely, and the effective sample size is n(m-1). The final prediction error would be

$$FPE_p = \frac{1}{n(m-1)} S_p \frac{n(m-1) + p}{n(m-1) - p}. \tag{73}$$

In the linear regression method of estimating autoregression coefficients, the sample size is reduced because time points are lost at the beginning of the series. Let n' be the actual length of the vectors in the regression equation. The most n' can be when fitting an autoregression of order p is n-p, but it may be less if the normal equations are calculated once for the largest order to be considered. If we let RSS be the residual sum of squares based on n' observations and assume a sample mean has been subtracted, then

$$FPE_p = \frac{1}{n'} RSS \frac{n' + p + 1}{n' - p - 1}. \tag{74}$$

When using the maximum entropy algorithm, the sample size reduces by one each time the order is increased by one, since the algorithm works entirely within the data span. In this case, it would seem that

$$\hat{\sigma}^2 = \frac{1}{n - 2p} S_p \tag{75}$$

would give an unbiased estimate of the one-step prediction variance. However, simulation studies (Jones, 1976) have shown that the estimate

$$\hat{\sigma}^2 = \frac{1}{n-p} S_p \tag{76}$$

is approximately unbiased if $p < n/2$. This is probably because the estimate of order p is not based entirely on a sample of size n-p but the estimates of lower order, which were based on longer data spans, also contributed. Therefore, for the maximum entropy algorithm,

$$FPE_p = \frac{1}{n} S_p \frac{n + p + 1}{n - p - 1}, \quad p < n/2. \tag{77}$$

Experience has shown that p should be restricted to this range. Note that the definition of S_p in Jones (1976) differs from S_p used here by a factor of n.

The FPE order selection criterion was extended (Akaike, 1973) to any maximum likelihood estimation situation, not just regression. In regression analysis, if the errors have independent Gaussian distributions with zero means and constant variances, then (except for an additive constant)

$$-2 \ln(\text{likelihood}) = n \ln(\text{RSS}). \tag{78}$$

Akaike's Information Criterion (AIC), when estimating p parameters, is

$$AIC_p = -2 \ln(\text{likelihood}) + 2p. \tag{79}$$

In the equations presented here, S_p plays the role of the residual sum of squares, so

$$AIC_p = n \ln(S_p) + 2p. \tag{80}$$

The model for which AIC_p is a minimum is selected. For large n, this procedure is asymptotically equivalent to the FPE criterion; specifically, except for a constant which does not depend on p,

$$AIC_p \simeq n \ln(FPE_p). \tag{81}$$

The AIC is related to chi-squared tests for testing differences in models based on changes in $-2 \ln(\text{likelihood})$ (Mood et al., 1974, p. 441). This relation gives a concept of whether two models are close competitors for being best. If a model has an AIC value within about two of the minimum AIC, it is a competitor and the model with fewer parameters would usually be selected. There has been some work (Shibata, 1976) showing that AIC may be an inconsistent estimate of the true order; that is, as the sample size increases, AIC may select too large an order autoregression. Choosing a positive constant c and saying that any model which falls within that distance of the minimum, gives a large probability that the true model is among the competitors (Duong, 1981). The AIC for competitors would be models for which

$$AIC_c = AIC_{min} + c. \tag{82}$$

This concept can be transformed back to FPE giving

$$FPE_c = FPE_{min} e^{c/n}. \tag{83}$$

A typical value would be c = 2.

5. AUTOREGRESSIVE-MOVING AVERAGE MODELS

Autoregressive-moving average (ARMA) models incorporate into predictions, not only past values of the data, but past values of the prediction residuals as well; namely,

$$x(t) = \sum_{k=1}^{p} \alpha_k x(t-k) + \sum_{k=1}^{q} \beta_k \epsilon(t-k) + \epsilon(t). \qquad (84)$$

The simplest possible situation would be $p = 1$ and $q = 1$, usually referred to as ARMA(1,1); that is,

$$x(t) = \alpha_1 x(t-1) + \beta_1 \epsilon(t-1) + \epsilon(t). \qquad (85)$$

This model says that there is information for prediction, not only in the previous observation, but in the size of the error made at the previous time point. The way this model is used in practice to make predictions is to replace the previous errors (which are unobservable) by prediction residuals. A one-step prediction would be

$$x^{(1)}(t+1) = \alpha_1 x(t) + \beta_1 [x(t) - x^{(1)}(t)]. \qquad (86)$$

It can be seen from back substitutions of the one-step predictions, that (86) is highly nonlinear in the parameters α_1 and β_1. In addition, there is a problem getting the prediction process started at $t = 1$. The parameters are much more difficult to estimate in the ARMA model than in the autoregressive model. However, by considering ARMA models, it is sometimes possible to obtain a more parsimonious model; that is, a model with fewer parameters which does as well or better than a model with more parameters.

The problem of estimating the parameters will be discussed only briefly here, since computer programs are available. We let

$$e(t) = x(t) - x^{(1)}(t) \qquad (87)$$

be residuals from one-step predictions, where

$$x^{(1)}(t) = \sum_{k=1}^{p} \alpha_k x(t-k) + \sum_{k=1}^{q} \beta_k e(t-k) \qquad (88)$$

are the one-step predictions. The estimates of the α_k's and β_k's can be obtained by nonlinear least squares, choosing the values which minimize

$$RSS = \sum_{t=1}^{n} e^2(t). \qquad (89)$$

Nonlinear optimization programs require initial guesses at the parameters and then carry out a systematic search for the minimum RSS. Recalling that the mean-value function has been removed from the data, the predictions could be started by assuming that

$$x(t) = 0, \quad e(t) = 0, \quad t \leq 0. \tag{90}$$

The method of starting the predictions is not important if the number of observations is large (say $n \geq 100$), but can make important differences for small sample sizes. Box and Jenkins (1976) often use a technique called "back forecasting" to obtain the necessary starting values of $x(t)$ and $e(t)$. This procedure starts at the end of the data using the best available α_k's and β_k's and forecasts backwards towards the beginning of the data and off the end to obtain the initial values. When programming the estimation using nonlinear optimization, it is only necessary to calculate RSS for given values of of α_k's and β_k's. The nonlinear optimization program then varies the α_k's and β_k's in its search for the minimum RSS.

One method of obtaining initial guesses at the α_k's and β_k's is to start with simple models and build up to more complicated models, setting the initial values of the new parameters in the model to zero at each stage. For example, since autoregressions are much easier to fit, an autoregressive model of order p can be fitted to the data by the methods of Section 3, and the moving average parameters added one at a time.

Another possibility, which is not commonly used, is to fit a "long" autoregression with the order chosen by some selection criterion as in Section 4. If the order of this autoregression is p', a search can be made among all models for which

$$p + q < p' \tag{91}$$

to see if a more parsimonious model can be found. The FPE or AIC procedures can still be used to select the best model. Since all order autoregressions have been tried up to p', there are $p'(p'-1)/2 - 1$ possible models left to try. This number could be reduced by only considering models for which $q \leq p$.

Initial guesses for the α_k's and β_k's can be obtained by generating one-step prediction residuals from the long autoregression and forming the regression equation

$$
\begin{bmatrix} x(p'+1) \\ x(p'+2) \\ \vdots \\ \\ x(n) \end{bmatrix} = \begin{bmatrix} x(p') & \cdots & x(p'-p+1) & e(p') & \cdots & e(p'-q+1) \\ x(p'+1) & \cdots & x(p'-p+2) & e(p'+1) & \cdots & e(p'-q+2) \\ \vdots & & \vdots & \vdots & & \vdots \\ x(n-1) & \cdots & x(n-p) & e(n-1) & \cdots & e(n-q) \end{bmatrix} \begin{bmatrix} \alpha_1 \\ \vdots \\ \alpha_p \\ \beta_1 \\ \vdots \\ \beta_q \end{bmatrix}
$$

$$
+ \begin{bmatrix} \varepsilon(p'+1) \\ \varepsilon(p'+2) \\ \vdots \\ \\ \varepsilon(n) \end{bmatrix} . \tag{92}
$$

The α_k's and β_k's calculated from this linear regression equation can be used as initial estimates in the nonlinear estimation.

When the model has been determined and the parameters estimated, the estimated spectrum can be calculated from

$$
S(f) = h_\sigma^2 \left| [1 + \sum_{k=1}^{q} \beta_k e^{2\pi i k h f}] / [1 - \sum_{k=1}^{p} \alpha_k e^{2\pi i k h f}] \right|^2 . \tag{93}
$$

This estimate can be plotted superimposed on a windowed estimate to give a better feeling for the shape of the true spectrum.

If the model which has been fit to the data is stationary [see (43)], predictions can be made any number of steps into the future, with the prediction variance converging to the variance of the process. As in (60)-(62), the process can be inverted into a one-sided moving average using the following recursion (Box and Jenkins, 1976, Eq. 5.2.3),

$$\psi_1 = \alpha_1 + \beta_1,$$

$$\psi_2 = \alpha_1\psi_1 + \alpha_2 + \beta_2,$$

$$\psi_3 = \alpha_1\psi_2 + \alpha_2\psi_1 + \beta_3, \tag{94}$$

$$\cdot$$
$$\cdot$$
$$\cdot$$

$$\psi_j = \alpha_1\psi_{j-1} + \cdots + \alpha_p\psi_{j-p} + \beta_j,$$

where

$$\psi_0 = 1,$$

$$\psi_j = 0, \quad j < 0, \tag{95}$$

$$\beta_j = 0, \quad j > q.$$

For $j > \max(p-1,q)$, the ψ_j's satisfy the difference equation

$$\psi_j = \alpha_1\psi_{j-1} + \alpha_2\psi_{j-2} + \cdots + \alpha_p\psi_{j-p}. \tag{96}$$

These calculations are very easy to implement on a computer because of their recursive nature.

The variance of a one-step prediction is σ^2, the variance of $\varepsilon(t)$, so an approximate one-step prediction 95% confidence interval is

$$x^{(1)}(t) \pm 2\sigma. \tag{97}$$

For a two-step prediction, the prediction error is

$$\varepsilon(t) + \psi_1\varepsilon(t-1), \tag{98}$$

with variance

$$V_2 = \sigma^2(1 + \psi_1^2). \tag{99}$$

A two-step prediction is

$$x^{(2)}(t) = \alpha_1 x^{(1)}(t-1) + \sum_{k=2}^{p} \alpha_k x(t-k) + \sum_{k=2}^{q} \beta_k e(t-k). \tag{100}$$

In this case, the previous one-step prediction is used for the auto-regressive part, but the corresponding prediction residual is unavailable for the moving average part. The two-step prediction confidence interval is

$$x^{(2)}(t) \pm 2(V_2)^{1/2}. \tag{101}$$

This procedure can be extended to predictions of any number of steps as in (62).

If the data are differenced before fitting models (a procedure not usually recommended in meteorology), it is important to be aware of the effect this has on the spectral density and model parameters. A simple first difference

$$y_t = x_t - x_{t-1} \tag{102}$$

has frequency response function

$$H(f) = 1 - e^{2\pi i h f} \tag{103}$$

and variance transfer function

$$|H(f)|^2 = 4\sin(2\pi h f). \tag{104}$$

If the original data are stationary, this is the function which multiplies the input spectrum to obtain the spectrum of the output of the differencing operation. This function converges to zero at zero frequency. If a first-order moving average is fit to the data, the numerator term of (93) is

$$1 + \beta_1 e^{2\pi i h f} \tag{105}$$

and the estimated value of β_1 will often be close to minus one. The moving average term is modeling the effect of differencing.

6. HYPOTHESIS TESTS

Since the coefficients of parametric models are estimated by least squares or maximum likelihood, standard statistical methods can be used to test hypotheses. These tests would usually be based on the "extra sum of squares" principle (Draper and Smith, 1981), or changes in -2 ln(likelihood). For example, testing whether two or more spectra are the same, or testing for stationarity by dividing a time series into several segments and testing whether the segments have the same spectrum, are equivalent. If there are m segments and segment i has n_i observations, then the total sample size is

$$n = \sum_{i=1}^{m} n_i. \tag{106}$$

Suppose that the same model with p parameters is fit independently to each segment by least squares, and that the residual sum of squares for segment i is denoted by RSS_i. This estimate would have $n_i - p$ degrees of freedom. If the data are now pooled and the model with p parameters is fit once to all the data, the residual sum of squares, RSS_c say, would have n-p degrees of freedom. This is sometimes referred to as the constrained model, since the parameters fit to each segment are constrained to be the same. The "extra sum of squares principle" provides an F-test for the null hypothesis that stochastic structure of the segments are the same. The residual sum of squares for the unconstrained model can be obtained by summing the residual sum of squares for the segments,

$$RSS_{uc} = \sum_{i=1}^{m} RSS_i \qquad (107)$$

and has n - mp degrees of freedom. The number of square terms entering into the sums RSS_c and RSS_{uc} should be the same, so loss of data due to initial conditions should be avoided. The use of the modified residual sum which occurs in maximum likelihood estimation [the special case of a first-order autoregression is shown in (22)], or "back forecasting" to obtain initial estimates, would give the correct number of square terms. The F-test is

$$F_{(m-1)p, n-mp} = \frac{(RSS_c - RSS_{uc})/(m-1)p}{RSS_{uc}/(n-mp)}. \qquad (108)$$

A large F-ratio rejects the hypothesis of stationarity, or that the segments have the same spectra.

When using maximum likelihood estimation for a Gaussian process, an asymptotically equivalent test based on the minimum values of

$$\ell = -2 \ln(\text{likelihood}), \qquad (109)$$

is obtained by fitting the model separately to the segments and then to the pooled data. Let ℓ_i be the value obtained for segment i, and ℓ_c the value for the constrained model fit to the pooled data. The unconstrained minimized value is

$$\ell_{uc} = \sum_{i=1}^{m} \ell_i, \qquad (110)$$

and the chi-squared test for stationarity, or that the segments have the same spectra, is

$$\chi^2_{(m-1)p} = \ell_c - \ell_{uc}, \qquad (111)$$

which has (m-1)p degrees of freedom. The degrees of freedom are the difference in the number of parameters fit to the constrained and unconstrained models.

A very simple test for white noise is to test the hypothesis of a zero-order autoregression against the alternative hypothesis of a first-order autoregression. This is equivalent to testing whether the lag-one serial correlation is zero. While this is a rather specific alternative hypothesis, it is very practical in many applications since serial correlation often shows up at lag one due to persistence in data. An easy way to carry out this test is to compute the correlation $\hat{\rho}$ and look the value up in tables of significance values of estimated correlation coefficients, or transform the estimate to a t-statistic,

$$t_{n-2} = \hat{\rho}[(n-2)/(1 - \hat{\rho}^2)]^{1/2}. \tag{112}$$

The degrees of freedom are the number of observations minus two (one for the sample mean and one for the estimated autoregression coefficient). For large n, this test statistic is approximately normally distributed, so a value of $|t_{n-2}| > 2$ would reject the hypothesis of white noise at approximately the 5% level.

An interesting variation on this test is to fit autoregressions of all orders up to some maximum m', chosen by the investigator using AIC to select the order. It has been shown by both theoretical arguments and simulations, that if the data are white noise, the procedure has about an 85% chance of selecting order zero. This is equivalent to testing the hypothesis of white noise at the 15% level against the alternative of an autoregression of any order.

7. MULTIVARIATE TIME SERIES

Stepwise fitting of autoregressions was generalized to multivariate time series by Robinson (1963) and Whittle (1963), and programmed and applied to meteorological time series by Jones (1964). The equations are given in these references and will not be repeated here. Equations for the extension of the AIC selection procedure are given in Jones (1974), which also gives examples of the use of multivariate autoregression for estimating spectra, coherence and phase. A multivariate extension of the maximum entropy algorithm was programmed by Jones (1978). Programs are also available for multivariate extensions of ARMA models (Akaike et al., 1975, 1976; Tiao and Box, 1979).

In multivariate prediction models, each time series is predicted from its own past and from the past of the other series. It is often of interest to pick out one of the series to be predicted, and explore relationships among that series, its past, and the past or past and present of other series. This is possible using a multivariate extension of the idea of subset autoregression introduced by McClave (1975). The leaps and bounds algorithm (Furnival and Wilson, 1974) can be used in these analyses.

As an example, consider the question of whether sunspot activity has any effect on drought. The data for this example were supplied to the author by J. Murray Mitchell of the National Oceanic and Atmospheric Administration (NOAA). The drought data are yearly areas of the western United States with Palmer index less than minus one (indicating drought) as reconstructed from tree-ring data (Fritts series). The sunspot data are yearly sunspot indices with the sign reversed on every other cycle because of the magnetic reversal which takes place on the sun. This series is referred to as the 22-year sunspot data, since there is one cycle approximately every 22 years, except during extended periods of low sunspot activity. Two time periods were studied, 1700-1970 and 1848-1970. In the latter, shorter period, the data are of higher quality.

We let D_t be the drought index for year t, and S_t be the sunspot index. From previous analysis of the two series as univariate time series, it was known that the drought series was fairly unstructured, showing only a first-order autoregressive structure with observational error (Jones, 1980). The sunspot data are relatively structured, and autoregressions of up to order nine are necessary to get good fits.

The full model considered for the leaps and bounds algorithm, after subtracting the mean of both series, is

$$D_t = \beta_1 D_{t-1} + \beta_2 S_t + \beta_3 S_{t-1} + \cdots + \beta_{11} S_{t-9} + \varepsilon_t. \qquad (113)$$

We note that it is necessary to include at least one lagged value of the variable to be predicted D_{t-1}, since the drought data are known to be serially correlated. This acts as a covariate which decreases the error variance and helps remove the serial correlation so that the tests of significance used in the analysis are meaningful. It is not of interest whether or not the coefficient β_1 is significantly different from zero. The null hypothesis is that

$$\beta_2 = \beta_3 = \cdots = \beta_{11} = 0.$$

The question of whether the term S_t is included on the right-hand side depends on the purpose of the analysis. If the purpose is prediction, this term should not be indluced. However, if the purpose is to look for relationships, including simultaneous relationships, this term can be included.

The result of the analysis using the 1700-1970 data, including only the single lagged value D_{t-1} in the model, is

$$D_t = \overline{D} + 0.34(D_{t-1} - \overline{D}), \qquad (114)$$

with $R^2 = 11.8\%$. In other words, 11.8% of the variation of the yearly drought index is explained by the previous year's drought

index. Using AIC to help pick the best model from the eleven independent variables gives the model

$$D_t = \overline{D} + 0.32(D_{t-1} - \overline{D}) + 0.024(S_t - \overline{S}), \qquad (115)$$

with $R^2 = 14.0\%$. The estimated coefficient 0.024 has a P-value of
0.01, indicating that there is one chance in a hundred of obtaining an
estimated coefficient this far or farther away from zero when the
true coefficient is actually zero (i.e., when no relationship exists
between the simultaneous sunspot index and the drought index). We
note that the percent variation explained increased by only 2.2% when
the sunspot index was included. This brings up the difference between
practical significance and statistical significance. This result
seems to be statistically significant, indicating that there may be a
relationship between sunspots and drought, but using the concurrent
sunspot index to predict drought would show very little skill. In
fact, it would be impossible to use (115) to make predictions, since
the yearly sunspot index would not be available until the year was
over and the drought conditions would already be known. The statistical significance indicates that there may be a relationship, but it
cannot be used for forecasting.

Using data from the years 1848-1970 gives the following equation
for drought alone:

$$D_t = \overline{D} + 0.40(D_{t-1} - \overline{D}), \qquad (116)$$

with $R^2 = 16.5\%$. No significant improvement in this model was
obtained by including the sunspot terms.

The final conclusion of this analysis is that there may be an
effect of sunspots on drought, but that this apparent effect could be
due to chance. Since it is not possible to duplicate the experiment
without waiting a couple of hundred years, the relationship is still
in the realm of speculation.

Another example concerns whether the aa-index magnetic data
(yearly means) helps to forecast the yearly sunspot index. In this
example, the eleven-year sunspot data were used without the sign
reversal on every other cycle. The magnetic data were obtained from
NOAA and were available for the years 1868-1976. Since both sunspots
and the magnetic index are highly structured series, the full model
considered by the leaps and bounds algorithm, after subtracting means,
was

$$S_t = \beta_1 S_{t-1} + \beta_2 S_{t-2} + \cdots + \beta_8 S_{t-8} + \beta_9 M_{t-1}$$

$$+ \beta_{10} M_{t-2} + \cdots + \beta_{16} M_{t-8} + \varepsilon_t, \qquad (117)$$

where M_t is the magnetic aa-index. We note that M_t is not included,
since this is a prediction model. The best subset of size two is the
following model:

$$S_t = \overline{S} + 1.03(S_{t-1} - \overline{S}) - 0.44(S_{t-3} - \overline{S}), \qquad (118)$$

with $R^2 = 83.8\%$. The best subset of size three is

$$S_t = \overline{S} + 1.04(S_{t-1} - \overline{S}) - 0.44(S_{t-3} - \overline{S}) + 1.15(M_{t-6} - \overline{M}), \qquad (119)$$

with $R^2 = 86.2\%$. The coefficient of the lag-six magnetic index term is significant with a P-value of 0.0001, indicating a strong statistical relationship between the magnetic aa-index and the sunspot index six years later. We note that this should not be interpreted as a cause-and-effect relationship, since there is feedback from the interaction of these two series.

The above examples involve expressions of input-output relationships between time series. Eq. (100) of Chapter 5,

$$y(t) = \sum_{k=0}^{\infty} w(k)x(t-k) + n(t), \qquad (120)$$

expresses an open loop relationship between two stationary time series. The term "open loop" means that the x(t) series is feeding information into the y(t) series, but the y(t) series is not feeding information back into the x(t) series. The noise series n(t) is a stationary series which is uncorrelated with the input series x(t). The weights w(k) are the impulse response function. While this is a linear system, direct estimation of the impulse response function by regressing the y(t) series on the present and past of the x(t) series is inefficient because of the serial correlation in the noise series. The basic assumption of regression analysis that the errors are uncorrelated is violated. Since the impulse response function may contain a large number of non-zero terms, a more parsimonious representation with fewer parameters may be desired. The approach to this problem taken by Box and Jenkins (1976) is to use nonlinear estimation techniques to fit ARMA models to the various series. Here the use of linear methods, which are much easier to compute is emphasized.

The time series can be modeled as follows (Akaike, 1968; Granger, 1969; Otomo et al., 1972; Granger and Newbold, 1977; Haugh and Box, 1977):

$$y_t = \sum_{k=1}^{\infty} a_k y_{t-k} + \sum_{k=0}^{\infty} b_k x_{t-k} + \varepsilon_t, \qquad (121)$$

where ε_t is an uncorrelated sequence of random variables with variance σ^2. Choosing maximum orders for the a_k's and b_k's by analyzing the univariate time series, subset regression programs can be used to estimate the significant coefficients. Letting

$$A(f) = 1 - \sum_{k=1}^{\infty} a_k e^{-2\pi ikhf},$$

$$\hspace{8cm} (122)$$

$$B(f) = \sum_{k=0}^{\infty} b_k e^{-2\pi ikhf}$$

gives a rational approximation to the frequency response function

$$H(f) = B(f)/A(f). \hspace{3cm} (123)$$

A by-product of this analysis is an estimate of the spectral density of the noise in (120)

$$S_n(f) = \sigma^2/|A(f)|^2. \hspace{3cm} (124)$$

In order to test for feedback, the model

$$x_t = \sum_{k=1}^{\infty} c_k x_{t-k} + \sum_{k=1}^{\infty} d_k y_{t-k} + n_t, \hspace{2cm} (125)$$

can be fitted to the data by again assuming finite orders of the autoregressions. We note here that the term y_t is not included in the model, since an instantaneous relationship between x_t and y_t could cause this term to be significant even if there is no feedback from y to x. It is not possible to determine the direction of instantaneous relationships from the data. This direction can only be determined on prior grounds. For example, it can be speculated that sunspots can affect the weather, but that the weather probably does not affect the sunspots. In (125), if all $d_k = 0$, there is no feedback from the y_t series to the x_t series, with the possible exception of instantaneous feedback from y_t to x_t which presumably has been ruled out on prior grounds.

The frequency response function of the model in (115) is

$$H(f) = 0.024/(1 - 0.32e^{-2\pi if}). \hspace{2cm} (126)$$

This expression can be inverted to give the impulse response function

$$H(f) = 0.024[1 + 0.032e^{-2\pi if} + (0.32)^2 e^{-4\pi if} + \ldots], \hspace{0.5cm} (127)$$

so that

$$W_k = 0.024(0.32)^k. \hspace{3cm} (128)$$

This inversion, however, is usually not necessary in practice, since the original form in (115) can be used to make predictions.

8. STATE SPACE RECURSIVE ESTIMATION

Kalman filtering (Kalman, 1960) is a popular term among engineers and others for models which can be formulated in a state space framework, with estimation and prediction being carried out recursively. Assuming knowledge of the model parameters and covariance matrices, the recursive estimation allows for real time calculations which keep up with the data. The concept of state space involves determining the minimum amount of information about the past and present needed to predict the future and incorporating this information into a vector. When model parameters are not known, for any given values of the parameters, the likelihood function can be calculated recursively and non-linear optimization programs used to find maximum likelihood estimates of the parameters. Because of the recursive form of the calculations, missing observations are handled very simply. By using continuous time models, unequally spaced data can also be handled.

The simplest possible example is a first-order autoregression

$$x(t+1) = ax(t) + \varepsilon(t). \tag{129}$$

Here $x(t)$ would be the state at time t with zero mean, and the parameter a plays the role in this scalar example of the state transition matrix. The state at time t is premultiplied by the state transition matrix to give the state at time $t+1$. The random input to the model $\varepsilon(t)$ is sometimes called the shock or plant noise and has variance σ^2. If the process is observed with error, the observation equation becomes

$$y(t) = x(t) + v(t), \tag{130}$$

where $v(t)$ is a white noise sequence with variance R. This model is equivalent to an ARMA(1,1) process. The notation $x(t+1|t)$ indicates the best estimate of $x(t+1)$ given observations up to time t. The variance of this estimated state, or in the general case, the covariance matrix, is

$$P(t+1|t) = E\{[x(t+1) - x(t+1|t)]^2\}. \tag{131}$$

To begin the recursion, it is necessary to specify the initial conditions. Before any data are collected, the best estimate of the state of a zero-mean process is zero; that is,

$$x(0|0) = 0. \tag{132}$$

The corresponding variance is the variance of the process

$$P(0|0) = \frac{\sigma^2}{1-a^2},$$ (133)

which is also the variance of a no-skill forecast. A general step of the recursion is as follows:

(a) Make a one-step forecast

$$x(t+1|t) = ax(t|t).$$ (134)

(b) Calculate its variance

$$P(t+1|t) = a^2 P(t|t) + \sigma^2.$$ (135)

(c) Predict the next observation by

$$y(t+1|t) = x(t+1|t).$$ (136)

(d) When the next observation is available, calculate the residual or "innovation"

$$I(t+1) = y(t+1) - y(t+1|t).$$ (137)

(e) The variance of the innovation is

$$V(t+1) = P(t+1|t) + R.$$ (138)

(f) If the errors are Gaussian, the contribution to -2 ln(likelihood) from this step of the recursion is

$$\ln[V(t+1)] + [I(t+1)]^2/V(t+1).$$ (139)

(g) The state is updated by

$$x(t+1|t+1) = x(t+1|t) + P(t+1|t)I(t+1)/V(t+1).$$ (140)

This equation can be reexpressed as a weighted average of the old estimate and the new observation, weighting inversely proportional to the variances, so that

$$x(t+1|t+1) = \left[\frac{x(t+1|t)}{P(t+1|t)} + \frac{y(t+1)}{R}\right] \Big/ \left[\frac{1}{P(t+1|t)} + \frac{1}{R}\right].$$ (141)

(h) Finally, the variance is updated by

$$P(t+1|t+1) = P(t+1|t) - [P(t+1|t)]^2/V(t+1).$$ (142)

The -2 ln(likelihood) function, for the assumed values of a, σ^2, and R, is calculated by summing (139) over all the data points. The nonlinear optimization program then varies the values of the parameters in search of the minimum of -2 ln(likelihood). When data points are missing, only steps (a) and (b), which make a prediction and calculate its variance, are calculated. The algorithm then returns to step (a). This procedure gives the exact likelihood even when observations are missing. It is possible to remove σ^2 from the nonlinear estimation problem by differentiating -2 ln(likelihood) with respect to σ^2, setting the result equal to zero, solving for σ^2 in terms of the other parameters and data, and substituting the result back into -2 ln(likelihood) (Jones, 1980).

The general form of the Kalman state space model consists of two equations, the first of which is the state equation

$$X(t+1) = \Phi(t)X(t) + U(t), \tag{143}$$

where $X(t)$ is the state vector, $\Phi(t)$ is the state transition matrix, and $U(t)$ is a vector of the random plant noise with covariance matrix $Q(t)$. The second equation is the observation equation

$$Y(t) = M(t)X(t) + v(t), \tag{144}$$

where $M(t)$ is a matrix indicating which linear combinations of the state vector are observed. We note that the observation vector $Y(t)$ is not necessarily the same length as the state vector, it can be longer or shorter. The random observational error vector $v(t)$ has covariance matrix denoted by $R(t)$. The algorithm is as follows:

(a) Make a one-step forecast by

$$X(t+1|t) = \Phi(t)X(t|t). \tag{145}$$

(b) Calculate its covariance by

$$P(t+1|t) = \Phi(t)P(t|t)[\Phi(t)]' + Q(t). \tag{146}$$

(c) Predict the next observation vector by

$$Y(t+1|t) = M(t+1)X(t+1|t). \tag{147}$$

(d) When the next observation is available, calculate the innovation vector

$$I(t+1) = Y(t+1) - Y(t+1|t). \tag{148}$$

(e) The covariance matrix of the innovation vector is

$$V(t+1) = M(t+1)P(t+1|t)[M(t+1)]' + R(t+1). \tag{149}$$

(f) The contribution to -2 ln(likelihood) is

$$\ln|V(t+1)| + [I(t+1)]'[V(t+1)]^{-1}I(t+1), \tag{150}$$

where $|V(t+1)|$ is the determinant of the covariance matrix.

(g) The state is updated by

$$X(t+1|t+1) = X(t+1|t) + P(t+1|t)[M(t+1)]'[V(t+1)]^{-1}I(t+1). \tag{151}$$

(h) Finally, the state covariance matrix is updated by

$$P(t+1|t+1) = P(t+1|t) - P(t+1|t)[M(t)]'[V(t+1)]^{-1}M(t)P(t+1|t). \tag{152}$$

The details of fitting a general autoregressive-moving average model with missing observations using a state space representation are given in Jones (1980).

For unequally spaced data, when there is no basic sampling interval but data can be observed at arbitrary time points, it is necessary to formulate the problem in terms of continuous time models. Again, the simplest possible example is a continuous time "red noise" process, which satisfies the first-order equation

$$\frac{dx(t)}{dt} + \alpha x(t) = \varepsilon(t), \tag{153}$$

where α is positive and $\varepsilon(t)$ is a continuous time "white noise" process with spectral density σ^2. The spectral density of $x(t)$ is

$$S(f) = \sigma^2/[\alpha^2 + (2\pi f)^2], \tag{154}$$

and the covariance function is

$$C(\tau) = \frac{\sigma^2}{2\alpha} e^{-\alpha|\tau|}. \tag{155}$$

If this process is sampled at equally spaced times with time interval h, the result is a first-order autoregression with covariance function

$$C_k = \frac{\sigma^2}{2\alpha} e^{-\alpha h|k|}. \tag{156}$$

The autoregression coefficient of the discrete time process is

$$\alpha_d = e^{-\alpha h}, \tag{157}$$

which depends on the sampling interval. Note that the spectral density of the discrete time process can not be obtained from (154) simply by substituting parameters. Because of aliasing, it is necessary to sum over all aliased frequencies to obtain the expression for the discrete time spectrum (2).

Since this is a Markov process, the state of the system at any time t is simply the value of the process at that time. The equation of state which predicts the process at an arbitrary time in the future, given the present, is

$$x(t+\tau) = e^{-\alpha\tau} x(t) + u(t). \tag{158}$$

The variance of the random input u(t) depends on the length of the step and has variance (Kalman and Bucy, 1961)

$$Q(\tau) = \frac{\sigma^2}{2\alpha} [1 - e^{-2\alpha\tau}]. \tag{159}$$

We note that this expression approaches the variance of the process as the prediction interval τ becomes large.

The recursion for unequally spaced data starts, for a zero mean process, with

$$x(0|0) = 0. \tag{160}$$

The initial variance is

$$P(0|0) = \frac{\sigma^2}{2\alpha}. \tag{161}$$

This variance depends on the two unknown parameters of the process, α and σ^2. As before, σ^2 can be removed from the likelihood function and calculated after α has been estimated by nonlinear optimization. For given values of α and σ^2, the likelihood is calculated from the Kalman recursion:

(a) Predict from the present observation to the next observation time by

$$x(t_{k+1}|t_k) = e^{-\alpha(t_{k+1} - t_k)} x(t_k|t_k). \tag{162}$$

(b) Calculate its variance by

$$P(t_{k+1}|t_k) = e^{-2\alpha(t_{k+1} - t_k)} P(t_k|t_k) + Q, \tag{163}$$

where Q is calculated from (159).

(c) When the next observation is available, the innovation is

$$I(t_{k+1}) = y(t_{k+1}) - x(t_{k+1}|t_k). \tag{164}$$

(d) It has variance

$$V(t_{k+1}) = P(t_{k+1}|t_k) + R.$$

The parameter R is the observational error variance, if any. This could represent the variance caused by roundoff error or could be actual observational error. If R is unknown, it can also be estimated by this nonlinear optimization procedure.

(e) The contribution to -2 ln(likelihood) for this step, if the errors are Gaussian, is

$$\ln[V(t_{k+1})] + [I(t_{k+1})]^2/V(t_{k+1}). \qquad (165)$$

(f) The updated estimate of the state is

$$x(t_{k+1}|t_{k+1}) = x(t_{k+1}|t_k) + P(t_{k+1}|t_k)I(t_{k+1})/V(t_{k+1}). \qquad (166)$$

(g) The variance is then updated by

$$P(t_{k+1}|t_{k+1}) = P(t_{k+1}|t_k) - [P(t_{k+1}|t_k)]^2/V(t_{k+1}). \qquad (167)$$

To remove σ^2 from the recursion, we note that it can be factored out of P, Q, and V, and if observational error is included in the model, R is replaced by R/σ^2. For n observations at times t_1, t_2, ..., t_n, -2 ln(likelihood) is

$$\ell(\alpha,\sigma^2) = \sum_{k=1}^{n} \{\ln[\sigma^2 V(t_k)] + [I(t_k)]^2/\sigma^2 V(t_k)\}. \qquad (168)$$

Differentiating with respect to σ^2 and setting the result equal to zero gives

$$\sigma^2 = \frac{1}{n} \sum_{k=1}^{n} [I(t_k)]^2/V(t_k). \qquad (169)$$

Substituting back into (168) and dropping constants,

$$\ell(\alpha) = n \ln(\sigma^2) + \sum_{k=1}^{n} \ln[V(t_k)]. \qquad (170)$$

In this case,

$$[I(t_k)]^2/V(t_k), \quad \ln[V(t_k)]$$

are accumulated separately during the recursion in which σ^2 has been set equal to one. The parameter σ^2 is then calculated at the end from (169) and substituted into (170) to give -2 ln(likelihood), which is to be minimized with respect to α. If observational error is included in the model and R is estimated from the data, it must be remembered that in this form it is R/σ^2 which is actually being estimated so the final result must be multiplied by σ^2 to obtain the observational error variance.

This scalar Markov process has been treated by Robinson (1977). Extensions to higher-order continuous time autoregressions using state space representations are given in Jones (1981), including a computer program for calculating the exact likelihood function from unequally spaced data with observational error.

REFERENCES

Akaike, H., 1968: On the use of a linear model for the identification of feedback systems. Annals of the Institute of Statistical Mathematics, 20, 425-439.

Akaike, H., 1969: Fitting autoregressions for predictions. Annals of the Institute of Statistical Mathematics, 21, 243-247.

Akaike, H., 1973: Information theory and an extension of the maximum likelihood principle. Second International Symposium on Information Theory (B. N. Petrow and F. Csaki, editors). Budapest, Akademia Kaido, pp. 267-281.

Akaike, H., E. Arahata, and T. Ozaki, 1975, 1976: TIMSAC-74, A time series analysis and control program package (1)-(2). Computer Science Monographs, Nos. 5-6. Tokyo, Institute of Statistical Mathematics.

Box, G. E. P., and G. M. Jenkins, 1976: Time Series Analysis Forecasting and Control, Revised Edition. San Francisco, Holden-Day.

Burg, J. P., 1967: Maximum entropy spectral analysis. Paper presented at 37th Annual International SEG Meeting, 31 October, Oklahoma City, Okla.

Draper, N. R., and H. Smith, 1981: Applied Regression Analysis, Second Edition. New York, Wiley.

Duong, Q. P., 1981: On the choice of the order of autoregressive models: a ranking and selection approach. Department of Statistics and Actuarial Science, TR-81-08. London, Ontario, University of Western Ontario.

Durbin, J., 1960: The fitting of time series models. Review of the International Statistical Institute, 28, 233-289.

Furnival, G. M., and R. W. Wilson, 1974: Regression by leaps and bounds. Technometrics, 16, 499-511.

Granger, C. W. J., 1969: Investigating causal relations by econometrical models and cross-spectral methods. Econometrics, 37, 424-438.

Granger, C. W. J., and P. Newbold, 1977: Forecasting Economic Time Series. New York, Academic Press.

Haugh, L. D., and G. E. P. Box, 1977: Identification of dynamic regression (distributed lag) models connecting two time series. Journal of the American Statistical Association, 72, 121-130.

Jones, R. H., 1964: Prediction of multivariate time series. Journal of Applied Meteorology, 3, 285-289.

Jones, R. H., and W. M. Brelsford, 1967: Time series with periodic structure. Biometrika, 54, 403-408.

Jones, R. H., 1974: Identification and autoregressive spectrum estimation. IEEE Transactions on Automatic Control, Ac-19, 894-897.

Jones, R. H., 1976: Autoregression order selection. Geophysics, 41, 771-773.

Jones, R. H., 1977: Spectrum estimation from unequally spaced data. Preprints Fifth Conference on Probability and Statistics in Atmospheric Sciences. Boston, American Meteorological Society, pp. 277-282.

Jones, R. H., 1978: Multivariate autoregression estimation using residuals. Applied Time Series Analysis (D. F. Findley, editor). New York, Academic Press.

Jones, R. H., 1980: Maximum likelihood fitting of ARMA models to time series with missing observations. Technometrics, 22, 389-395.

Jones, R. H., 1981: Fitting a continuous time autoregression to discrete data. Applied Time Series Analysis, II (D. F. Findley, editor). New York, Academic Press.

Kalman, R. E., 1960: A new approach to linear filtering and prediction problems. Transactions of the ASME (Journal of Basic Engineering), 82D, 34-35.

Kalman, R. E., and R. S. Bucy, 1961: New results in linear filtering and prediction theory. Transactions of the ASME (Journal of Basic Engineering), 83D, 95-108.

Levinson, H., 1947: The Wiener RMS (root mean square) error criterion in filter design and prediction. Journal of Mathematics and Physics, 25, 261-278.

McClave, J., 1975: Subset autoregression. Technometrics, 17, 213-220.

Mood, A. M., F. A. Graybill, and D. C. Boes, 1974: Introduction to the Theory of Statistics (third edition). New York, McGraw-Hill.

Otomo, T., T. Nakagawa, and H. Akaike, 1972: Statistical approach to computer control of cement rotary kilns. Automatica, 8, 35-48.

Robinson, E. A., 1957: Predictive decomposition of seismic traces. Geophysics, 12, 767-778.

Robinson, E. A., 1963: Mathematical development of discrete filters for the detection of nuclear explosions. Journal of Geophysical Research, 68, 5559-5567.

Robinson, P. M., 1977: Estimation of a time series model from unequally spaced data. Stochastic Processes and Their Applications, 6, 9-24.

Shibata, R., 1976: Selection of the order of an autoregressive model by Akaike's information criterion. Biometrika, 63, 117-126.

Siddiqui, M. M., 1958: On the inversion of the sample covariance matrix in a stationary autoregressive process. Annals of Mathematical Statistics, 20, 585-588.

Tiao, G. C., and G. E. P. Box, 1979: An introduction to applied multiple time series analysis. Technical Report No. 582, Department of Statistics. Madison, Wis., University of Wisconsin.

Whittle, P., 1963: On the fitting of multivariate autoregressions, and the approximate canonical factorization of a spectral density matrix. Biometrika, 50, 129-134.

7
Probabilistic Models

Richard W. Katz

1. INTRODUCTION

Meteorological quantities (temperature, precipitation, etc.)
exhibit variations both in time and space. On the other hand, a
substantial degree of similarity usually is present between observations of a meteorological quantity taken either: (a) at the same
location separated by short periods of time; or (b) at the same time
at different locations separated by small distances. To represent
such meteorological quantities in a realistic manner, it is necessary
to consider probabilistic models that allow for both variability and
dependence. Such models are useful for characterizing meteorological
processes in terms of a few meaningful parameters, and they are also
necessary to make valid statistical inferences about meteorological
data.

In work published in 1852 (see Stigler, 1975), Quetelet proposed
what is apparently the first probabilistic model ever fit to meteorological observations. He examined runs of consecutive rainy days
(and of consecutive dry days) at Brussels for the time period
1833-1850 and discovered that rainy weather (and dry weather) tends
to persist. To take into account this dependence, he proposed a
model that is a special case of what now is called a two-state first-
order Markov chain. In later work published in 1854 (see Stigler,
1975), Quetelet also found that cold weather (and warm weather) tends
to persist.

More recently, Gabriel and Neumann (1962) proposed a two-state
first-order Markov chain to fit daily precipitation occurrences at
Tel Aviv. This probabilistic model can be viewed as a generalization
of the one considered by Quetelet over one hundred years earlier.
Since Gabriel and Neumann's work, more complex models have been suggested for fitting meteorological observations. Higher than first-
order Markov chains, for instance, have been employed to fit daily
precipitation occurrences (Chin, 1977; Gates and Tong, 1976). The
Markov chain model for daily precipitation occurrence also has been

generalized to bivariate processes that represent both daily precipitation occurrences and amounts (Todorovic and Woolhiser, 1975; Katz, 1977a, 1977b). Further, point processes have been employed to model precipitation simultaneously over both time and space (Waymire and Gupta, 1981a, 1981b, 1981c).

In this chapter we review some of the simplest probabilistic models that have found applications in meteorology. Some specific examples of probabilistic models are discussed in Section 2, while probability theory for stochastic processes, necessary to make statistical inferences about dependent meteorological observations, is introduced in Section 3. Finally, Section 4 consists of some methods of statistical inference for probabilistic models, including the demonstration of their application to a specific set of meteorological data.

2. EXAMPLES

In this section some examples of probabilistic models are given. Basic properties of these models are reviewed and applications to meteorology are described. The reader is advised to consult, for example, Feller (1968), Karlin and Taylor (1975), Parzen (1962), or Ross (1970), for a more complete treatment of applied probability theory and stochastic processes.

2.1 First-Order Markov Chains

2.1.1 Properties. A Markov chain is conceptually one of the simplest probabilistic models and has found considerable application in meteorology. We consider a stochastic process $\{J_n: n = 1, 2, \ldots\}$ with s states (say, for convenience, 0, 1, ..., s-1). For instance, J_n might represent the state of the weather (e.g., dry or wet) on the nth day at a particular location. The J_n-process is said to be a first-order Markov chain if

$$\Pr\{J_{n+1} = j \mid J_1 = i_1, \ldots, J_{n-1} = i_{n-1}, J_n = i\}$$
$$= \Pr\{J_{n+1} = j \mid J_n = i\} \tag{1}$$

for all states $i_1, \ldots, i_{n-1}, i, j,$ and for n = 1, 2, ... In other words, given the past states of the process $J_1, \ldots, J_{n-1}$, and the present state of the process J_n, the conditional distribution of the future state J_{n+1} is independent of the past states. This condition (1) is referred to as the Markovian property.

The probabilities specified by (1) are called <u>transition</u> <u>probabilities</u>. If these probabilities are independent of n, then the Markov chain is said to have stationary transition probabilities. In meteorological applications, the assumption of stationary transition probabilities is often reasonable when dealing with daily observations over a time period of a month or, perhaps, a season. When data for longer periods of time are modeled, seasonal cycles would be present in the transition probabilities. When the data to be modeled are recorded more frequently than daily (e.g., hourly), then diurnal cycles would be present in the transition probabilities. Only chains with stationary transition probabilities will be considered. In this case, we denote a transition probability by

$$P_{ij} = Pr\{J_{n+1} = j | J_n = i\}$$

and let $\underline{P} = (P_{ij})$ denote the s by s matrix of transition probabilities. Here

$$\sum_{j=0}^{s-1} P_{ij} = 1, \quad i = 0, 1, \ldots, s-1,$$

because of the definition of the transition probabilities. Such a square matrix $\underline{P}$, having non-negative elements and row sums equal one, is called a <u>stochastic matrix</u>. Finally, we let $\lambda_i = Pr\{J_1 = i\}$, $i = 0, 1, \ldots, s-1$, denote the initial probability distribution of the process, and set $\underline{\lambda} = (\lambda_i)$, the row vector (s elements) of initial probabilities. Here

$$\sum_{i=0}^{s-1} \lambda_i = 1.$$

The transition matrix $\underline{P}$ and the initial probability vector $\underline{\lambda}$ completely specify the probabilistic structure of the Markov chain. For example,

$$Pr\{J_1 = i_1, J_2 = i_2, \ldots, J_{n-1} = i_{n-1}, J_n = i_n\}$$

$$= \lambda_{i_1} P_{i_1 i_2} \cdots P_{i_{n-1} i_n}.$$

The k-step transition probability that the process goes from state i to state j in k steps is denoted by

$$P_{ij}^{(k)} = Pr\{J_{n+k} = j | J_n = i\}, \quad k = 0, 1, \ldots .$$

It is easily verified that the k-step transition probabilities are elements of the matrix $\underline{P}^k$; that is, the transition probability matrix

$\underline{P}$ multiplied by itself k times. The state j is accessible from the state i if there is some $k \geq 0$ for which $P_{ij}^{(k)} > 0$. The Markov chain is said to be irreducible if all states are accessible from each other. In particular, if $P_{ij} > 0$ for all states i and j, (as is generally the case in meteorological applications), then the chain is clearly irreducible. The state i is said to have period d if $P_{ii}^{(k)} = 0$ except when k = d, 2d, ... and d is the greatest integer with this property. The Markov chain is said to be aperiodic if all states have period d = 1. Again, since $P_{ij} > 0$ for all states i and j in most meteorological applications, the chain to be used as a model would be aperiodic.

Only irreducible aperiodic Markov chains will be considered. In this case, a unique stationary distribution exists, namely

$$\pi_j = \lim_{k \to \infty} P_{ij}^{(k)}, \quad \sum_{j=0}^{s-1} \pi_j = 1, \quad \pi_j > 0,$$

j = 0, 1, ..., s-1. We set $\underline{\pi} = (\pi_i)$, the row vector (s elements) of stationary probabilities. The vector $\underline{\pi}$ is the solution of the system of equations $\underline{\pi} = \underline{\pi}P$, and

$$\lim_{n \to \infty} Pr\{J_n = i\} = \pi_i, \quad i = 0, 1, ..., s-1.$$

Thus, π_i can be thought of as the long-run probability of the occurrence of the state of weather i. Further, if the initial distribution of the Markov chain is taken to be the stationary distribution (i.e., $\lambda = \underline{\pi}$), then the Markov chain is a stationary stochastic process (i.e., all joint distributions of the process are time invariant) with

$$Pr\{J_n = i\} = \pi_i, \quad n = 1, 2, ...$$

For simplicity, we will ordinarily assume that $\lambda = \underline{\pi}$, rather than conditioning on the state of the process prior to the start of the record. In this case, π_i also can be thought of as the unconditional probability of the occurrence of the state of weather i on the nth day.

A common misinterpretation of the Markovian property (1) is the belief that J_{n+1} must be independent of J_1, ..., J_{n-1}; that is, the future is independent of the past. This belief is erroneous, since J_{n+1} is dependent on J_n, J_n is dependent on J_{n-1}, and so on, making J_{n+1} dependent on all of the previous states of the process. If the J_n-process is stationary (i.e., if $\underline{\lambda} = \underline{\pi}$), the nature of this dependence can be seen by examining the form of the autocorrelation function

$$\rho_k = \text{Cov}(J_n, J_{n+k})/\text{Var}(J_n).$$

In the two-state case for a stationary Markov chain (i.e., taking s = 2 and $\underline{\lambda} = \underline{\pi}$),

$$\rho_k = (P_{11} - P_{01})^k, \quad k = 0, 1, \ldots . \tag{2}$$

We observe that the autocorrelation function (2) is nonzero for all lags (taking $P_{11} \neq P_{01}$, since otherwise the J_n's are independent random variables), but decreases to zero at a geometric rate as the lag k increases. In general, for a stationary first-order Markov chain with a finite number of states, the autocorrelation function is a mixture of exponentials (e.g., Lloyd, 1974).

2.1.2 Application. The most common meteorological application is the two-state (i.e., s = 2) first-order Markov chain model for the occurrence of precipitation (Gabriel and Neumann, 1962). We consider the sequence of daily precipitation occurrences at a given meteorological station and define

$$J_n = \begin{cases} 1 & \text{if precipitation occurs on nth day,} \\ 0 & \text{otherwise.} \end{cases}$$

If $J_n = 1$ it is said to be a _wet day_, whereas if $J_n = 0$ it is said to be a _dry day_.

Thinking in terms of this application to daily precipitation occurrences, we mention a few special properties of two-state first-order Markov chains. The assumption that the chain is irreducible and aperiodic corresponds to requiring that all four of the transition probabilities P_{ij}, i = 0, 1, j = 0, 1, are positive. In this case, the stationary probability of a wet day is given by

$$\pi_1 = P_{01}/(P_{01} + P_{10}) \tag{3}$$

(note that $\pi_0 = 1 - \pi_1$). The k-step transition probabilities $P_{ij}^{(k)}$ can be expressed in terms of the stationary probabilities of wet and dry days, π_0 and π_1, and the first-order autocorrelation coefficient $\rho_1 = P_{11} - P_{01}$ [see (2)]. Specifically, the conditional probability of a wet day k days from now given today is wet can be expressed as

$$P_{11}^{(k)} = \pi_1 + \rho_1^k \pi_0 \tag{4}$$

(note that $P_{10}^{(k)} = 1 - P_{11}^{(k)}$), whereas the conditional probability of a dry day k days from now given today is dry can be expressed as

$$P_{00}^{(k)} = \pi_0 + \rho_1^k \pi_1 \tag{5}$$

(note that $P_{01}^{(k)} = 1 - P_{00}^{(k)}$). The second terms on the right-hand side of (4) and (5) specify the differences between these conditional probabilities and the corresponding stationary probabilities.

To provide an example of parameter values for the first-order Markov chain model for daily precipitation occurrences, we present the parameter estimates that will be obtained in Section 4 for a set of precipitation data during February at State College, Pennsylvania. The conditional probability that tomorrow will be wet given today is wet is $P_{11} = 0.441$ (note that $P_{10} = 1 - P_{11}$), whereas the conditional probability that tomorrow will be dry given today is dry is $P_{00} = 0.658$ (note that $P_{01} = 1 - P_{00}$). The transition probability P_{11} can be compared with the stationary probability of a wet day $\pi_1 = 0.380$ [by (3)]. In particular, the conditional probability that tomorrow will be wet given today is wet is 0.061 greater than the stationary (or unconditional) probability of a wet day. Likewise, the conditional probability that tomorrow will be dry given today is dry is 0.038 greater than the stationary probability of a dry day.

It is evident that both wet and dry weather tend to persist, with the extent of this dependence being characterized by the first-order autocorrelation coefficient (sometimes referred to as the "persistence parameter")

$$\rho_1 = P_{11} - P_{01} = 0.099.$$

The persistence parameter ρ_1 ordinarily will be positive when modeling daily precipitaiton occurrences, resulting in an autocorrelation function (2) that is positive for all lags. Consequently, the second terms on the right-hand side of (4) and (5), whose interpretations were discussed earlier, also will be positive.

In applying this model to precipitation data, it must be assumed that the Markovian property (1) holds. For instance, given that today is a wet day, the conditional probability of tomorrow being wet should not depend on whether the past days were dry or wet. This assumption that the Markovian property holds will be tested for the State College precipitation data in Section 4.

2.2 Higher-Order Markov Chains

The Markovian property (1) can be easily generalized to the situation where the conditional distribution of the next state of the process J_{n+1}, given the previous states J_1, ..., J_n, depends on only the m most recent states J_n, J_{n-1}, ..., J_{n-m+1}, $m \geq 1$. Specifically, a <u>mth-order Markov chain</u> satisfies

$$\Pr\{J_{n+1} = j_{n+1} | J_1 = j_1, \ldots, J_{n-m+1} = j_{n-m+1}, \ldots, J_n = j_n\}$$

$$= \Pr\{J_{n+1} = j_{n+1} | J_{n-m+1} = j_{n-m+1}, \ldots, J_n = j_n\}$$

for all states j_1, ..., j_{n+1} and for $n = m, m+1, \ldots$. All of the terminology and properties for first-order Markov chains (Section 2.1.1) can be easily extended to higher-order chains. For instance, a transition probability $P_{i_1 i_2 \cdots i_{m+1}}$ for a mth-order chain is given by

$$P_{i_1 i_2 \cdots i_{m+1}} = \Pr\{J_{n+1} = i_{m+1} | J_{n-m+1} = i_1, \ldots, J_n = i_m\}.$$

Higher-order Markov chains also have been applied to meteorological observations (e.g., Chin, 1977). In particular, Gates and Tong (1976) have claimed that, on the basis of a certain statistical procedure, the original Tel Aviv precipitation occurrence data which Gabriel and Neumann (1962) modeled using a first-order Markov chain is fit significantly better by a second-order chain. Statistical methods for determining the appropriate order of Markov chain to fit to meteorological observations will be discussed in Section 4.

2.3 Markov Processes

The Markovian property (1) also may be extended to a more general stochastic process $\{X_n: n = 1, 2, \ldots\}$, where X_n is a real-valued random variable. In this case, the Markovian property can be expressed as

$$\Pr\{X_{n+1} \leq y | X_1 = x_1, \ldots, X_{n-1} = x_{n-1}, X_n = x\}$$

$$= \Pr\{X_{n+1} \leq y | X_n = x\} \tag{6}$$

for all possible values of x_1, ..., x_{n-1}, x, y. A process satisfying condition (6) is called a <u>Markov process</u>. A first-order Markov chain is a special case of a Markov process, since (6) reduces to (1) when X_n assumes only a finite number of values.

2.3.1 <u>Random Walks</u>. One example of a Markov process is called a random <u>walk</u> and consists of a sequence of random variables $\{X_n : n = 1, 2, \ldots\}$ with

$$X_{n+1} = X_n + \varepsilon_{n+1}, \quad n = 1, 2, \ldots . \tag{7}$$

Here the ε_n's are independent and identically distributed random variables with zero expected value.

Coin tossing is a classical example of a random walk. In this case, ε_n represents the outcome of the nth toss of a fair coin and is assigned the value one for a head and minus one for a tail; that is,

$$\Pr\{\varepsilon_n = 1\} = \Pr\{\varepsilon_n = -1\} = 1/2.$$

Now by (7), X_n is the difference between the total number of heads and the total number of tails in n tosses. The behavior of this particular random walk, and random walks in general, is counter-intuitive. Even when the total number of coin tosses is large, it is quite likely that either X_n will be greater than zero for virtually all the time (i.e., for most values of n) or less than zero for virtually all the time (Feller, 1968, Chapter III). An individual sample realization of the process would very likely exhibit large fluctuations suggestive of "trends" or "cycles."

For this reason, it has been suggested by several researchers (e.g., Curry, 1962; Jacchia, 1975) that the apparent cyclic behavior and other large fluctuations characteristic of long-term climatic time series would not be at all surprising if, in fact, such series were generated by a random walk. Here X_n could represent, for instance, the mean annual temperature for a particular meteorological station expressed in the form of deviations from a long-term mean. The possibility that a climatic series could be represented as a sum of random deviations has significant implications for meteorological researchers. Major fluctuations in climatic time series then might be properly viewed as unpredictable, and it might not make any sense to refer to such behavior as "climatic changes" or as "cycles."

2.3.2 <u>Autoregressive Processes</u>. Another example of a Markov process is a <u>first-order autoregressive process</u>. It consists of a sequence of random variables $\{X_n : n = 1, 2, \ldots\}$, with expected value $E(X_n) = \mu$ and variance $\mathrm{Var}(X_n) = \sigma^2$, satisfying

$$X_{n+1} - \mu = \phi(X_n - \mu) + \varepsilon_{n+1}, \quad n = 1, 2, \ldots . \tag{8}$$

Here the ε_n's are independent and identically distributed random variables with zero expected value and the parameter ϕ is the first-order autocorrelation coefficient. This process is sometimes simply referred to as a Markov process or a "red noise" process.

If the parameter ϕ in (8) satisfies the condition $|\phi| < 1$, then the X_n-process is stationary. Ordinarily, ϕ has to be estimated from a sample of meteorological observations. This parameter is positive in most meteorological applications, meaning that deviations from the mean tend to persist over time [see (8)]. To make statistical inferences about the process, some assumption needs to be made about the specific distribution of ε_n, and it usually is assumed that ε_n has a normal distribution.

The condition (8) for a first-order autoregressive process can easily be extended to the case of higher-order dependence. For a mth-order autoregressive process, the next state of the process X_{n+1} would be expressed as a linear combination of the m most recent states of the process, X_n, X_{n-1}, ..., X_{n-m+1}, and an error term. Much of time series analysis deals with the modeling of data using such processes (see Chapter 6). First-order autoregressive processes, in particular, have been fit to numerous meteorological time series, including temperature and pressure (e.g., Jenkinson, 1957; Klein, 1951).

2.4 Generalizations of Markov Processes

To fit daily precipitation amounts, a generalization of the two-state first-order Markov chain model for daily precipitation occurrences (Section 2.1.1) is necessary. One such model, involving a random number of random variables, was used by Todorovic and Woolhiser (1975). In this model, J_k indicates the occurrence of precipitation on the kth day and it is assumed, as in Section 2.1.1, that the J_k-process constitutes a first-order Markov chain. The amount of precipitation (or intensity) on the jth wet day is denoted by X_j, and the X_j's are required to be independent and identically distributed. We shall refer to this particular type of probabilistic model as a random i.i.d. process.

The total number of wet days in a n-day time period is a random variable, say N; that is,

$$N = \sum_{k=1}^{n} J_k. \tag{9}$$

Methods for computing the distribution of N are discussed in Section 3. If a random i.i.d. process is used to model daily precipitation amounts, then the total amount of precipitation, S_N say, for a n-day time period can be represented as a random sum of independent and identically distributed random variables; that is,

$$S_N = \sum_{j=1}^{N} X_j. \tag{10}$$

In this case, S_N has what is commonly called a "generalized distribution," being dependent on both the distribution of N and the common distribution of the X_j's. The expected value and variance for the random sum S_N are given by

$$E(S_N) = E(N)E(X), \tag{11}$$

$$Var(S_N) = E(N)Var(X) + Var(N)[E(X)]^2 \tag{12}$$

(Feller, 1968, p. 301), where E(X) denotes the common expected value of the X_j's and Var(X) denotes the common variance of the X_j's. Expressions for the expected value E(N) and variance Var(N) of the total number of wet days N will be given in Section 3. Katz (1977a, 1977b) has considered a generalization of the random i.i.d. process, called a chain-dependent process, for fitting daily precipitation amounts.

3. PROBABILITY THEORY

Probabilistic models are necessary to take into account the dependence of meteorological observations. The standard statistical distribution theory for independent random variables does not apply to such models. In this section we review some of the probability theory necessary to make statistical inferences about dependent meteorological data. Methods for computing exact probability distributions associated with some specific probabilistic models first are discussed.

3.1 Exact Distributions

The distributions of certain functions of stochastic processes, such as the sum (or average) and the extremes (e.g., maximum or minimum) of the process, are commonly of interest. Because of the dependency structure of probabilistic models, exact expressions for such probability distributions may be either quite complicated or even impossible to derive. Nevertheless, simple methods, such as recurrence relations, are still sometimes available for computing these probability distributions. We discuss one specific example in detail, the distribution of the total number of wet days when the daily occurrence of precipitation is represented by a two-state first-order Markov chain.

3.1.1 Number of Wet Days. With the two-state first-order Markov chain model for the daily occurrence of precipitation (Section 2.1.2), the distribution of the number of wet days N in a n-day time period is often of interest (Section 2.4). If the occurrence of precipitation could be assumed to be independent from day to day, then the exact distribution of the number of wet days would have a simple expression, namely that for the binomial distribution. Because of the dependence of the Markov chain, the exact formula for this distribution is much more complicated than that for the binomial distribution, but has been derived by Gabriel (1959). Alternatively, the distribution of the number of wet days can be computed with relative ease using a recurrence relation approach.

It is convenient, in this case, to denote by J_0 (instead of J_1) the initial state of the process immediately preceding the n-day time period. We define probabilities

$$W(j;n) = Pr\{N = j\}, \quad W_0(j;n) = Pr\{N = j|J_0 = 0\},$$

$$W_1(j;n) = Pr\{N = j|J_0 = 1\}, \quad j = 0, 1, \ldots, n.$$

Conditioning on whether the initial day J_0 is dry or wet,

$$W(j;n) = \lambda_0 W_0(j;n) + \lambda_1 W_1(j;n). \tag{13}$$

A recurrence relation for $W_0(j;n)$ and $W_1(j;n)$ is obtained by conditioning on J_1 (see Katz, 1974 for further details):

$$W_0(j;n) = P_{00}W_0(j;n-1) + P_{01}W_1(j-1;n-1), \tag{14}$$

$$W_1(j;n) = P_{10}W_0(j;n-1) + P_{11}W_1(j-1;n-1). \tag{15}$$

Here $j = 0, 1, \ldots, n$, and $n = 1, 2, \ldots$, with initial conditions $W_0(0;0) = W_1(0;0) = 1$ and the conventions that $W_0(n;n-1) = W_1(-1;n-1) = 0$. For a given two by two matrix of transition probabilities $\underline{P}$ and initial probability vector $\underline{\lambda}$, (14) and (15) can be used to compute recursively $W_0(j;n)$ and $W_1(j;n)$, letting $n = 1$, 2, ... and $j = 0, 1, \ldots, n$. Then $W(j;n)$ can be determined from $W_0(j;n)$ and $W_1(j;n)$ using (13).

Using the State College, Pennsylvania Markov chain parameters for February of $P_{11} = 0.441$, $P_{00} = 0.658$, and $\lambda_1 = \pi_1 = 0.380$ (see

Section 2.1.2), the exact probability distribution of the number of wet days N in a week (i.e., n = 7) was computed by means of the recurrence relations (14) and (15), as well as (13). These probabilities are listed in Table 1, along with the corresponding binomial probabilities based on the assumption of independence, and these two probability distributions are shown in Figure 1. It is evident that the effect of the dependence is to make the probability distribution of N more variable, giving higher probabilities to relatively extreme events (e.g., N = 0 or N = 7).

3.1.2 <u>Total and Maximum Precipitation</u>. The total amount of precipitation and the maximum daily amount of precipitation in a n-day time period are two statistics whose distributions are sometimes required. If a random i.i.d. process or a chain-dependent process is fit to daily precipitation amounts (Section 2.4), formulas for the exact distributions of these statistics are, in general, impossible to derive. Nevertheless, like the distribution of the number of wet days, recurrence relations can be obtained for computing these distributions (Katz, 1977b).

3.2 Asymptotic Distribution of Sums

Much of statistical inference relies upon asymptotic distribution theory; namely, the limiting distribution of specific functions of a stochastic process as the number of observations of the process tends to infinity. Further, much of this asymptotic theory is predicated upon the assumption that the stochastic process consists of a sequence of independent random variables. The central limit theorem, establishing that under very general conditions the sum of a sequence of random variables, suitably normalized, has a normal distribution asymptotically, is generally proven under this assumption of independence. We have seen, however, that dependent sequences of random variables are required to adequately model meteorological variables. It is important to note that the central limit theorem still holds for many dependent processes used to model meteorological data.

3.2.1 <u>Central Limit Theorem</u>. Let $\{X_j: \quad j = 1, 2, \ldots\}$ be a stationary stochastic process with finite expected value $\mu = E(X_j)$, finite non-zero variance $\sigma^2 = Var(X_j)$, and autocorrelation function

$$\rho_k = Cov(X_j, X_{j+k})/\sigma^2, \quad k = 1, 2, \ldots .$$

We set

$$S_n = \sum_{j=1}^{n} X_j .$$

Table 1. Distribution of total number of wet days in a week for State College parameters (P_{11} = 0.441, P_{00} = 0.658, λ_1 = π_1 = 0.380).

j	$W_0(j;7)$	$W_1(j;7)$	$W(j;7)$	Binomial[a]
0	0.053	0.045	0.050	0.035
1	0.169	0.154	0.164	0.151
2	0.263	0.254	0.260	0.278
3	0.255	0.260	0.257	0.284
4	0.166	0.178	0.170	0.174
5	0.072	0.081	0.075	0.064
6	0.019	0.023	0.021	0.013
7	0.003	0.003	0.003	0.001

[a] P_{11} = P_{01} = λ_1 = π_1 = 0.380.

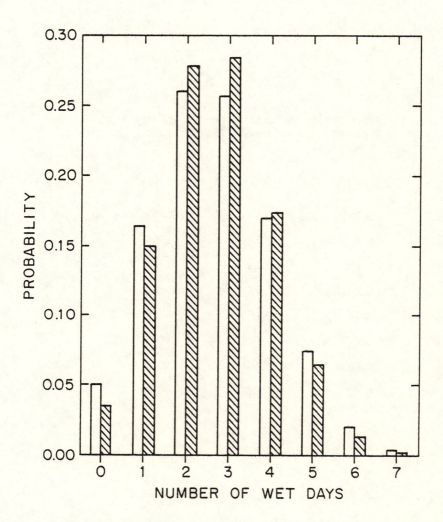

Figure 1. Distribution W(j;7) of total number of wet days in a week
for State College parameters based on Markov chain model
(solid lines) with binomial distribution (dashed lines)
included for comparison.

Then, under certain assumptions concerning the dependency structure of the X_j-process too technical to be stated here (e.g., Billingsley, 1968, p. 174),

$$(S_n - n\mu)/n^{1/2}V$$

has asymptotically a normal distribution with zero expected value and unit variance. Here

$$V^2 = \sigma^2(1 + 2 \sum_{k=1}^{\infty} \rho_k). \tag{16}$$

Roughly speaking, the central limit theorem for stationary stochastic processes requires that the process possess a certain type of asymptotic independence (i.e., observations separated by a large amount of time should be nearly independent). These conditions, in particular, will ensure that the variance normalizing factor (16) is a well-defined finite quantity. We note that this central limit theorem is identical to that for independent and identically distributed random variables, except for a change in the variance normalizing factor (16) to take into account the correlation structure of the stochastic process.

3.2.2 Autoregressive Processes. The central limit theorem holds for first-order autoregressive processes (Section 2.3.2) under the stationarity condition that $|\phi| < 1$. The variance normalizing factor (16) reduces in this case to

$$V^2 = \sigma^2(1 + \phi)/(1 - \phi). \tag{17}$$

As mentioned earlier, the parameter ϕ is typically greater than zero for meteorological data. So, by (17), the variance of the sum of the process will be greater than that for an independent process. The central limit theorem also holds for higher-order autoregressive processes, with a slightly more complicated expression for the variance normalizing factor. In the case of autoregressive processes of order possibly greater than one, this central limit theorem has been applied to perform tests of significance for general circulation model climate experiments (Katz, 1982).

One common problem is that of estimating the variance of time averages of meteorological data. If a meteorological variable can be represented as a first-order autoregressive process (8), then the variance of the average of n consecutive observations of the process is approximately V^2/n, where V^2 is given by (17). With the first-order autocorrelation coefficient ϕ replaced by an estimate, such an expression or slightly more complex expressions have been employed in practice to estimate the variance of meteorological time averages (e.g., Madden, 1979, 1981).

3.2.3 <u>Number of Wet Days</u>. The central limit theorem can be applied
to the two-state first-order Markov chain model for the daily
occurrence of precipitation (Section 2.1.1) to establish that the
number of wet days N in a n-day time period has asymptotically a normal distribution. Using the representation of N as the sum of the
J_k's (9), the expected value and variance of J_k (taking $\underline{\lambda} = \underline{\pi}$) are
given by

$$\mu = \pi_1, \qquad \sigma^2 = \pi_1(1 - \pi_1).$$

Using (2), the variance normalizing factor (16) reduces to

$$V^2 = \pi_1(1 - \pi_1)(1 + \rho_1)/(1 - \rho_1), \tag{18}$$

where $\rho_1 = P_{11} - P_{01}$. In particular, the expected value and variance of N (even if $\underline{\lambda} \neq \underline{\pi}$) are approximately

$$E(N) \simeq n\pi_1, \tag{19}$$

$$Var(N) \simeq n\pi_1(1 - \pi_1)(1 + \rho_1)/(1 - \rho_1). \tag{20}$$

As mentioned earlier, ρ_1 is generally positive, and the variance of
the number of wet days is, by (20), greater than that for an independent process. Recurrence relations should be employed for small
samples to compute the exact distribution of the number of wet days
(Section 3.1.1), whereas for large samples the normal approximation
just discussed can be employed instead to reduce the required computations. For instance, Gabriel (1959) has verified that, if n =
31, the normal approximation is very close to the exact distribution
when the Markov chain parameters are equal to the Tel Aviv parameter
estimates. We note that the normal approximation is valid even if
$\lambda \neq \pi$.

3.2.4 <u>Total Precipitation</u>. The central limit theorem can be applied
to the random i.i.d. process or the chain-dependent process for daily
precipitation amounts (Section 2.4) to establish that the distribution of the total amount of precipitation in a n-day time period
should be approximately normal (Katz, 1977a). Empirical results
suggest that, because of the high degree of positive skewness of the
distribution of daily precipitation intensities, the Gaussian
approximation may not work well unless the time period n is quite
large, say n greater than 30 (Katz, 1977b). The central limit
theorem for a chain-dependent process has been applied by Klugman and
Klugman (1981), obtaining an analysis of variance procedure to test
for climatic change in daily precipitation data. If daily precipitation amounts are represented by a random i.i.d. process, then the
approximate expressions (19) and (20) for the expected value and
variance of the number of wet days N can be substituted into (11) and

(12), yielding the following approximate expressions for the expected value and variance of the total amount of precipitation S_N [see (10)] in a n-day time period:

$$E(S_N) \simeq n\pi_1 E(X),$$

$$Var(S_N) \simeq n\pi_1 Var(X) + n\pi_1(1 - \pi_1) \frac{1 + \rho_1}{1 - \rho_1} [E(X)]^2.$$

3.3 Extreme Value Theory

The distribution of extremes (e.g., maximum or minimum) is sometimes required in meteorological applications of probabilistic models. Like the case of sums just discussed, an asymptotic theory for the distribution of extreme values of a stochastic process is well established. In particular, under very general conditions the maximum of a sequence of random variables, suitably normalized, has one of the three so-called "extreme value distributions" as its limiting distribution. As with the central limit theorem, this extreme value theorem holds for many dependent processes.

3.3.1 Extreme Value Theorem.

We let $\{X_j: j = 1, 2, \ldots\}$ denote a stationary stochastic process and set

$$M_n = \max_{1 \leq j \leq n} X_j.$$

Then, under certain assumptions concerning the dependency structure of the X_j-process (e.g., Loynes, 1965; Leadbetter, 1974), there exist normalizing constants $\{(a_n, b_n): n = 1, 2, \ldots\}$, with $a_n > 0$, such that

$$\lim_{n \to \infty} Pr\{(M_n - b_n)/a_n \leq x\} = G(x).$$

Here G is one of the three possible extreme value distribution functions:

(a) Type I with

$$G(x) = \exp[-\exp(-x)], \qquad -\infty < x < \infty; \qquad (21)$$

(b) Type II with

$$G(x) = \exp[-(x^{-\alpha})], \qquad \alpha > 0, \quad x > 0; \qquad (22)$$

(c) Type III with

$$G(x) = \exp[-(-x)^{\alpha}], \qquad \alpha > 0, \quad x < 0. \tag{23}$$

Criteria for choosing among the three possible extreme value distributions (21)-(23) and methods for determining the normalizing constants are available (e.g., Gumbel, 1958). Unlike the central limit theorem, the normalizing constants in the extreme value theorem are identical to those for the independent case. The extreme value theorem for stationary stochastic processes holds under dependency conditions quite similar to those for the central limit theorem.

3.3.2 Maximum Precipitation. The extreme value theorem can be applied to the random i.i.d. process or the chain-dependent process for daily precipitation amounts (Section 2.4). If daily precipitation amounts are represented by a random i.i.d. process, then the maximum amount of daily precipitation, M_N say, in a n-day time period can be expressed as the maximum of a random number of random variables; that is,

$$M_N = \max_{1 \le j \le N} X_j.$$

Suppose it further is assumed that daily precipitation intensity X_j has, for instance, a gamma distribution, say with probability density function

$$F_1(x) = [\beta\Gamma(\nu)]^{-1}(x/\beta)^{\nu-1}\exp(-x/\beta),$$

$x > 0$, $\nu > 0$, $\beta > 0$. Then, as a special case of an extreme value theorem for chain-dependent processes (Katz, 1977a), the distribution of M_N, suitably normalized, converges to the Type I extreme value distribution (21) as n tends to infinity. Specifically,

$$\lim_{n\to\infty} \Pr\{nF'(u_n)(M_N - u_n) \le x\} = \exp[-\exp(-x)],$$

where the distribution function F is given by

$$F(x) = \pi_0 + \pi_1 F_1(x)$$

and u_n is the solution to

$$1 - F(u_n) = 1/n.$$

Empirical results suggest that the Type I extreme value approximation for the distribution of the maximum amount of daily precipitation works quite well even when the time period n is quite small, say n = 10 or 20 (Katz, 1977b). The Type I extreme value approximation is still in force if other assumptions are made about the distribution of daily precipitation intensity; for instance, if lognormality is assumed.

4. STATISTICAL INFERENCE

To apply probabilistic models to meteorological data, questions of statistical estimation and inference need to be considered. The parameters of the model, such as the transition probabilities of a Markov chain, are generally unknown and have to be estimated from the available meteorological observations. Often the particular form of model, such as the order of a Markov chain, also needs to be determined. Although standard statistical texts give a thorough treatment of statistical inference for independent random variables, many of the corresponding procedures for stochastic processes are found only in technical scientific articles.

In this section we discuss statistical inference for Markov chains. For this class of probabilistic models, making statistical inferences is relatively straightforward. We consider the two-state Markov chain model for the occurrence of precipitation (Section 2.1.2). One particular set of data, daily precipitation occurrences at State College, Pennsylvania, is used for illustrative purposes.

4.1 Parameter Estimation

We present methods for estimating the parameters of both first-order and higher-order Markov chains. In this section it is assumed that the appropriate order has already been determined (Section 4.2 describes methods of order selection). To simplify matters, the case of a first-order Markov chain is considered first.

4.1.1 <u>First-Order Markov Chains</u>. Suppose that a sample consisting of n consecutive observations, J_1, J_2, ..., J_n, from a first-order Markov chain is available. To completely specify the Markov chain, the transition probabilities need to be estimated. Under very general conditions (Billingsley, 1961), the maximum likelihood estimate of the transition probability P_{ij}, denoted by $\hat{P}_{ij}$, can be calculated in the following manner. Let n_{ij} denote the number of transitions from state i to state j that occur in the sample. We call n_{ij} a <u>transition count</u>, and denote the transition count matrix by $\underline{N} = (n_{ij})$. Then

$$\hat{P}_{ij} = \frac{n_{ij}}{n_{i \cdot}}, \quad n_{i \cdot} > 0, \tag{24}$$

with

$$n_{i.} = \sum_{j=0}^{s-1} n_{ij}.$$

Intuitively, $\hat{P}_{ij}$ is simply the relative frequency of transitions from state i to state j as compared to all transitions from state i to any possible state.

Frequently, when dealing with meteorological data, it is necessary to analyze many records of observations of a fixed length (e.g., daily precipitation occurrences for a single month over many years), instead of a single long record of observations. For this situation, the transition probabilities can be estimated in a similar fashion to (24) (Anderson and Goodman, 1957). We assume that L records are available, and let $n_{ij}^{(\ell)}$ denote the transition count for the ℓth record. If we redefine n_{ij} to be

$$n_{ij} = \sum_{\ell=1}^{L} n_{ij}^{(\ell)}, \tag{25}$$

then the maximum likelihood estimates of the transition probabilities are still given by (24) for this more general case.

The two-state Markov chain model for the occurrence of precipitation will now be applied to State College daily precipitation data for 1-28 February 1930-1969 (a total of 1120 = 28 x 40 observations). The data are restricted to a single month so that the assumption of stationary transition probabilities will hold at least approximately. Using (25), the transition count matrix for the State College observations is

$$\underline{N} = \begin{bmatrix} 442 & 230 \\ 228 & 180 \end{bmatrix},$$

and the corresponding matrix of transition probability estimates, denoted by $\underline{P}$, is

$$\hat{\underline{P}} = \begin{bmatrix} 0.658 & 0.342 \\ 0.559 & 0.441 \end{bmatrix}.$$

4.1.2 Higher-Order Markov Chains. Parameter estimation for a general kth-order Markov chain (Section 2.2) is now discussed. The expression (24) for the maximum likelihood estimate of a transition probability for a first-order Markov chain can easily be extended to the case of a higher-order chain. Given a sample of n observations,

J_1, J_2, ..., J_n, from a kth-order Markov chain, let $n_{i_1 \cdots i_{k+1}}$
denote the number of transitions (or transition count) from state
i_1 to state i_2 ... to state i_{k+1} that occur in this sample. The
maximum likelihood estimate of the transition probability
$P_{i_1 \cdots i_{k+1}}$, denoted by $\hat{P}_{i_1 \cdots i_{k+1}}$, is given by

$$\hat{P}_{i_1 \cdots i_{k+1}} = \frac{n_{i_1 \cdots i_{k+1}}}{n_{i_1 \cdots i_k \cdot}}, \quad n_{i_1 \cdots i_k \cdot} > 0, \quad (26)$$

with

$$n_{i_1 \cdots i_k \cdot} = \sum_{i_{k+1}=0}^{s-1} n_{i_1 \cdots i_{k+1}}$$

(Billingsley, 1961). Again, it is straightforward to extend (26) to
the case of several records, as pointed out earlier for first-order
chains [see (25)]. The next section includes the determination of
which order Markov chain is most appropriate for the State College
data.

4.2 Model Selection

In this section we describe methods for determining the appropriate form of probabilistic model to fit to a given set of meteorological observations. When fitting Markov chains to daily precipitation occurrences, several assumptions need to be verified including
stationarity (both day to day and year to year) and first-order
dependence. The specific problem of choosing the order of a Markov
chain is considered in detail.

4.2.1 Hypothesis Tests.

Several procedures are available for
selecting Markov chain order. One commonly used approach is to rely
upon hypothesis tests, generally either the likelihood ratio or chi-
squared goodness-of-fit testing procedures (Anderson and Goodman,
1957). Suppose we are given a sample of n consecutive observations,
J_1, J_2, ..., J_n, from a kth-order Markov chain. Here k = 0, 1, ...
with a zero-order chain corresponding to the case of independence.

The likelihood ratio test of the null hypothesis of kth-order
Markov chain versus the alternative hypothesis of mth-order (m a
fixed integer greater than k) proceeds as follows. Denoted by
$M_k(J_1, J_2, \ldots, J_n)$, the maximized likelihood function for a kth-
order Markov chain (ignoring the first few terms) is

$$M_k(J_1, J_2, \ldots, J_n) = \prod_{i_1,\ldots,i_{k+1}} \hat{P}_{i_1 \cdots i_{k+1}}^{n_{i_1 \cdots i_{k+1}}}. \tag{27}$$

Here $\hat{P}_{i_1 \cdots i_{k+1}}$ is the maximum likelihood estimate of the transition probability $P_{i_1 \cdots i_{k+1}}$ and is computed using (26). The likelihood ratio test statistic for kth-order versus mth-order Markov chain is

$$n_{k,m} = -2 \ln \lambda_{k,m} \tag{28}$$

where

$$\lambda_{k,m} = \frac{M_k(J_1, J_2, \ldots, J_n)}{M_m(J_1, J_2, \ldots, J_n)}. \tag{29}$$

Under the null hypothesis that the Markov chain is kth order, the likelihood ratio test statistic (28) has asymptotically a chi-squared distribution with degrees of freedom (df) as follows:

$$df = (s^m - s^k)(s - 1) \tag{30}$$

(Anderson and Goodman, 1957).

With a revised definition of the transition count $n_{i_1 \cdots i_{k+1}}$ (Section 4.1), the likelihood ratio test for Markov chain order can be applied to meteorological observations consisting of several records. The identical formulas (27)-(30) still hold in this case. The chi-squared goodness-of-fit test for Markov chain order, closely related to the likelihood ratio test, could also be used to test the hypothesis of kth-order versus mth-order chain (Anderson and Goodman, 1957).

We now demonstrate the application of the likelihood ratio test, employing the same State College precipitation occurrence observations considered in Section 4.1.1. The test of zero-order versus first-order Markov chain results in a likelihood ratio test statistic value of $n_{0,1} = 10.49$. The chi-squared approximation with one df [setting $s = 2$ in (30)] yields an observed significance level (or probability value) of about 0.001 for this test statistic, indicating overwhelming evidence that daily precipitation occurrences are not independent. The test of first-order versus second-order Markov chain results in a likelihood ratio test statistic value of $n_{1,2} = 4.03$, with an associated observed significance level of about 0.134. Consequently, the evidence is not strong enough to conclude that a higher than first-order model is necessary.

4.2.2 <u>Model Selection Criteria</u>. Multiple hypothesis testing approaches to model selection may be undesirable for several reasons. For instance, in the State College precipitation occurrence example just discussed, we tested two hypotheses, both zero-order versus first-order Markov chain and first-order versus second-order chain. This multiplicity of hypothesis tests required to choose the order of a Markov chain makes it very difficult, if not impossible, to determine the overall significance level for the procedure. Such approaches, moreover, are undesirable on philosophical grounds, since they violate some of the fundamental tenets on which hypothesis testing is based (i.e., the questionable manner in which the "null hypothesis" and "alternative hypothesis" are specified and then revised).

As an alternative to such hypothesis testing procedures, several model selection criteria have recently been proposed as techniques for determining Markov chain order and for other model building problems. The basic principle underlying these model selection criteria is to achieve a parsimonious approach to model building, balancing the desire for a better fitting model against the desire for a model with as few parameters as possible. One such criterion, called <u>Akaike's Information Criterion</u> (AIC) (e.g., Akaike, 1974) is based upon heuristic arguments employing an information-theoretic concept known as the Kullback-Leibler mean information. Tong (1975) has extended these same heuristics to the specific problem of Markov chain order selection, while Gates and Tong (1976) have applied this procedure to precipitation occurrence data.

To apply the AIC technique, we need to assume that the order of the Markov chain is less than some fixed upper bound m. The AIC procedure chooses the order k (k = 0, 1, ..., m-1) for which the quantity

$$AIC(k) = \eta_{k,m} - 2(s^m - s^k)(s - 1) \tag{31}$$

is smallest. The first term on the right-hand side of (31), namely the likelihood ratio test statistic, is a goodness-of-fit measure for the kth-order Markov chain; whereas the second term, namely twice the degrees of freedom, is the penalty function to achieve parsimony.

Another procedure, called the <u>Bayesian Information Criterion</u> (BIC), has also been proposed for model selection (Schwarz, 1978). The BIC procedure is identical to the AIC procedure except for a change in the form of penalty function. Katz (1981) has extended the BIC technique to the specific problem of Markov chain order selection and has applied this technique to precipitation occurrence data (Katz, 1979). The BIC procedure chooses the order k (k = 0, 1, ..., m - 1) for which the quantity

$$BIC(k) = \eta_{k,m} - (s^m - s^k)(s - 1)(\ln n) \tag{32}$$

Table 2. Markov chain order selection procedures applied to State College precipitation data.

Order k	df	$n_{k,m}$	AIC(k)	BIC(k)	BP(k)
0	31	33.67	-28.33	-183.98	0.1491
1	30	23.18	-36.82	-187.45[a]	0.8452[b]
2	28	19.15	-36.85[a]	-177.44	0.0057
3	24	17.63	-30.37	-150.88	0.0000
4	16	9.59	-22.41	-102.75	0.0000

[a]Denotes minimum.

[b]Denotes maximum.

is smallest. Being based on a Bayesian argument, BIC(k) is a transformation of the approximate posterior probability, BP(k) say, of a
kth-order Markov chain; specifically,

$$BP(k) \propto \exp[-(1/2)BIC(k)]. \tag{33}$$

The BIC technique corresponds to selecting the model which is most
likely a posteriori (i.e., the model for which BP(k) is largest).
Which particular model selection criterion is most appropriate,
whether on the basis of statistical optimality or on the basis of
superiority in practice when dealing with meteorological data, has
not yet been established.

We now demonstrate the application of the AIC and BIC procedures
to the State College precipitation occurrence observations. The
maximum possible order is assumed, for convenience, to be less than
five (i.e., m = 5). Using (31)-(33), the statistics for the AIC and
BIC techniques are given in Table 2. We observe that AIC(k) is
smallest for k = 2, so that the AIC procedure chooses a second-order
Markov chain as the most appropriate model to be fitted to the State
College data, thus differing from the results of the hypothesis
testing approach. On the other hand, the BIC procedure chooses a
first-order Markov chain to fit the State College data [with the
approximate posterior probability of a second-order chain being only
0.0057 (see Table 2)], in agreement with the hypothesis testing
approach.

ACKNOWLEDGMENT

This work was supported in part by the National Science Foundation (Atmospheric Sciences Division) under grant ATM80-04680.

REFERENCES

Akaike, H., 1974: A new look at the statistical model identifica-
    tion. IEEE Transactions on Automatic Control, 19, 716-723.

Anderson, T. W., and L. A. Goodman, 1957: Statistical inference
    about Markov chains. Annals of Mathematical Statistics, 28,
    89-110.

Billingsley, P., 1961: Statistical methods in Markov chains. Annals
    of Mathematical Statistics, 32, 12-40.

Billingsley, P., 1968: Convergence of Probability Measures. New
    York, Wiley.

Chin, E. H., 1977: Modeling daily precipitation occurrence process
    with Markov chain. Water Resources Research, 13, 949-956.

Curry, L., 1962: Climatic change as a random series. Annals of the Association of American Geographers, 52, 21-31.

Feller, W., 1968: An Introduction to Probability Theory and Its Applications, Volume I, Third Edition. New York, Wiley.

Gabriel, K. R., 1959: The distribution of the number of successes in a sequence of dependent trials. Biometrika, 46, 454-460.

Gabriel, K. R., and J. Neumann, 1962: A Markov chain model for daily rainfall occurrence at Tel Aviv. Quarterly Journal of the Royal Meteorological Society, 88, 90-95.

Gates, P., and H. Tong, 1976: On Markov chain modeling to some weather data. Journal of Applied Meteorology, 15, 1145-1151.

Gumbel, E. J., 1958: Statistics of Extremes. New York, Columbia University Press.

Jacchia, L. G., 1975: Some thoughts about randomness. Sky and Telescope, 50, 371-374.

Jenkinson, A. F., 1957: Relations between standard deviations of daily, 5-day, 10-day, and 30-day mean temperatures. Meteorological Magazine, 86, 169-176.

Karlin, S., and H. M. Taylor, 1975: A First Course in Stochastic Processes, Second Edition. New York, Academic Press.

Katz, R. W., 1974: Computing probabilities associated with the Markov chain model for precipitation. Journal of Applied Meteorology, 13, 953-954.

Katz, R. W., 1977a: An application of chain-dependent processes to meteorology. Journal of Applied Probability, 14, 598-603.

Katz, R. W., 1977b: Precipitation as a chain-dependent process. Journal of Applied Meteorology, 16, 671-676.

Katz, R. W., 1979: Estimating the order of a Markov chain: another look at the Tel Aviv rainfall data. Preprints Sixth Conference on Probability and Statistics in Atmospheric Sciences, American Meteorological Society, Boston, pp. 217-221.

Katz, R. W., 1981: On some criteria for estimating the order of a Markov chain. Technometrics, 23, 243-249.

Katz, R. W., 1982: Statistical evaluation of climate experiments with general circulation models: a parametric time series modeling approach. Journal of the Atmospheric Sciences, 39 (in press).

Klein, W. H., 1951: A hemisphere study of daily pressure variability at sea level and aloft. Journal of Meteorology, 8, 332-346.

Klugman, M. R., and S. A. Klugman, 1981: A method for determining change in precipitation data. Journal of Applied Meteorology, 20, 1506-1509.

Leadbetter, M. R., 1974: On extreme values in stationary sequences. Zeitschrift für Wahrscheinlichkeitstheorie und Verwandte Gebiete, 28, 289-303.

Lloyd, E. H., 1974: What is, and what is not, a Markov chain? Journal of Hydrology, 22, 1-28.

Loynes, R. M., 1965: Extreme values in uniformly mixing stationary stochastic processes. Annals of Mathematical Statistics, 36, 993-999.

Madden, R. A., 1979: A simple approximation for the variance of meteorological time averages. Journal of Applied Meteorology, 18, 703-706.

Madden, R. A., 1981: A quantitative approach to long-range prediction. Journal of Geophysical Research, 86, 9817-9825.

Parzen, E., 1962: Stochastic Processes. San Francisco, Holden-Day.

Ross, S. M., 1970: Applied Probability Models with Optimization Applications. San Francisco, Holden-Day.

Schwarz, G., 1978: Estimating the dimension of a model. Annals of Statistics, 6, 461-464.

Stigler, S. M., 1975: The transition from point to distribution estimation. Bulletin of the International Statistical Institute, XLVI-2, 332-340.

Todorovic, P., and D. A. Woolhiser, 1975: A stochastic model of n-day precipitation. Journal of Applied Meteorology, 14, 17-24.

Tong, H., 1975: Determination of the order of a Markov chain by Akaike's information criterion. Journal of Applied Probability, 12, 488-497.

Waymire, E., and V. K. Gupta, 1981a: The mathematical structure of rainfall representations, 1, A review of the stochastic rainfall models. Water Resources Research, 17, 1261-1272.

Waymire, E., and V. K. Gupta, 1981b: The mathematical structure of rainfall representations, 2, A review of the theory of point processes. Water Resources Research, 17, 1273-1285.

Waymire, E., and V. K. Gupta, 1981c: The mathematical structure of rainfall representations, 3, Some applications of point process theory to rainfall processes. Water Resources Research, 17, 1287-1294.

8
Statistical Weather Forecasting

Harry R. Glahn

1. INTRODUCTION

Statistical weather forecasting, in its broadest sense, has undoubtedly been practiced for thousands of years. All that is necessary is for someone to collect some data, someone to process it, and someone to use the results to make a forecast. Ancient man, seeing a dark cloud approaching and thinking that rain was likely, would be practicing statistical weather forecasting even if he had no knowledge of the physical processes involved. However, in this chapter we use the term statistical weather forecasting to mean forecasting through the use of a formal statistical analysis of the data, with the results of that analysis being clearly stated.

Statistical forecasting is a branch of objective weather forecasting, the other branch being numerical weather prediction. Allen and Vernon (1951) have defined an objective forecast as "... a forecast which does not depend for its accuracy upon the forecasting experience or the subjective judgment of the meteorologist using it. Strictly speaking, an objective system is one which can produce one and only one forecast from a specific set of data." (Subjective judgment is, of course, used in the development of the system.) Occasionally, these restrictions are relaxed slightly or some subjectivity may enter into the definition of "a specific set of data." For instance, an observation of temperature at a certain location may be needed as input to an objective scheme, and this observation may not be available for some reason. It may, then, have to be estimated from other data. Even though this estimate is made subjectively and requires skill on the part of the meteorologist, the forecast would probably still be called objective.

Some statistical techniques are very simple, whereas other procedures are more complicated. Various forms of scatter diagrams and histograms fall into the first category. Discriminant analysis and logit analysis are examples of the latter category.

In the early years of operational numerical weather prediction, competition rather than cooperation dominated the relationship between those individuals engaged in developing statistical models and those researchers concerned with developing numerical models. Each group thought that its approach was the best way to proceed and that the other branch of objective weather prediction was not necessary. Even though the barriers between the two groups have not yet vanished, each group has become much more tolerant of the other group's viewpoint. Statistical modelers now use the results from (rather than compete with) numerical models, and numerical modelers recognize the usefulness of properly applied statistical procedures.

In this chapter, we will review the three general methods of application of statistical models and describe the statistical techniques that have been applied to weather prediction. Emphasis will be placed on those techniques that have been used operationally. Other discussions of statistical models employed in objective weather forecasting can be found in Allen and Vernon (1951), Gringorten (1955), Panofsky and Brier (1958), U. S. Navy (1963), Glahn (1965), and Miller (1977).

2. METHODS OF APPLICATION

2.1 Classical Method

Before the days of numerical models, statistical techniques necessarily incorporated the time lag. That is, if one wanted to develop a scheme for forecasting the maximum (max) temperature for tomorrow, the input would consist only of observational data available at the time that the forecast was to be made. This situation can be expressed as

$$\hat{Y}_t = f_1(\underline{X}_0), \tag{1}$$

where $\hat{Y}_t$ is the estimate (forecast) of the predictand (dependent variable) Y at time t and $\underline{X}_0$ is a vector of observational data (independent variables) at time 0. (The observations are not necessarily all made at time 0 but must be available at that time.) This technique has become known as the "classical" approach for lack of a better name (Klein, 1969). In application, the input is the same as in development.

2.2 Perfect Prog Method

As numerical models were implemented and improved, it was recognized that their output must be exploited to the greatest possible extent. However, these models did not predict many of the weather variables with which users were concerned - for instance, max temperature. This situation led to the development of the perfect prog (prog for prognostic) technique (Klein et al., 1959).

A concurrent relationship between the predictand variable and the predictor variables is developed, which can be expressed as

$$\hat{Y}_0 = f_2(\underline{X}_0), \tag{2}$$

where $\hat{Y}_0$ is the estimate of the predictand Y at time 0 and $\underline{X}_0$ is a vector of observations of variables that can be predicted by numerical models. The time relationship need not be exactly concurrent, but it is much more nearly so than in the classical technique. Even though $\hat{Y}_0$ is an estimate, it is not a "forecast" in the sense of "looking ahead"; it is more appropriately called a "specification."

In application, $\hat{\underline{X}}_t$ is inserted into Eq. (2) to provide a forecast $\hat{Y}_t$:

$$\hat{Y}_t = f_2(\hat{\underline{X}}_t). \tag{3}$$

The vector $\hat{\underline{X}}_t$ is obtained from numerical model output. This approach assumes that the model output is "perfect" (hence, the name "perfect prog").

2.3 Model Output Statistics Method

Although the perfect prog technique makes use of numerical model output, it is not necessarily true that the statistical relationship between Y and $\underline{X}$ at time 0 is the best relationship for time t when $\underline{X}_t$ is estimated by numerical models as in Eq. (3). In order to overcome this problem, the model output statistics (MOS) technique was developed (Glahn and Lowry, 1972). In this approach, a sample of model output is collected and a statistical relationship is developed, which can be expressed as

$$\hat{Y}_t = f_3(\hat{\underline{X}}_t), \tag{4}$$

where $\hat{Y}_t$ is the estimate of the predictand Y at time t and $\hat{\underline{X}}_t$ is a vector of forecasts from numerical models. The numerical model predictions $\underline{X}_t$ need not be limited to time t but could be valid either before or after time t; however, the projection times of the different variables will usually be grouped around t. In application, Eq. (4) is used as developed.

2.4 Comparison of Classical, Perfect Prog, and MOS Techniques

Table 1 summarizes the development and application aspects of the three techniques. Since the classical technique does not depend on numerical models, it is most useful for very short-range forecasting. The strength of most numerical models lies in predicting events several hours to a few days in advance. For predictions of up to 4 hours, say, simple statistical models and even persistence may be quite good in comparison to numerical models or statistical forecasts derived from them. The classical technique is relatively simple to use, observations are usually abundant for model development, and there is no dependence on a numerical model to complicate the application.

For many purposes, the perfect prog technique gives quite good results. Since Eq. (2) is based entirely on observations (or simple calculations made from them), a large data sample usually can be obtained to ensure a stable relationship. (A stable relationship is one that will give similar results on dependent and independent data.) The availability of observations also may allow useful stratifications of the data. That is, different relationships can be developed for different months of the year, hours of the day, etc. In addition, as numerical models become more accurate, forecasts based on Eq. (3) will improve even without redevelopment of the functional relationship f_2.

For medium range forecasting, MOS is the best technique if (a) a sufficient sample of model output can be obtained for development and (b) the model does not undergo major changes. Use of MOS usually requires more planning than the other techniques because the model output desired may not be saved without special arrangements. The major disadvantage is that a relationship f_3 developed for one model may not hold for another model. Therefore, if the operational model is changed substantially, a new relationship should be developed. This redevelopment can be done only after the new model has been used for a long enough period to obtain an adequate data sample. At the time that this chapter was written, changes in the National Weather Service (NWS) models being employed by the National Meteorological Center (NMC) have not presented serious problems in MOS applications.

Changes in numerical models that might materially affect MOS applications could be any of three types: (a) the model produces the same output variables, but the overall skill is higher; (b) the model produces the same output variables and the overall skill is about the same, but the error characteristics are different; or (c) the model produces different output variables with or without an increase in skill. (We assume that a new model would not be implemented if the skill level was below that of the old model.) In the first case, use of the new model would probably decrease the skill of MOS forecasts slightly (without redevelopment) unless the model skill was increased

considerably. If the skill did increase markedly, then the MOS skill would probably also improve. For a new model with equal skill but different error characteristics, the MOS skill would undoubtedly decrease. If the new model didn't produce the same output variables, the old variables would have to be estimated (by interpolation or other computations from the new variables). A decrease in MOS skill would likely result unless the new model was considerably more skillful.

Even with these potential problems, it is likely that MOS will be used operationally more than perfect prog and will produce better forecasts for many years to come. At some point, some of the applications may shift to perfect prog. However, when the predictand is a dichotomous event (e.g., precipitation/no precipitation) and the statistical relationship estimates the probability of that event, MOS will always be superior to perfect prog. MOS incorporates the inaccuracies of the numerical model, and as the skill becomes small for large projections (i.e., long lead times) the estimated probability will approach climatology. Perfect prog will not give this result; that is, the possible range of predictions in the perfect prog approach is just as great for long-time projections (say 5 days) as for short-time projections (say 12 hours), unless the numerical model itself becomes much smoother with time and perhaps approaches climatology. Therefore, perfect prog probabilities will not be reliable (i.e., will not correspond to observed relative frequencies). Figure 1 shows schematically the relationship between MOS and perfect prog probabilities as a function of projection.

Although relatively little experience to date has been obtained in applying Eq. (4) to a model other than that on which it was developed, the evidence available suggests that the decrease in skill is minimal. Major operational models share many of the same characteristic errors - incorrect phase of systems at long projections, missed cyclogenesis, etc. The forecasts prepared by two models frequently look more like one another than either looks like reality. That is, major error characteristics for different models are similar. As long as this situation exists, relationships developed on one model can be applied to another model without major loss in skill.

All of the statistical models presented in the following sections, such as scatter diagrams or regression, can be used with any of the techniques discussed above. In the following sections, an estimate of a predictand may at times be called a "forecast" even though no time projection is actually involved.

3. HISTOGRAMS

Perhaps the simplest statistical model one might apply is the histogram. Figure 2 shows the relative frequency of frozen precipitation at Salt Lake City as a function of the 1000-500 mb thickness

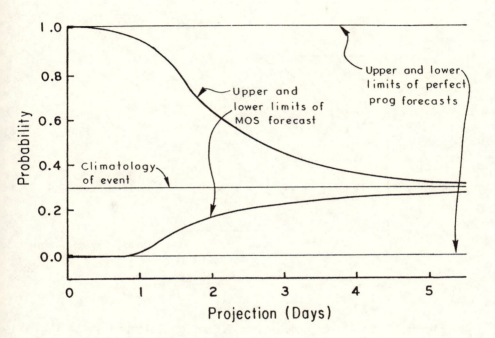

Figure 1.   Schematic diagram of upper and lower limits of MOS and
perfect prog forecasts as a function of projection.

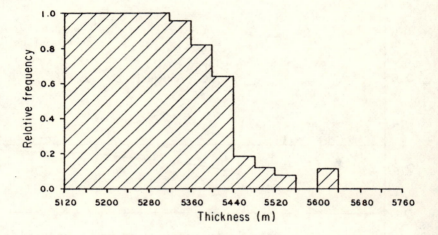

Figure 2. Relative frequency of frozen precipitation at Salt Lake City, Utah, as a function of forecast 1000-500 mb thickness (after Glahn and Bocchieri, 1975).

forecast by the primitive equation (PE) model (Shuman and Hovermale, 1968). The relative frequencies have been calculated for 40 m intervals. The interval must be wide enough to encompass several cases in the region of greatest concern and yet be small enough to give sufficient detail to be useful. The intervals need not be the same width.

The histogram can be applied directly. However, Figure 2 poses a question: "Should the relative frequency for each thickness band be used exactly as plotted?" This question really has two parts, one involving smoothing and one involving interpolation. Is there any reason to believe that the relative frequency should be higher for the 5600-5640 m band than for the bands on either side? If not, smoothing is suggested. Also, should one use 19% for 5441 m and jump to 64% for 5439 m? If not, interpolation is suggested. In any case, judicious use of histograms can produce a useful objective tool, and a computer is not required for its development or use.

4. SCATTER DIAGRAMS

Another model rivaling the histogram in its simplicity is the scatter diagram. It is primarily a noncomputer technique and was used as early as the beginning of this century by Besson (1905). The technique has also been called graphical regression and was studied in detail by Brier (1946). It has been used extensively by the U.S. Weather Bureau, now the National Weather Service, since the mid-1940's and several papers appeared in the Monthly Weather Review circa 1950 illustrating the use of this model. A typical paper from this period is that by Thompson (1950).

In its simplest form, coordinate axes are established on a diagram such that the ordinate represents the dependent variable or predictand and the abscissa represents a single independent variable or predictor. Points are then plotted on this diagram depicting the available data sample. Finally a line can be drawn by eye which seems to fit the data points. In application, a forecast of the predictand is found by reading the ordinate value of the line at the abscissa value of the predictor. Such a completed diagram is shown in Figure 3.

Usually, however, one wants to use two or more predictors. In this case, the coordinate axes should be the values of the pair of predictors, and the predictand values are plotted at the points on the diagram representing the data sample. An analysis is then made of the plotted data. The analysis is subjective and will depend on the skill of the analyst. The analyst must be careful not to "over analyze" the data, especially in regions where few data points exist. In general, the analysis should be rather smooth and, in case of doubt, known physical relationships may furnish a key to correct analysis.

298

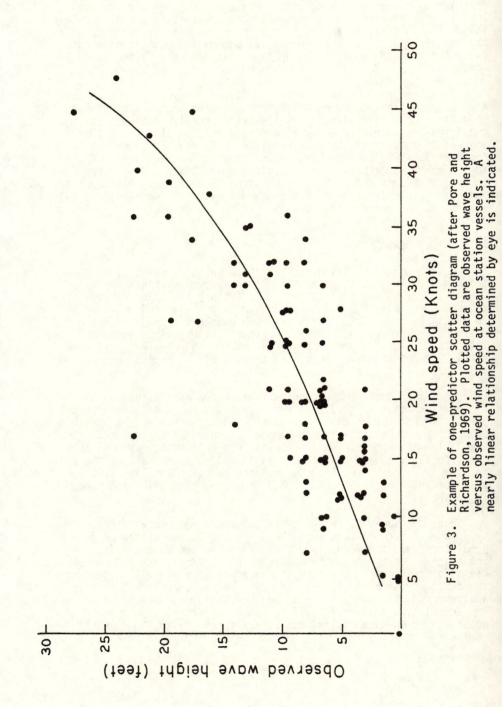

Figure 3. Example of one-predictor scatter diagram (after Pore and Richardson, 1969). Plotted data are observed wave height versus observed wind speed at ocean station vessels. A nearly linear relationship determined by eye is indicated.

Figure 4 shows an example of a two-predictor scatter diagram
taken from Thompson (1950). Precipitation amount can be forecast
with Figure 4 and the observed values of the two predictors, x_1
(700 mb height at Oakland) and x_2 (San Francisco minus Los Angeles
sea level pressure difference). If one wishes to use more than two
predictors, other diagrams can be plotted and analyzed. For
instance, predictors 3 and 4 can be combined on a diagram similar to
Figure 4. The predictand value "forecast" from Figure 4 can be
called predictor x_5. Similarly, the predictand value estimated from
x_3 and x_4 can be called x_6. Then x_5 and x_6 can be the coordinates
on another scatter diagram in which the actual predictand values are
plotted as a function of x_5 and x_6. (A variation of this procedure
is to plot deviations between the preliminary estimates x_5 and x_6
and the actual values Y.) An analysis of these values will then
define estimates of Y given values of x_5 and x_6. Thompson (1950)
presents an example of a six-predictor scatter diagram procedure.

The scatter diagram model is very simple in principle, yet it
allows for any degree of complexity that the data warrant. Its success will depend on the analyst's ability to choose meaningful predictors, as will the success of any technique. Thompson (1950)
offered the following comments concerning the analysis: "While the
meteorological relationships brought out by the primary graphical
combination of each pair of variables may ... be discussed from a
physical standpoint, and thereby the reasonableness of the isograms
checked, very little can be said about the secondary combinations.
Here the complexity of the joint relationships, as well as the probable effect of other variables not considered in the integration,
defeats any attempt to supply a theoretical or physical justification
for the distribution of the isograms. Consequently the construction
of these charts must depend almost entirely upon an analysis of the
data."

Scatter diagram analysis is very useful when resources are
limited and only small amounts of data are available. It does not
lend itself easily to processing by electronic computer, and the
method itself implies hand analysis [although some individuals,
including Freeman (1961), have attempted to automate the process].
For this reason, other techniques are usually to be preferred when
samples consisting of several thousand cases and a computer are
available. Also, no reliable significance test exists for the
scatter diagram procedure to determine whether added predictors will
lead to increases in forecast accuracy on new data. Therefore, a
forecasting system based on scatter diagrams should always be tested
on new data if at all possible.

300

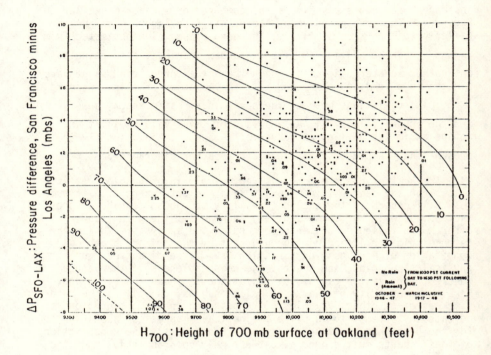

Figure 4. Precipitation amount plotted as a function of 700 mb
height at Oakland versus the San Francisco minus Los
Angeles sea level pressure difference. Isolines representing rainfall amount have been adjusted to a scale of
0 to 100 for input to another diagram (after Thompson,
1950).

5. REGRESSION

It has become increasingly clear during the last 20 years that the use of large samples is very desirable in the solution of meteorological prediction problems. Three reasons can be cited for this conclusion. First, the autocorrelation of many meteorological variables does not approach zero unless observations are taken quite some time apart. Therefore, the number of degrees of freedom for such data is much less than the sample size. Second, many variables usually can be found that possess a relationship to the predictand, and one frequently wants to include - or at least test the desirability of including - a large number of these potential predictors. In the application of most models, this process uses up many degrees of freedom. Third, the distribution of the predictand is frequently highly skewed, so that the very weather situations that are most important to predict occur very infrequently. A large sample is necessary to include a representative number of such situations.

The use of large samples and the inclusion and testing of many predictors necessitates the use of an electronic computer and a model that lends itself to computer application. Linear regression is such a model.

5.1 Simple Linear Regression

The simple linear regression model is of the form

$$Y = \alpha + \beta X + \varepsilon,$$

where Y is the predictand, X is a predictor, α and β are parameters, and ε is the error term. The predicted value of Y is $\hat{Y}$, where

$$\hat{Y} = a + bX,$$

in which a and b are estimates of the parameters.

In this model, the sum of squares of the observed errors (i.e., the e_i's) is minimized over a dependent sample of size n:

$$\min \sum_{i=1}^{n} e_i^2 = \min \sum_{i=1}^{n} (Y_i - \hat{Y}_i)^2 = \min \sum_{i=1}^{n} (Y_i - a - bX_i)^2.$$

Taking partial derivatives of $\sum_i e_i^2$ with respect to a and b and setting each derivative equal to zero yields the so-called normal equations:

$$an + b \sum_{i=1}^{n} X_i - \sum_{i=1}^{n} Y_i = 0,$$

$$a \sum_{i=1}^{n} X_i + b \sum_{i=1}^{n} X_i^2 - \sum_{i=1}^{n} X_i Y_i = 0.$$

Solving for a and b gives

$$b = \frac{n \sum_{i=1}^{n} X_i Y_i - \sum_{i=1}^{n} X_i \sum_{i=1}^{n} Y_i}{n \sum_{i=1}^{n} X_i^2 - (\sum_{i=1}^{n} X_i)^2}$$

$$a = \frac{1}{n} \sum_{i=1}^{n} Y_i - \frac{b}{n} \sum_{i=1}^{n} X_i.$$

This one-predictor model is useful for illustrative purposes and can be applied to situations such as that shown in Figure 3. Usually, however, several predictors need to be considered. As a result, one is led to multiple linear regression.

5.2   Multiple Linear Regression

If we form matrices from data samples of size n for the predictand and p predictors, then

$$\underline{x} = \underline{X} - \overline{\underline{X}}$$

and

$$\underline{y} = \underline{Y} - \overline{\underline{Y}}$$

are nxp and nx1 matrices, respectively, in which each column is the deviation from the mean of the corresponding original variable. Variance-covariance matrices can be calculated as follows:

$$\underline{S}_{11} = \frac{1}{n} \underline{x}'\underline{x},$$

$$\underline{S}_{12} = \underline{S}_{21}' = \frac{1}{n} \underline{x}'\underline{y},$$

$$\underline{S}_{22} = \frac{1}{n} \underline{y}'\underline{y} = \hat{\sigma}_Y^2,$$

where a prime denotes a matrix transpose. The multiple regression equation, derived in a manner analagous to the one-predictor case in Section 5.1, is

$$\hat{Y} = \underline{X} \; \underline{S}_{11}^{-1} \; \underline{S}_{12} - \overline{\underline{X}} \; \underline{S}_{11}^{-1} \; \underline{S}_{12} + \overline{Y}. \tag{5}$$

Associated with Eq. (5) are a reduction of variance (RV) and a multiple correlation coefficient (R) that are defined as follows:

$$RV = R^2 = \frac{\frac{1}{n} \sum_{i=1}^{n} (Y_i - \overline{Y})^2 - \frac{1}{n} \sum_{i=1}^{n} (Y_i - \hat{Y}_i)^2}{\frac{1}{n} \sum_{i=1}^{n} (Y_i - \overline{Y})^2},$$

$$= \frac{\text{variance of } Y - \text{error variance of } \hat{Y}}{\text{variance of } Y}.$$

These quantities are easily calculated from the variance-covariance matrices:

$$RV = R^2 = \frac{\underline{S}_{21} \; \underline{S}_{11}^{-1} \; \underline{S}_{12}}{\underline{S}_{22}} = \frac{\underline{S}_{21} \; \underline{S}_{11}^{-1} \; \underline{S}_{12}}{\hat{\sigma}_Y^2}.$$

Regression analysis has been used extensively in meteorology for many years. However, it wasn't until the age of electronic computers that large data samples and many predictors could be handled easily. Regression, as a mathematical model, can be used no matter what the joint distribution of predictand and predictors, except that no predictor may be an exact linear function of one or more other predictors. In that case, the inverse of $\underline{S}_{11}$ could not be determined;

that is, it would be singular. For real meteorological data and sample sizes much larger than the number of parameters (i.e., $p + 1$), this problem seldom arises.

Under certain conditions, analysis of variance can be used for testing the significance of the reduction of variance, R^2, and of individual terms in the equation. The conditions are that the sample be drawn randomly from a multivariate normal population and that no preselection of variables be made using the same sample on which the regression equation is developed. The F value corresponding to R^2 is calculated according to line 2 in Table 2. This calculated value can then be compared to the tabled F value for a desired α level (probability of Type I error) with p and n-p-1 degrees of freedom.

Table 2. Analysis of variance table for reduction of variance associated with multiple linear regression model.[a]

Source	Sum of squares[b]	Degrees of freedom	Mean square[b]	F value
Total	1	$n-1$		
Regression equation - p predictors	R_p^2	p	$\dfrac{R_p^2}{p}$	$\dfrac{R_p^2 (n-p-1)}{(1-R_p^2)\, p}$
pth predictor in regression equation	$R_p^2-R_{p-1}^2$	1	$R_p^2-R_{p-1}^2$	$\dfrac{(R_p^2-R_{p-1}^2)(n-p-1)}{1-R_p^2}$
Residual	$(1-R_p^2)$	$n-p-1$	$\dfrac{1-R_p^2}{n-p-1}$	

[a]Patterned after Panofsky and Brier (1958).

[b]All entries in these columns should be multiplied by $n\hat{\sigma}_Y^2$. This factor cancels out in computing F.

Suppose R^2 is significant in a particular situation, and one wonders whether a <u>particular</u> predictor is really adding any predictive information over and above the other p-1 predictors. Consider that predictor as the last or pth predictor. Then the appropriate F value is shown on line 3 of Table 2 and has 1 and n-p-1 degrees of freedom. This test is valid, under the conditions stated above, provided that the <u>choice</u> of which predictor to test is not based on an analysis of the data sample. This topic will be discussed further in Section 5.3.

The model discussed in this section <u>is</u> a linear model, and other models may be more appropriate. However, if the population distribution is multivariate normal, then this model is the best model that can be found. If a researcher wants to "screen" out from a much larger set of possible predictors the p predictors to include in the equation, then the approach described in the next section can be used.

5.3 Screening Regression

Screening regression, as the term is usually defined in meteorology, combines multiple linear regression with an objective method of selecting a "good" set of predictors to use in the equation from a larger set of m potential predictors. Since regression finds the solution that minimizes the estimated error variance on the dependent sample, it is logical to choose a set of predictors that would be better than any other set for reducing this error variance. However, if p predictors were to be picked from a set of m predictors, then the number of combinations for even a moderate value of m is quite large (unless p is very small or approaches m). Specifically, the number of such combinations is

$$C_p^m = \frac{m!}{p!(m-p)!}.$$

It is usually not feasible to compute all these combinations, so some shortcut must be taken to find a "good" set that may not be the "best" set.

Screening can be done in one of several ways. The simplest is what may be called forward selection. This procedure consists of first selecting the one predictor from the total set of m predictors under consideration that reduces the variance of the predictand more than any other possible predictor, then choosing the predictor that together with the first one selected reduces the variance more than any other such combination of two predictors, and continuing the selection procedure on a "one at a time" basis until the additional reduction of variance afforded by any predictor is very small. This procedure insures that the first predictor selected is the best single predictor, but it does not insure that the first two chosen are the best pair, etc. This stepwise selection was discussed as early as 1940 by Wherry (1940) and was introduced into the meteorological literature by Miller (1958) following some unpublished work by Bryan (1944).

The question arises as to how many predictors to choose. One might be tempted, after selecting p-1 predictors, to test the additional reduction of variance given by the pth predictor using the F value computed in the third line of Table 2. This procedure was suggested by Lubin and Summerfield (1951) and is sometimes done. However, it must be considered only as a stopping criterion. That is, one must not attach any particular significance level to it. The reason for this limitation is that the test is being performed on the next best predictor and not on a predictor that has been selected at random.

Miller (1958) has suggested a modification to the standard F test that compensates for the testing of the best of several remaining potential predictors. Instead of using the critical value $F_{(1-\alpha)}$ at each selection step, he suggests using the value $F^*_{(1-\alpha)} = F_{1-\alpha/(m-p+1)}$ at the pth selection step with some desired probability of Type I error α. This criterion is rather harsh, since it assumes (approximately) that the (m-p+1) tests that could be performed at the pth selection are independent and, in the absence of additional complications, tends to lead to the selection of too few predictors if some accepted value of α such as 0.05 is used (Zurndorfer and Glahn, 1977). Additional complications could include highly nonnormal distributions or non-zero autocorrelation for the predictors. The test proposed by Miller (op. cit.) will also tend to compensate for nonzero autocorrelations. However, it should be remembered that, because of all the complications, this test is primarily a stopping procedure and no exact level of significance should be attached to it.

The decision as to the exact number of predictors to select is many times overemphasized. The mean square error for independent data is usually not very sensitive to the number of predictors in the equation within rather broad limits. Figure 5 shows schematically the kind of results that have been obtained from experiments (e.g., see Bocchieri and Glahn, 1972). R^2 on dependent data always increases with the addition of another predictor but tends to "level out" so that little is to be gained in terms of the mean square error of the predictand by including more than, say, 12 terms. The mean square error on test data need not decrease monotonically. A small test sample will frequently cause the mean square error curve to be irregular. Also, for a large number of predictors, the R^2 test sample curve will usually turn downward. However, a broad, flat maximum will generally be found where, for practical purposes, the

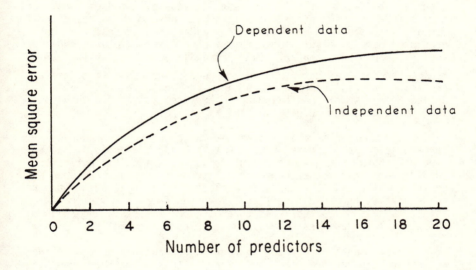

Figure 5.   Schematic diagram of mean square error as a function of
number of predictors.

predictions are of equal quality. For this reason, almost any "practical" stopping procedure is quite adequate, such as (a) when the added reduction of variance of the next predictor is less than 0.005, (b) when 12 predictors have been selected, (c) when the reduction in mean square error is less than, say, 0.05°F for temperature or 0.2 mph for wind speed, or (d) Miller's $F_{(1-\alpha)}$.

Another version of screening regression is to find the reduction of variance for all m predictors, and then start eliminating predictors one by one until some stopping criterion is met. This backward elimination procedure also does not yield the unique best set, and significance testing for it has been inadequately studied. However, some simple stopping procedure similar to those methods described for forward selection can be used. A complication could occur if one predictor were an exact linear function of a set of other predictors. Then $\underline{S}_{11}$ would be singular. This possibility is very unlikely with real meteorological data, unless one were to actually formulate a predictor from a linear relationship. For instance, one could include only two of the following three predictors: 500 mb height, 1000 mb height, and 1000-500 mb thickness.

Still another algorithm combines the above two procedures. Forward selection is done with an F test being performed at each step. When the added reduction of variance is insufficient to be judged significant, the procedure is stopped. However, between each selection step, all the variables selected up to that point (and not subsequently discarded) are tested for significance. The least significant is discarded if it does not meet the test, and again all those remaining are tested until none is discarded.

Forward selection screening regression has been used more than any other computer-oriented model for statistical weather prediction. Many studies were made at the Travelers Research Center, Inc. in the late 1950's and early 1960's using this procedure in combination with the classical approach. For instance, Veigas et al. (1958) produced an objective method for predicting the behavior of hurricanes in the western Atlantic and Gulf of Mexico that was subsequently used operationally by the National Weather Service. More recently, the Techniques Development Laboratory (TDL) of the National Weather Service has developed many operational products based on this model. Most of these products involve the MOS technique, but some forecasts are based on perfect prog and classical procedures. These present-day uses of regression analysis are discussed in Section 11.

5.4 Regression Estimation of Event Probabilities

The screening regression model can be used when the predictand is binary. For instance, the predictand might take on the value of 1 when an event occurred and 0 when it didn't. The regression equation, then, can be thought of as yielding the probability (or relative frequency) of the event for realistic combinations of predictor values. This approach was used by Mook (1948) and Lund (1955), but no extensive application was made of it until Miller (1964) and others began using it at the Travelers Research Center in the 1960's. They dubbed it REEP for regression estimation of event probabilities. Miller (op. cit.) realized that the probabilistic model also held for multiple categories. For instance, if ceiling height is divided into five mutually exclusive and exhaustive categories and each category is used to define a binary predictand, then the set of five regression equations (all with the same predictors) will give a set of probabilities, p_i, where $\sum_i p_i = 1$ (i = 1, ..., 5).

An important property of this model is that it minimizes the P-score defined by Brier (1950), which has certain desirable characteristics (Brier, op. cit.; Murphy, 1974) and which is frequently used in probabilistic forecast verification (see Chapter 10). Unfortunately, the individual p_i's are not constrained to the zero-one interval. Selection of predictors can be made by choosing next the predictor that contributed most to the R^2 of any one of the categories. Significance tests based on assumptions of normality are not appropriate. Experience has shown that, generally, a larger sample is required to obtain stable results when the predictand is binary than when it is continuous.

5.5 Binary Predictors

The screening regression model can be used when one or more of the predictors are binary. All predictors were binary in the first applications of REEP. An early reference to the use of binary variables in regression is Suits (1957), and Neter and Wasserman (1974) give a good discussion of the subject.

A binary variable (sometimes called an indicator or dummy variable) can arise naturally. For example, an observation may be made in this format, as in the case of rain or no rain. In addition, a continuous variable can also be "dummied"; this process is usually carried out in one of the two ways indicated in Table 3. (Variables are never really continuous, since only discrete values are used in practice. However, temperature measured to the nearest degree is quasi-continuous and will be considered to be continuous in this chapter.) In transformation 1, each dummy variable indicates whether or not the original variable has a value corresponding to its particular defining interval. In transformation 2, each dummy variable indicates whether or not the original variable has a value less than the upper limit of its particular defining interval. Note that any one of the four dummy variables for transformation 1 is redundant with the other three and, although all four could be screened (a

Table 3. Two methods of transforming a continuous variable into binary variables. In any particular column, ones and zeroes can be interchanged without affecting predictive capability.

Original variable category	Binary variable transformation 1				Binary variable transformation 2		
	1	2	3	4	1	2	3
1	1	0	0	0	0	0	0
2	0	1	0	0	1	0	0
3	0	0	1	0	1	1	0
4	0	0	0	1	1	1	1

fourth would never be selected), only three (any three) can be
included in a regression equation. For transformation 2, only three
meaningful variables are possible; the fourth would always have the
same value. Dummy variable No. 1 corresponds to No. 1, and No. 4 to
No. 3, for transformations 1 and 2, respectively. However, Nos. 2
and 3 for transformation 1 have no match in transformation 2.

Any combination of three dummy variables for transformation 1
will give the same reduction of variance as the three dummy variables
for transformation 2. However, No. 2 for transformation 2 may be
better than any single predictor for transformation 1. When several
dummy variables are created from a continuous variable, careful consideration should be given to which transformation to use. A predictor such as No. 2 for transformation 1 treats a certain category of
the original variable one way and the categories on both sides of it
another way. This procedure may be appropriate if the predictand is
continuous and is a quadratic function of the original (undummied)
predictor or if the predictand is binary and was dummied by transformation 1. For instance, suppose a binary predictand represents
ceiling height from 1000 to 1900 feet at 1200 GMT. Then a binary
predictor representing that same ceiling interval at 0900 GMT would
be a good predictor. However, if the predictand is continuous, then
a few binary predictors defined by transformation 2 will usually
yield better results than the same number of predictors defined by
transformation 1.

Although dummy predictors have the potential of accommodating a
nonlinear relationship between the predictand and predictors, some
information is lost since all values within a defining interval are
treated the same way (unless each value is represented by a different
dummy variable). Also, it takes several binary predictors to provide
about the same information as one continuous predictor. The number
of binary variables to be defined for a given predictor is usually
quite arbitrary as is the interval to associate with each variable.
Traditional statistical significance tests are even less applicable
in this case, due both to the binary nature of the variables and to
the unknown number of degrees of freedom used in choosing, say, five
out of eight possible dummy variables created from a single predictor.

The use of all binary variables permits much more efficient computer use, since one value need occupy only one bit rather than a
complete word and, in addition, faster logical rather than arithmetic
operations can be used in obtaining sums of squares and cross products. However, realization of this advantage usually requires considerable programming effort and use of such a program would have to
be rather extensive before the effort would be worthwhile.

5.6 Computed Predictors

Although regression as presented above is a linear model, non-linear relationships can be incorporated through special computations. For instance, divergence is not observed but can be calculated from wind observations. In addition, a predictor can be "linearized" in various ways. That is, it can be transformed in such a way that it has a more nearly linear relationship with the predictand than did the original variable. Consider again the histogram example shown in Figure 2. If we want to use several predictors, including 1000-500 mb thickness, then we could include a transformed or computed predictor representing the heights of the bars in Figure 2. The transformation could be exactly as indicated in Figure 2, or a curve could be fit that would undoubtedly provide a variable that would be more robust on independent data.

5.7 Orthogonal Predictors

We may have a problem in which we want to distill most of the linear predictive information from a large set of variables without using a large number of degrees of freedom. For instance, 1000-500 mb thickness values may be available at each of 25 stations surrounding a station for which we wish to predict max temperature. The predictors are highly correlated and we wouldn't want to include all 25 in a regression equation. We could screen the 25 and select, say, five. Another alternative is to transform the 25 variables into another set of variables that are more efficient in terms of retaining the large scale predictive information and discarding the small scale "noise." Orthogonal functions can be used for this purpose.

Assume that pressure values are available at each of m points and at each of n times. At each point the mean over time can be found and the deviations from the means put into an nxm matrix $\underline{P}$. The element P_{ii} on the diagonal of $\frac{1}{n} (\underline{P}'\underline{P})$ is the variance of the pressure at the ith point and the element P_{ij} is the covariance of the pressures at the ith and jth points. The time series of k new variables, represented by the nxk matrix $\underline{U}$ can be found by

$$\underline{U} = \underline{P} \ \underline{T},$$

where $\underline{T}$ is an mxk matrix of coefficients of k functions at m points. The matrix $\underline{T}$ is an efficient transformation matrix if the columns are orthogonal. Also, the total variance of the columns of $\underline{U}$ is equal to the total variance of the columns of $\underline{P}$ if the columns of $\underline{T}$ are orthonormal and k = m. Then

$$\underline{T}'\underline{T} = \underline{I}$$

($\underline{I}$ is a kxk identity matrix) and

$$tr(\underline{U}'\underline{U}) = tr(\underline{P}'\underline{P}),$$

where tr represents the trace of a matrix. The original pressure deviations at the m points can be approximately reproduced from the new functions by

$$\hat{\underline{P}} = \underline{U}\ \underline{T}',$$

and, if k = m,

$$\hat{\underline{P}} = \underline{P}.$$

Several authors, including Wadsworth (1948), White et al. (1958), and Jorgensen (1959), have adapted orthonormal Tschebyscheff functions for this use. The m points occur in a rectangular array and functions of degree zero through r and zero through s are used in the two dimensions, respectively. The columns of $\underline{T}$ are then made up of cross products of two functions. The function composed of the function of degree zero in both directions represents the mean of the m points. Functions composed of one function of degree zero and one function of degree not zero represent patterns which vary in only one direction. In general, the low degree functions represent large scale features of the map, whereas the high degree functions represent small scale features.

Much of the variance of pressure, and of many other meteorological variables, is explained by large scale components. On the other hand, it is the very small scale components that contain most of the observational error and are the least predictable. If very small scale features in the pressure map furnish much predictive information for other variables, then it is usually not beneficial to represent those features in terms of orthogonal functions.

Even though the transforming functions - the columns in $\underline{T}$ - are orthogonal, the new variables - the columns in $\underline{U}$ - are not necessarily orthogonal. Therefore, $\underline{T}$ is not as efficient as it might be, and the regression constants which relate a predictand to the new variables must be determined by considering the covariances between those new variables as well as the variances. That is, it is necessary to examine the complete matrix

$$\frac{1}{n}\ (\underline{U}'\underline{U}).$$

The most efficient way of representing the linear information in a set of data is through principal components. These functions were introduced into meteorology by Lorenz (1956) who called them Empirical Orthogonal Functions (EOFs). EOFs have been used to study meteorological data and for predictive purposes by several researchers, including Gilman (1957), White et al. (1958), Glahn (1962), and Grimmer (1963). In addition to

$$\underline{T}'\underline{T} = \underline{I},$$

the condition

$$\frac{1}{n} (\underline{U}'\underline{U}) = \underline{D}$$

is also imposed here, where $\underline{D}$ is an mxm diagonal matrix. Since

$$\underline{U} = \underline{P}\ \underline{T},$$

substitution can be made to yield

$$\frac{1}{n} (\underline{T}'\underline{P}'\underline{P}\ \underline{T}) = \underline{D}.$$

The matrix

$$\frac{1}{n} (\underline{P}'\underline{P}) = \underline{R}$$

is the covariance matrix of the original variables. Therefore,

$$\underline{T}'\underline{R}\ \underline{T} = \underline{D}.$$

The columns of $\underline{T}$ are the characteristic vectors and the corresponding diagonal elements of $\underline{D}$ are the roots of the matrix $\underline{R}$. The k columns of $\underline{T}$ which correspond respectively to the k largest diagonal elements of $\underline{D}$ explain a larger fraction of the total variance of the original variables, $\frac{1}{n} \text{tr}(\underline{P}'\underline{P})$, than any other k linear combinations of those variables.

Regression estimates $\hat{\underline{y}}$ of a predictand time series $\underline{y}$ (in terms of deviations from the mean) can be found by

$$\hat{\underline{y}} = \underline{U}\ \underline{A},$$

where $\underline{A}$ is the kx1 vector of regression coefficients corresponding to k of the new variables. The vector $\underline{A}$ is now easily determined from

$$\underline{A} = \frac{1}{n} (\underline{D}^{-1}\underline{U}\ \underline{y}),$$

since the inversion of $\underline{D}$ is trivial.

It is worth noting that if the original variables are normalized to unit variance, the characteristic vectors and roots will not in general be the same as with the nonnormalized variables. For instance, the pressure at a point in middle latitudes has a larger variance than it does in low latitudes. If points from both regions are included without normalization, the columns of $\underline{U}$ having the largest variances will be dominated by the middle latitude pressures.

An advantage in the use of EOF's is that the points do not have to be in any organized pattern spatially. Therefore, observations taken at stations can be used directly rather than requiring the field to be specified in terms of a grid before the orthogonal functions are applied.

5.8 Normalized Variables

Although regression can be applied to variables with practically any distribution, it is the optimum model if the variables have a multivariate normal distribution. Boehm (1976) suggests that each variable used in the analysis should be "transnormalized." That is, some method such as histogram analysis or curve fitting should be applied to each variable separately, both predictand and predictors, so that the resulting transformed variable will have a (near) normal distribution. Boehm (op. cit.) uses the term transnormalized to highlight the fact that this transformation is not just subtraction of the mean and division by the standard deviation. It is the same transformation discussed by Panofsky and Brier (1958, p. 41). After the regression analysis is performed on the normalized variables, predictions of the normalized dependent variable can be made. Then these values must be transformed back to the original variable.

6. DISCRIMINANT ANALYSIS

Certain meteorological variables do not lend themselves well to prediction by linear regression due to their nonnumerical nature, highly nonnormal distribution, or nonlinear relationships to the predictors. In such cases, multiple discriminant analysis (MDA) provides a useful tool that has been applied extensively to weather forecasting problems by Miller (1962) and others at the Travelers Research Center, Inc.

Discriminant analysis was conceived by Fisher (1936) and first brought into the literature by Barnard (1935). MDA refers specifically to the Fisher analysis on more than two predictand groups. Barnard (op. cit.) used the analysis on 4 groups, but she considered only one discriminant function. Hotelling (1935) and others (Fisher, op. cit.; Brier, 1940) evidently appreciated the possibility of more than one function and mentioned the determinantal equation involved, but the burden of calculations forbade extensive use of MDA until the computational scheme of Bryan (1950) or electronic computers became available.

For a given problem, there is a maximum of p or G-1 (whichever is smaller) discriminant functions, where p is the number of predictors and G the number of groups. These functions are mutually uncorrelated and are found through the solution of the equations:

$$(\underline{W}^{-1}\underline{B} - \lambda_j\underline{I})\underline{V}_j = \underline{0}, \qquad [j = 1, \ldots, \min(p, G-1)]$$

where $\underline{W}$ and $\underline{B}$ are respectively the matrices of within and between groups sums of squares of the predictors, $\underline{I}$ is the identity matrix, the $\underline{V}_j$'s (eigenvectors) are the coefficients in the discriminant functions, and the λ_j's are the roots (eigenvalues) of the determinantal equation

$$|\underline{W}^{-1}\underline{B} - \lambda\underline{I}| = 0.$$

In the special case involving only two groups, only one function is possible and its coefficients are proportional to those derived by regression. Therefore, for this special case, the two analyses are equivalent.

A significance test, which is a generalization of Mahalanobis' D^2, for the predictand-predictors relationship based on large sample theory has been developed by Rao (1952) and uses the statistic

$$V_{pG} = n(\text{tr } \underline{W}^{-1}\underline{B}) = n \sum_{j=1}^{G-1} \lambda_j,$$

where n is the sample size and it is assumed that G-1 roots exist. V_{pG} is distributed as χ^2 with p(G-1) degrees of freedom provided that the predictors are multivariate normal within each group and that the covariance matrices for each group are identical.

The importance of each discriminant function $\underline{V}_j$ is indicated by its associated root λ_j. Since the number of discriminant functions may be less than the minimum of p and G-1, the significance of each root can be tested by an approximate procedure due to Bartlett (1934). For each nonzero root, the test statistic

$$[n - (1/2)(p + G)] \ln (1 + \lambda_j)$$

is computed. This statistic is approximately distributed as χ^2 with p+G-2j degrees of freedom.

The selection of variables for MDA by screening has also been described by Miller (1962). In the same way that the selection for regression maximizes the F-statistic, the selection for MDA maximizes

V_{pG}. At each step, after p-1 predictors have been selected, the quantity

$$V_{pG} - V_{(p-1)G} = n[tr(\underline{W}_p^{-1} \underline{B}_p) - tr(\underline{W}_{p-1}^{-1} \underline{B}_{p-1})]$$

is evaluated for each remaining possible predictor and the largest value indicates the preferred variable. This statistic is considered to be distributed as χ^2 with G-1 degrees of freedom (Rao, 1952), and Miller (1962) suggests, as with regression, that the critical value $\chi^2_{1-\alpha/(m-p+1)}$ be used since the selection is not random.

MDA as used in prediction can be considered to be a linear transformation from a p-dimensional predictor space to a G-1 dimensional discriminant space (assuming G-1 $\leq$ p), such that the sample points plotted in the discriminant space exhibit as much clustering according to predictand category and as little dispersion from their respective cluster centers as possible. There will be, then, G regions in the discriminant space, one for each predictand category.

Let us consider the effect of the relationship between one of the predictors and the predictand on this transformation and the desirability of using this predictor. Three possibilities can be mentioned:

(a) It may be that a particular value of the predictor will indicate only one predictand category, and that category will be indicated by no other value of the predictor. This situation is the most desirable. It does not matter how the predictand categories are arranged on the predictor scale; group 1 could be between group 4 and group 5 just as well as anywhere else (see Figure 6). Thus, this type of nonlinearity is accommodated by MDA, since no numerical scale is associated with the predictand categories themselves.

(b) It may be that a low predictor value will indicate two different predictand categories. In this case the regions in the discriminant space containing the points representing these two predictand categories will be superimposed (unless the effect of other predictors can spread them apart), and the predictor will be worthless in discriminating between these two predictand categories. However, the predictor could still be very useful in separating these two predictand categories from all the rest and could also distinguish between those remaining categories. No transformation of the predictor before its inclusion in the analysis would be useful in separating the superimposed groups. See groups 3 and 4 in Figure 6 for an example of this situation.

318

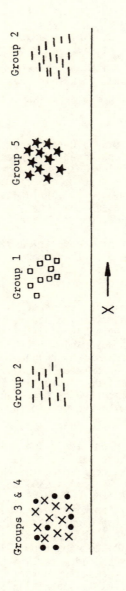

Figure 6. Hypothetical relationships of predictand groups 1 through
5 to the predictor x. See text for explanation.

(c) It could happen that both low and high values of the predictor would indicate the same predictand category. This
situation is very undesirable since the predictor would
tend to spread the points in the discriminant space representing this predictand category and these points might
group themselves into two distinct regions (see group 2 in
Figure 6). This type of nonlinearity is not accommodated
by MDA, and a nonlinear transformation of the predictor is
indicated if it is to be used. In a screening procedure
where the raw variable is a possible predictor, it will
probably be overlooked as its nonlinear relationship to the
predictand will not be recognized.

Since the criterion for selecting the variables and for determining the functions themselves is to maximize the between to within
groups variance ratio, groups with many cases will highly influence
the results. This state of affairs is generally detrimental, except
in those cases in which the costs of misclassification are all equal
and the concern is _only_ for the number of correct forecasts. Miller
(1962) has attempted to counteract the large group effect in predictor selection by making the size of all groups equal to that of the
smallest group during the selection process and then using the complete sample to determine the discriminant functions and probabilities of misclassification.

Discriminant analysis can be considered to be complete when the
functions and their associated roots have been found. However, the
problem of how to use these functions in probabilistic prediction of
the predictand groups still remains unspecified. Miller (1962) used
Bayes' theorem to find the a posteriori group probabilities from the
a priori group probabilities, which are estimated from the sample and
the assumed multivariate normal distribution of the discriminant
functions within each group. For the sample of meteorological data
considered by Miller (op. cit.), the multivariate normal assumption
proved to be untenable.

If the data do not appear to justify the use of a completely
parametric model, discriminant analysis can be used to map the p-
dimensional space into a G-1 or less dimensional space, and the conditional probability distributions can be determined by other means.
If the dimensionality of the discriminant space is only 1 or 2, a
scatter diagram can be employed to determine how the G groups are
distributed within this space.

Miller (1962) found a nonparametric method described by Fix and
Hodges (1951) to be useful in determining the a posteriori probabilities directly from the discriminant function values. At any point Y'
within the discriminant space, the probability of each group can be
estimated by the relative frequency of that group occurring in the k
sample points closest to Y'. The value of k should be relatively
large but small compared to the sample size. The k closest points
can be defined in terms of the Euclidean distance. This procedure is

essentially a smoothing process, and precautions must be taken to ensure that the importance of the discriminant functions is taken into account in this process. Specifically, an arbitrary metric has been used to transform the discriminant space into a space in which each function has zero mean and variance equal to λ_j/λ_1. The distance between a point Y' and a sample point Y in this space can be defined as

$$D = \left[\frac{\lambda_1}{\lambda_1}\left(\frac{y_1 - y_1'}{\hat{\sigma}_{y_1}}\right)^2 + \frac{\lambda_2}{\lambda_1}\left(\frac{y_2 - y_2'}{\hat{\sigma}_{y_2}}\right)^2 + \ldots + \frac{\lambda_j}{\lambda_1}\left(\frac{y_j - y_j'}{\hat{\sigma}_{y_j}}\right)^2\right]^{1/2},$$

where $\hat{\sigma}_{y_j}$ is the standard deviation of the jth discriminant function. This procedure usually produces extreme smoothing over the least important functions but retains the predictive information in the more important functions.

The most extensive use and thorough testing of the MDA technique in meteorology has been in the short range prediction of visibility and ceiling height undertaken at the Travelers Research Center, Inc. (e.g., see Enger et al., 1964).

7. CANONICAL CORRELATION

Canonical correlation, first developed by Hotelling (1936), is a technique for finding orthogonal relationships between two sets of variables. Consider a situation involving n observations of each of p variables X_i (i = 1, 2, ..., p) and of q variables Y_i (i = 1, 2, ..., q). These observations represent points in a p+q dimensional space and can be arranged in the nxp matrix $\underline{X}$ and the nxq matrix $\underline{Y}$. The variables have means $\overline{X}_i$ and $\overline{Y}_i$, respectively, and deviations from the mean are given by $x_i = X_i - \overline{X}_i$ and $y_i = Y_i - \overline{Y}_i$. New variables $\underline{x}\,\underline{A}_i$ and $\underline{y}\,\underline{B}_i$ (i = 1, 2, ..., r), where r is less than or equal to the smaller of p and q, can be formed such that their means are zero and

$$\underline{A}'\underline{x}'\underline{x}\,\underline{A} = n\,\underline{I}, \tag{6}$$

$$\underline{B}'\underline{y}'\underline{y}\,\underline{B} = n\,\underline{I}, \tag{7}$$

$$\underline{A}'\underline{x}'\underline{y}\,\underline{B} = n\,\underline{\Lambda}, \tag{8}$$

where $\underline{I}$ is an rxr identity matrix,

$$\underline{\Lambda} = \begin{bmatrix} \lambda_1 & & & \\ & \lambda_2 & & 0 \\ & & \ddots & \\ & 0 & & \ddots \\ & & & & \lambda_r \end{bmatrix}, \tag{9}$$

and $\lambda_1 \leq \lambda_2 \leq \cdots \leq \lambda_r$. Eqs. (6) and (7) state that the variance of each of the new variables is unity and each is uncorrelated with all others in its respective set. Eqs. (8) and (9), together with Eqs. (6) and (7), state that each $\underline{x}\,\underline{A}_i$ is uncorrelated with each $\underline{y}\,\underline{B}_j$ except when $i = j$ and then the correlation is λ_i.

It can be shown [for instance, see Anderson (1958)] that the $\underline{A}_i$ ($i = 1, 2, \ldots, r$) can be found from

$$(\underline{S}_{11}^{-1}\,\underline{S}_{12}\,\underline{S}_{22}^{-1}\,\underline{S}_{21} - \lambda_i^2\,\underline{I})\underline{A}_i = \underline{0}$$

(providing $\underline{S}_{11}$ and $\underline{S}_{22}$ are not singular), where the λ_i satisfy the determinantal equation

$$|\underline{S}_{11}^{-1}\,\underline{S}_{12}\,\underline{S}_{22}^{-1}\,\underline{S}_{21} - \lambda^2\,\underline{I}| = 0$$

and where

$$\underline{S}_{11} = \frac{1}{n}\,\underline{x}'\underline{x},$$

$$\underline{S}_{12} = \underline{S}_{21}' = \frac{1}{n}\,\underline{x}'\underline{y},$$

and

$$\underline{S}_{22} = \frac{1}{n}\,\underline{y}'\underline{y}$$

are the variance-covariance matrices. Then the $\underline{B}_i$ can be found from

$$\underline{B} = \underline{S}_{22}^{-1}\,\underline{S}_{21}\,\underline{A}\,\underline{\Lambda}^{-1}$$

Alternatively, we could use

$$(\underline{S}_{22}^{-1}\,\underline{S}_{21}\,\underline{S}_{11}^{-1}\,\underline{S}_{12} - \lambda_i^2\,\underline{I})\,\underline{B}_i = \underline{0},$$

$$|\underline{S}_{22}^{-1}\,\underline{S}_{21}\,\underline{S}_{11}^{-1}\,\underline{S}_{12} - \lambda^2\,\underline{I}| = 0,$$

and

$$\underline{A} = \underline{S}_{11}^{-1}\,\underline{S}_{12}\,\underline{B}\,\underline{\Lambda}^{-1}.$$

The latter equations are to be preferred if q < p, because the matrix that must be diagonalized is then of a lesser dimension.

The "first" pair of functions, defined by the first column of each $\underline{A}$ and $\underline{B}$, have as large a correlation λ_1 as any other possible pair of functions, each composed of a linear combination of the original variables. Also, the "second" function pair has as large a correlation λ_2 as any other possible pair of functions, each being composed of a linear combination of the original variables and each being uncorrelated with both members of the first pair.

Either set of new variables can be predicted in a least-squares sense by the new variables in the other set. The prediction equations are

$$(\underline{y}\,\hat{}\,\underline{B}) = \underline{x}\ \underline{A}\ \underline{\Lambda}$$

and

$$(\underline{x}\,\hat{}\,\underline{A}) = \underline{y}\ \underline{B}\ \underline{\Lambda}.$$

In addition, the original variables in one set can be predicted in a least-squares sense by the new variables in the other set; for example, by

$$\hat{\underline{y}} = \underline{x}\ \underline{A}\ \underline{\Lambda}\ \underline{B}'\underline{S}_{22}. \tag{10}$$

In the case that r = q, Eq. (10) can be written as

$$\hat{\underline{y}} = \underline{x}\ \underline{A}\ \underline{\Lambda}\ \underline{B}^{-1}. \tag{11}$$

Similar equations can be written for predicting $\underline{x}$.

Eq. (10) represents the prediction equation for each of the $\hat{y}_i$ in terms of all of the x_i. One may want to relate one set of variables to the other set but include only a portion of the correlations in $\underline{\Lambda}$, perhaps those k correlations that are judged to be significantly different from zero. An equation corresponding to Eq. (10) can be written as

$$\hat{\hat{\underline{y}}} = \underline{x}\ \underline{A}\ \underline{\Lambda}^*\underline{B}'\underline{S}_{22}, \tag{12}$$

where the rxr matrix $\underline{\Lambda}^*$ has only k nonzero elements, the others having been set equal to zero. Eq. (12) has the effect of including a contribution only from those k columns of $\underline{A}$ and k rows of $\underline{B}'$ corresponding to the k nonzero correlations.

The prediction equations can be expressed in terms of the original variables X_i and Y_i. For instance, Eq. (11) becomes

$$\underline{Y} = \underline{X} \, \underline{A} \, \underline{\Lambda} \, \underline{B}^{-1} - \overline{\underline{X}} \, \underline{A} \, \underline{\Lambda} \, \underline{B}^{-1} + \overline{\underline{Y}}. \tag{13}$$

Discriminant analysis and multiple regression, including REEP, are special cases of canonical correlation. For instance, when $q = 1$ Eq. (13) is the same as the least squares regression equation for a single predictand [see Eq. (5)]. In addition, if the predictands represent group membership in the same manner as for REEP, then Eq. (13) is the same as the set of REEP equations (see Glahn, 1968). Therefore, canonical correlation has little to offer in a purely predictive sense over the simpler regression or discriminant analysis (this statement is not meant to imply that canonical correlation is not useful in studying relationships between sets of variables). One possibility for canonical correlation in prediction does exist, and that involves the use of Eq. (12). As stated previously, defining EOFs on a set of predictors and using only those functions that explain a non-trivial portion of the predictor variance as new predictors in a regression equation filters out predictor "noise" (of course, one must be careful to ensure that it really is noise and not good predictive information). Eq. (12) seems to provide a way of filtering noise out of both the predictor and predictand sets and could provide more stable prediction equations. However, we know of no case where this possibility has been investigated. For further discussion of canonical correlation and an example using meteorological data, see Glahn (1968).

8. LOGIT MODEL

The logit model (Brelsford and Jones, 1967; Jones, 1968) provides a means of fitting a sigmoid or S-shaped curve to data when the dependent variable is binary and the independent variable is continuous. From this model, the probability of the binary variable Y having the value of one can be expressed as follows:

$$P(Y = 1|X) = \frac{1}{1 + \exp(\alpha + \beta X)} \tag{14}$$

The model can also be extended to several independent variables and to several, rather than two, categories of a dependent variable. Determination of the parameters [α and β in Eq. (14)] is usually more difficult than determination of coefficients in a regression equation. Iterative procedures can be used or, if each specific value of X in the sample is repeated and the relative frequency of the dependent event is neither zero or one, then a more direct method of solution can be used for the following linearized form of Eq. (14):

$$\ln(\frac{1 - p}{p}) = \alpha + \beta X.$$

Usually, in meteorological applications, several predictors are to be included, and this method of solution would require enough replications of each combination of predictor values to estimate the relative frequency of the predictand for that combination. We know of no meteorological application where this method of solution for multiple predictors has been used. For a discussion of this method, see Neter and Wasserman (1974).

9. MAP TYPING

    The concept of weather types arose early in the history of meteorology. The aim is to define a partition of weather maps (or sequences of maps), so that the differences between the maps (or sequences) of one type are small compared to the differences between maps (or sequences) of another type. Once a set of weather types has been defined, it can be used in various ways to forecast specific weather elements. Early work was done by Bowie and Weightman (1914) who stratified storms by their movement. Average tracks, expected direction, and average speeds were then computed. An historic development of weather types was done by Irving P. Krick at the California Institute of Technology (1943), leading to the identification of the so-called CIT types.

    The determination of map types can be accomplished in many ways. Initially, the methods employed were largely subjective, and even in the application phase, the user had to "decide" what type existed on a given day. More recently, with the advent of the electronic computer and the desire to process large quantities of data, objective methods of classification have been developed. One such method that has been rather extensively applied was developed by Lund (1963). The example he used to explain the method involved the classification of wintertime sea level pressure maps over the northeastern U.S. The steps involved in this method are:

Step 1. Correlate the sea level pressures on each map with the corresponding pressures on all of the other maps in the sample. That is, if each of 500 maps had 25 values of pressure (which could be reported values at stations or at grid points arrived at by an analysis of station values), then each of the 500 maps would be correlated with 499 other maps, the computations of the correlation coefficient involving 25 pairs of values.

Step 2. Select the map which has the most correlation coefficients ≥ 0.7 and designate it as Type A.

Step 3. Remove all of the cases that are correlated ≥ 0.7 with the Type A map, and select from the remaining maps the one with the most correlations ≥ 0.7. Designate this map as Type B.

Step 4. Remove Type B cases, and repeat the process until only
a few cases with correlations $\geq$ 0.7 remain.

In application, a map is classified according to the type with which
it correlates most highly. Of course, the 0.7 correlation criterion
stated above can be modified as desired.

A problem in the use of map types is that a particular map may
not classify well into any of the defined types. In the Lund method,
after the definition of several types, a few cases generally will
remain that are not very similar to any other map in the sample.

Many times only one variable, such as sea level pressure, is
used to define the types. However, the evolution of weather systems
and the correspondence to predictand variables depend on more than
that one element. Other variables can be included in the definition
of types, but finding "good" types - that is, maps which resemble
others in terms of all the considered variables - is then more difficult.

A possible forecast aid employing map types is to define the
conditional precipitation probability at a station given that a particular map type exists. This procedure could involve a lag relationship, in which case the application would probably be to existing
maps. On the other hand, it could be a concurrent relationship, in
which case the application would probably be to numerical forecasts
of the variable(s) used to define the types. This latter approach
would be a perfect prog application. Augulis (1969) describes a
forecast aid developed along these lines; it is still in use in the
western U.S.

10. ANALOGUES

The term analogue can be defined as follows: "In synoptic
meteorology, a past large-scale synoptic weather pattern which
resembles a given (usually current) situation in its essential
characteristics. The use of analogues as an aid in forecasting is
based upon the assumption that two similar synoptic weather patterns
will retain similarity through at least a short period of further
development" (American Meteorological Society, 1959). Analogues were
investigated comprehensively by Wadsworth (1948) and their use has
been discussed realistically by Willett (1951).

Selecting an analogue is much like selecting a weather type -
the idea is to choose one or more maps which are very similar to
other maps or to a particular map. Generally, map types of, say, sea
level pressure or 500 mb height are employed to forecast other
variables such as temperature or precipitation at specific points.
However, analogues of, say, sea level pressure and/or 500 mb height
may be used to forecast future states of those same variables over

the areas for which the analogues are defined. In these days of
numerical prediction models, analogues appear to be of very limited
use.

11. PRESENT STATUS

The most concentrated effort today in statistical weather forecasting is at the Techniques Development Laboratory (TDL) of the U.S.
National Weather Service. TDL's objective systems, implemented by the
National Meteorological Center, produce about 600,000 forecasts daily
from about 90,000 regression and logit equations (as of 1 April 1981).
These forecasts are disseminated by teletypewriter and facsimile to
civilian and military weather stations and to non-government users
throughout the United States (Glahn, 1976). The elements being forecast include probability of precipitation, precipitation type, precipitation amount (Bermowitz, 1975), surface wind at land stations
(Carter, 1975) and at marine stations and over the Great Lakes (Feit
and Pore, 1978), surface temperature and dew point (Dallavalle et al.,
1980), severe convective weather (Reap and Foster, 1979; Charba,
1979), cloud amount (Carter and Glahn, 1976), ceiling height and visibility (Globokar, 1974), storm surge (Pore, 1976; Richardson and Pore,
1969), and beach erosion. Some of these current applications are
discussed briefly in this section.

11.1 Probability of Precipitation

MOS probability of precipitation (PoP) forecasts have been produced operationally by the REEP model for several years. This statistical product replaced the subjectively produced NMC product in
January of 1972. The developmental sample was divided into 2 seasons - April through September, the summer season, and October
through March, the winter season. The event is defined to be 0.01
inches or more of measurable liquid equivalent precipitation in a
12 h period at a point, represented by a station rain gauge.
Separate equations were developed for the 12-24, 36-48, and 48-60 h
projections (i.e., lead times) and for each of the initial data times
of 0000 and 1200 GMT. Data for several stations within a region were
pooled, and one set of equations was developed that applied to all
stations within that region. Details of the evolution of the PoP
forecasting system are given by Lowry and Glahn (1976).

In terms of the P-score, the MOS PoP's improved upon the climatological relative frequency (defined by month and by station) by about
48%, 33%, and 34% for the 12-24, 24-36, and 36-38 h projections,
respectively, for the 1979-80 winter. The corresponding 1979 summer
improvements were about 29%, 22%, and 19%. Using the MOS PoP's as
guidance, the local forecasters were able to improve upon them by
8.1%, 1.4%, and 2.3% for the three periods, respectively, during the
1979-80 winter.

11.2 Precipitation Type

TDL's system for predicting the conditional probability of precipitation type (PoPT), conditional on the occurrence of precipitation, gives forecasts for three categories: frozen (snow or ice pellets), freezing (freezing rain or drizzle), and liquid (rain or mixed types) (Bocchieri, 1979). The PoPT system evolved from the conditional probability of frozen precipitation (PoF) system (Glahn and Bocchieri, 1975; Bocchieri and Glahn, 1976), which had been operational since November 1972. In PoF, explicit probability forecasts of freezing precipitation were not available. In the PoPT system, one logit equation was developed for each initial data time and each projection. Although data from about 200 stations were used in the development of each equation, the predictors were defined to be departures from 50% values. As an example, consider the 850 mb temperature as a predictor. For each station, the value that specifies a 50% conditional probability of frozen precipitation was found empirically. (This value was actually found by determining a one predictor logit equation for each station.) Then, the 850 mb temperature minus the unique 50% station value was used as a predictor in the multipredictor logit equation.

Heidke skill scores for the 1979-80 winter guidance forecasts were 0.88, 0.86, and 0.84 for 18, 30, and 42 h forecasts, respectively. These scores were computed only for those cases when the local PoP forecasts were greater than or equal to 30%.

11.3 Surface Wind

MOS surface wind forecasts for stations throughout the conterminous United States have been produced since May 1973 (Carter, 1975). Three regression equations are determined for each station for each projection - one for the U component, one for the V component, and one for speed. All three equations have the same predictors to ensure greater consistency between the three forecasts. Forecasts of the U and V components are used to determine direction. A separate equation is used for speed because speeds determined from regression estimates of the U and V components are biased toward zero (Glahn, 1970).

Verification of the MOS forecasts for the 1973-74 through 1979-80 winter shows a definite improving trend. Mean absolute errors in direction for the 1979-80 winter were 26, 30, and 35 degrees for the 18, 30, and 42 h projections, respectively. Corresponding skill scores for speed were 0.35, 0.34, and 0.26. The speed forecasts have been inflated (Klein et al., 1959) since 1975 in order to make a larger number of forecasts of strong winds.

11.4 Surface Temperature

Statistical forecasts of maximum and minimum temperature have
been made and disseminated operationally by the National Weather
Service since 1965 -longer than any other weather element. Initially
the forecasts were made by the perfect prog technique (Klein and
Lewis, 1970), but the MOS approach was adopted in August 1973 after
considerable testing showed that MOS furnished better forecasts
(Annett et al., 1972; Klein and Hammons, 1975). The forecasts are
made from regression equations developed for individual stations, one
for each of the initial data times and for each projection. A continuing evaluation has shown that MOS improves on the perfect prog
forecasts by about 0.5°F in mean absolute error at 24 and 36 h projections. These statistical forecasts have shown a consistent improvement since 1973. The mean absolute error for 24 h maximum temperature forecasts was 3.5°F for the 1979-80 winter period. The forecasters are able to improve on the guidance by a few tenths of a
degree Fahrenheit.

11.5 Extratropical Storm Surge

Storm surge is defined to be the piling up of water on the shore
due to meteorological conditions. TDL's statistical systems forecast
this surge at specific points on the Atlantic (Pore, 1976) and Great
Lakes (Richardson and Pore, 1969) coasts due to extratropical
storms. The perfect prog technique is used to develop regression
equations that relate the surge to concurrent values of sea level
pressure at grid points surrounding the forecast points. Since a
very good physical basis exists for the dependence of surge on
pressure gradients, the forecasts are quite good, and their skill
depends mainly on the skill of the numerical model used to provide
the pressure forecasts (Pore, 1972). Surge forecasts became operational for Buffalo and Toledo on Lake Erie in October 1969 and for
Atlantic coastal stations in October 1971.

11.6 Thunderstorms and Severe Convective Weather

Medium-range (24 h projection) probability forecasts of thunderstorms and severe convective weather have been operationally available since the spring of 1972 (Reap and Foster, 1979). In addition,
short-range (2-6 h projection) probability forecasts of the same
variables were implemented in the spring of 1974 (Charba, 1977,
1979). The medium-range forecasts are provided by REEP equations
developed by the MOS technique. The predictand is defined by radar
echoes within a specified time period and within an area approximately 75x75 km. These forecasts of severe convective weather are
conditional probabilities. That is, given a thunderstorm within the
defined area and time period, the forecast specifies the probability
of the occurrence of severe weather. It is interesting to note that
reliable forecasts of 30 to 40% can be made even though the climatological relative frequency is only 6%.

The short-range forecasts are also provided by REEP equations, but the equations contain more predictors derived from recent observations (of surface atmospheric variables and radar echoes) than from model output. Therefore, this technique is a blend of the classical and MOS approaches. In addition, the severe storm probabilities as well as the thunderstorm probabilities are unconditional. Reliable probabilities approaching 100% are forecast for both predictands, although for severe storms the climatological frequency for a 4 h period is only about 2%.

12. FUTURE OF STATISTICAL WEATHER FORECASTING

Stochastic-dynamic prediction is a term used to describe models that combine statistics and dynamics and produce output in probability form. These models show some promise and may be the models of the future. However, they require considerably more in the way of computer resources than conventional numerical models, and much more research is required before they can compete with present operational models. Also, like present models, they do not produce forecasts of many weather elements for which forecasts are required - ceiling height, cloud amount, minimum temperature, etc. So it is likely that MOS will be used for many years to translate numerical model forecasts into other needed products. The perfect prog technique may find increased use for medium-range projections if numerical models become accurate enough so that the perfect prog assumption is reasonably satisfied.

More efficient methods of processing large quantities of data, better statistical models, and better use of present models will help to improve and to extend the application of statistical forecasting in the future.

REFERENCES

Allen, R. A., and E. M. Vernon, 1951: Objective weather forecasting. Compendium of Meteorology (T. F. Malone, Ed.). Boston, Mass., American Meteorological Society, pp. 796-801.

American Meteorological Society, 1959: Glossary of Meteorology. Boston, Mass, AMS.

Annett, J. R., H. R. Glahn, and D. A. Lowry, 1972: The use of model output statistics (MOS) to estimate daily maximum temperatures. Silver Spring, Md., NOAA, National Weather Service, Technical Memorandum NWS TDL-45, 14 pp.

Anderson, T. W., 1958: An Introduction to Multivariate Statistical Analysis. New York, John Wiley and Sons.

Augulis, R. P., 1969: Precipitation probabilities in the Western Region associated with winter 500 mb map types. Salt Lake City, Utah, ESSA, National Weather Service, Technical Memorandum WBTM WR 45-1, 91 pp.

Barnard, M., 1935: The secular variations of skull characters in four series of Egyptian skulls. Annals of Eugenics, 6, 352-371.

Bartlett, M. S., 1934: The vector representation of a sample. Proceedings of the Cambridge Philosophical Society, 30, 327-340.

Bermowitz, R. J., 1975: An application of model output statistics to forecasting quantitative precipitation. Monthly Weather Review, 103, 149-153.

Besson, L., 1905: Essai de prevision methodique du temps. Observatorie Municipal de Monsouris, Annals, 6, 473-495.

Bocchieri, J. R., 1979: A new operational system for forecasting precipitation type. Monthly Weather Review, 107, 637-649.

Bocchieri, J. R., and H. R. Glahn, 1972: Use of model output statistics for predicting ceiling height. Monthly Weather Review, 100, 869-879.

Bocchieri, J. R., and H. R. Glahn, 1976: Verification and further development of an operational model for forecasting the probability of frozen precipitation. Monthly Weather Review, 104, 691-701.

Boehm, A. R., 1976: Transnormalized regression probability. Scott Air Force Base, Ill., USAF, Air Weather Service, Technical Report 75-259, 52 pp.

Bowie, E. H., and R. H. Weightman, 1914: Types of storms of the United States and their average movements. Monthly Weather Review, Washington Supplement No. 1, 147 pp.

Brelsford, W. M., and R. H. Jones, 1967: Estimating probabilities. Monthly Weather Review, 95, 570-576.

Brier, G. W., 1940: The discriminant function. Washington, D.C., George Washington University, M.A. Thesis, 34 pp.

Brier, G. W., 1946: A study of quantitative precipitation forecasting in the TVA basin. Washington, D.C., U.S. Weather Bureau, Research Paper No. 26, 40 pp.

Brier, G. W., 1950: Verification of forecasts expressed in terms of probability. Monthly Weather Review, 79, 1-3.

Bryan, J. G., 1944: Special techniques in multiple regression. Cambridge, Massachusetts Institute of Technology, unpublished manuscript, 17 pp.

Bryan, J. G., 1950: A method for the exact determination of the characteristic equation and latent vectors of a matrix with applications to the discriminant function for more than two groups. Cambridge, Mass., Harvard University, Ed. D. Dissertation, 290 pp.

California Institute of Technology, 1943: Synoptic weather types of North America. Pasadena, Calif., Department of Meteorology, Report, 237 pp.

Carter, G. M., 1975: Automated prediction of surface wind from numerical model output. Monthly Weather Review, 103, 866-873.

Carter, G. M., and H. R. Glahn, 1976: Objective prediction of cloud amount based on model output statistics. Monthly Weather Review, 105, 1565-1572.

Charba, J. P., 1977: Operational system for predicting thunderstorms two to six hours in advance. Silver Spring, Md., NOAA, National Weather Service, Technical Memorandum NWS TDL-64, 24 pp.

Charba, J. P., 1979: Two to six hour severe local storm probabilities: an operational forecasting system. Monthly Weather Review, 107, 268-282.

Dallavalle, J. P., J. S. Jensenius, Jr., and W. H. Klein, 1980: Improved surface temperature guidance from the limited-area fine mesh model. Preprints, Eighth Conference on Weather Forecasting and Analysis (Denver). Boston, Mass., American Meteorological Society, pp. 1-8.

Enger, I., J. A. Russo, Jr., and E. L. Sorenson, 1964: A statistical approach to 2-7 hr prediction of ceiling and visibility, volumes I and II. Hartford, Conn., Travelers Research Center, Inc., Contract No. CWB-10704, 48 pp. and 195 pp., respectively.

Feit, D. M., and N. A. Pore, 1978: Objective wind forecasting and verification on the Great Lakes. Journal of Great Lakes Research, 4, 10-18.

Fisher, R. A., 1936: The use of multiple measurements in taxonomic problems. Annals of Eugenics, 7, Part II, 179-188.

Fix, C., and J. L. Hodges, Jr., 1951: Discriminatory analysis, nonparametric discrimination: consistency properties. Randolph Field, USAF, School of Aviation Medicine, Report No. 4.

Freeman, M. H., 1961: A graphical method of objective forecasting derived by statistical techniques. Quarterly Journal of the Royal Meteorological Society, 87, 393-400.

Gilman, D. L., 1957: Empirical orthogonal functions applied to thirty-day forecasting. Cambridge, Massachusetts Institute of Technology, Department of Meteorology, Contract No. AF19 (604)-1283, Scientific Report No. 1, 129 pp.

Glahn, H. R., 1962: An experiment in forecasting rainfall probabilities by objective methods. Monthly Weather Review, 90, 59-67.

Glahn, H. R., 1965: Objective weather forecasting by statistical methods. The Statistician, 15, 111-142.

Glahn, H. R., 1968: Canonical correlation and its relationship to discriminant analysis and multiple regression. Journal of Atmospheric Sciences, 25, 23-31.

Glahn, H. R., 1970: A method for predicting surface winds. Silver Spring, Md., ESSA, National Weather Service, Technical Memorandum WBTM TDL 29, 18 pp.

Glahn, H. R., 1976: Progress in the automation of public weather forecasts. Monthly Weather Review, 104, 1505-1512.

Glahn, H. R., and J. R. Bocchieri, 1975: Objective estimation of the conditional probability of frozen precipitation. Monthly Weather Review, 103, 3-15.

Glahn, H. R., and D. A. Lowry, 1972: The use of model output statistics (MOS) in objective weather forecasting. Journal of Applied Meteorology, 11, 1203-1211.

Globokar, F. T., 1974: Computerized ceiling and visibility forecasts. Preprints, Fifth Conference on Weather Forecasting and Analysis (St. Louis). Boston, Mass., American Meteorological Society, pp. 228-233.

Grimmer, M., 1963: The space-filtering of monthly surface temperature anomaly data in terms of pattern, using empirical orthogonal functions. Quarterly Journal of the Royal Meteorological Society, 89, 395-408.

Gringorten, I. I., 1955: Methods of objective weather forecasting. Advances in Geophysics, Vol. II. New York, Academic Press, Inc., pp. 57-92.

Hotelling, H., 1935: The most predictable criterion. Journal of Educational Psychology, 26, 139-142.

Hotelling, H., 1936: Relations between two sets of variates. Biometrika, 28, 321-377.

Jones, R. H., 1968: A nonlinear model for estimating probabilities of k events. Monthly Weather Review, 96, 383-384.

Jorgensen, D. L., 1959: Prediction of hurricane motion with use of orthogonal polynomials. Journal of Meteorology, 16, 21-29.

Klein, W. H., 1969: The computer's role in weather forecasting. Weatherwise, 22, 195-218.

Klein, W. H., and G. A. Hammons, 1975: Maximum/minimum temperature forecasts based on model output statistics. Monthly Weather Review, 103, 796-806.

Klein, W. H., B. M. Lewis, and I. Enger, 1959: Objective prediction of five-day mean temperature during winter. Journal of Meteorology, 16, 672-682.

Klein, W. H., and F. Lewis, 1970: Computer forecasts of maximum and minimum temperatures. Journal of Applied Meteorology, 9, 350-359.

Lorenz, E. N., 1956: Empirical orthogonal functions and statistical weather prediction. Cambridge, Massachusetts Institute of Technology, Department of Meteorology, Scientific Report No. 1, 49 pp.

Lowry, D. A., and H. R. Glahn, 1976: An operational model for forecasting probability of precipitation - PEATMOS POP. Monthly Weather Review, 104, 221-232.

Lubin, A., and A. Summerfield, 1951: A square root method of selecting a minimum set of variables in multiple regression: I. The method. Psychometrika, 16, 271-284.

Lund, I. A., 1955: Estimating the probability of a future event from dichotomously classified predictors. Bulletin of the American Meteorological Society, 36, 325-328.

Lund, I. A., 1963: Map-pattern classification by statistical methods. Journal of Applied Meteorology, 2, 56-65.

Miller, R. G., 1958: The screening procedure. In Studies in Statistical Weather Prediction (B. Shorr, Ed.). Hartford, Conn., Travelers Research Center, Inc., Contract No. AF19 (604)-1590, Final Report, pp. 86-95.

Miller, R. G., 1962: Statistical prediction by discriminant analysis. Meteorological Monographs, 4, No. 25, 54 pp.

Miller, R. G., 1964: Regression estimation of event probabilities. Hartford, Conn., Travelers Research Center, Contract Cwb-10704, Technical Report No. 1, 153 pp.

Miller, R. G., Ed., 1977: Selected topics in statistical meteorology. Scott Air Force Base, Ill., USAF, Air Weather Service, AWS-TR-77-273, 164 pp.

Mook, C. P., 1948: An objective method of forecasting thunderstorms for Washington, D.C., in May. Washington, D.C., U.S. Weather Bureau, unpublished manuscript.

Murphy, A. H., 1974: A sample skill score for probability forecasts. Monthly Weather Review, 102, 48-55.

Neter, J., and W. Wasserman, 1974: Applied Linear Statistical Models. Homewood, Ill., Richard D. Irwin, Inc.

Panofsky, H. A., and G. W. Brier, 1958: Some Applications of Statistics to Meteorology. University Park, Pennsylvania State University, College of Mineral Industries.

Pore, N. A., 1972: Marine conditions and automated forecasts for the Atlantic coastal storm of February 18-20, 1972. Monthly Weather Review, 101, 363-370.

Pore, N. A., 1976: Automated forecasting of extratropical storm surges. Proceedings, Fifteenth Coastal Engineering Conference (Honolulu), Vol. 1, pp. 906-913.

Pore, N. A., and W. S. Richardson, 1969: Second interim report on sea and swell forecasting. Silver Spring, Md., ESSA, National Weather Service, Technical Memorandum WBTM TDL 17, 17 pp.

Rao, C. R., 1952: Advanced Statistical Methods in Biometric Research. New York, John Wiley and Sons.

Reap, R. M., and D. S. Foster, 1979: Automated 12-36 hour probability forecasts of thunderstorms and severe local storms. Journal of Applied Meteorology, 18, 1304-1315.

Richardson, W. S., and N. A. Pore, 1969: A Lake Erie storm surge forecasting technique. Silver Spring, Md., ESSA, National Weather Service, Technical Memorandum WBTM TDL 24, 23 pp.

Shuman, F. G., and J. B. Hovermale, 1968: An operational six-layer primitive equation model. Journal of Applied Meteorology, 7, 525-547.

Suits, D. B., 1957: Use of dummy variables in regression equations. Journal of the American Statistical Association, 52, 548-551.

Thompson, J. C., 1950: A numerical method for forecasting rainfall in the Los Angeles area. Monthly Weather Review, 78, 113-124.

U.S. Navy, 1963: A historical survey of statistical weather prediction. Norfolk, Va., U.S. Navy Research Facility, NWRF 41-1263-087, 25 pp.

Veigas, K. W., R. G. Miller, and G. M. Howe, 1958: Probabilistic prediction of hurricane movement by synoptic climatology. In Studies in Statistical Weather Prediction (B. Shorr, Ed.). Hartford, Conn., Travelers Research Center, Inc., Contract No. AF19 (604)-1590, Final Report, pp. 154-202.

Wadsworth, G. P., 1948: Short range and extended forecasting by statistical methods. Washington, D.C., U.S. Air Force, Air Weather Service, Technical Report No. 105-38, 202 pp.

Wherry, R. J., 1940: Occupational Counseling Techniques (W. H. Stead and C. L. Changle, Eds.). New York, American Book Company, pp. 245-250.

White, R. M., D. S. Cooley, R. C. Derby, and F. A. Seaver, 1958: The development of efficient linear statistical operators for the prediction of sea level pressure. Journal of Meteorology, 15, 426-434.

Willett, H. C., 1951: The forecast problem. Compendium of Meteorology (T. F. Malone, Ed.). Boston, Mass., American Meteorological Society, pp. 731-746.

Zurndorfer, E. A., and H. R. Glahn, 1977: Significance testing of regression equations developed by screening regression. Preprints, Fifth Conference on Probability and Statistics in the Atmospheric Sciences (Las Vegas). Boston, Mass., American Meteorological Society, pp. 95-100.

9
Probabilistic Weather Forecasting

Allan H. Murphy

1. INTRODUCTION

Weather forecasts are inherently uncertain. This uncertainty is
present whether the forecasts are based on "objective" numerical
(i.e., physical/dynamical) and/or statistical models, on the subjective judgments of human forecasters, or on a combination of objective
and subjective procedures. Uncertainty in the weather forecasting
process arises from several sources. First, observational systems and
networks in meteorology provide, at best, an incomplete description of
the state of the atmosphere at any point in time. Second, the models
used to describe the behavior and evolution of the atmosphere - and
its attendant weather - necessarily involve a variety of assumptions
and approximations concerning the relevant physical and dynamical processes. Third, the procedures used to derive forecasts from these
models generally entail additional simplifications or approximations.
These factors all make significant contributions to the uncertainty
that currently resides in weather forecasts.

The importance of quantifying the uncertainty in weather forecasts and of communicating this information to users of forecasts was
first recognized more than seventy-five years ago (Cooke, 1906a,b).
In the intervening period, a variety of studies and experiments in
probabilistic weather forecasting have been undertaken (see Sections 3
and 4; see also Murphy and Winkler, 1984). This work has included the
development and application of statistical and numerical-statistical
methods of probability forecasting as well as the formulation of
probability forecasts by weather forecasters on an operational and
experimental basis. A significant event in the historical development
of probabilistic weather forecasting occurred in 1965, when the
National Weather Service (NWS) in the United States initiated a nationwide operational program of subjective precipitation probability
forecasting. Moreover, for more than a decade now, NWS forecasters
have received objective precipitation probability forecasts as guidance in the formulation of their subjective forecasts. In recent
years, subjective and objective probability forecasts of other
meteorological events have been prepared on an experimental and/or

operational basis in the U.S. and probabilistic weather forecasting
has been initiated in several other countries.

This chapter is concerned primarily with methods of formulating
probability forecasts and with the results of operational and experimental probability forecasting programs in meteorology. First, the
definition and interpretation of probability forecasts and the motivation for probability forecasting are discussed briefly in Section 2.
Section 3 treats objective probability forecasts, including methods of
preparing such forecasts as well as some results of current objective
probability forecasting programs. Subjective probability forecasts
are considered in Section 4, with subsections containing an introduction to methods of probability assessment and aggregation and the
results of operational and experimental subjective probability forecasting programs. Section 5 examines some issues related to the
communication of uncertainty in weather forecasts provided to the
general public and specific users.

2. PROBABILITY FORECASTS: DEFINITION, INTERPRETATION, AND MOTIVATION

A probability forecast can be considered to be a discrete or
continuous probability distribution defined over the values of the
predictand of interest. It will be assumed that such distributions
satisfy the usual probability axioms (e.g., Lindley, 1965, pp. 6-14).
In the case of a discrete predictand with N possible values or
classes, a probability forecast consists of a set of N probabilities
determining a discrete distribution. Since the probabilities must sum
to one, only N-1 probabilities need to be specified. A probability
forecast for a continuous predictand consists, in general, of a probability density or distribution function. Such a function might be
specified in parametric form (e.g., by indicating that the forecast
follows a normal or beta distribution) or graphically (by drawing the
density or distribution function by hand). Probabilistic weather
forecasts for continuous predictands generally are described in terms
of certain summary measures of the underlying distributions (e.g.,
means, variances, credible intervals). In this chapter, we will be
concerned primarily with probability forecasts for discrete
predictands.

As indicated in Section 1, probability forecasts can be formulated on the basis of objective models or procedures or on the basis
of subjective judgments of forecasters. In the case of the former,
the probabilities generally are given a relative frequency interpretation and are considered to be estimates of the "true" (but unknown)
probabilities of the events of interest. On the other hand, when
probability forecasts are formulated on a subjective basis the probabilities can be interpreted as expressions of a forecaster's subjective degrees of belief concerning the likelihood of occurrence of the
relevant events. In this framework it is not necessary (or even

appropriate) to postulate the existence of true probabilities and, as a result, such forecasts frequently are referred to as probability assessments (as opposed to probability estimates). It is important to recognize that a sound theoretical basis exists for both the objective (i.e., relative frequency) and subjective (i.e., personal) interpretations of probability (e.g., Lindley, 1965; Winkler, 1972). For the purposes of this chapter, the differences between these two interpretations of probabilities will not be of major importance.

Two principal reasons exist for formulating and expressing weather forecasts in probabilistic terms. First, as noted in Section 1, uncertainties are inherent in the forecasting process, whether this process is based on objective procedures or subjective judgments. The existence of these uncertainties implies that categorical or deterministic forecasts will seldom if ever be warranted. In other words, such forecasts will seldom accurately reflect the true "state of knowledge" of the forecasting system (regardless of whether the system consists of an objective model or a human forecaster). Only a forecast expressed in probabilistic terms can provide an appropriate description of this state of knowledge. Second, in the absence of perfect forecasts, users require estimates of the likelihood of occurrence of the relevant events in order to make optimal decisions in uncertain situations. It is relatively easy to demonstrate that the value of probabilistic forecasts generally is equal to or greater than the value of categorical forecasts for all users of such forecasts (e.g., Thompson, 1962; Murphy, 1977; Krzysztofowicz, 1983). Thus, a strong motivation exists for formulating and expressing weather forecasts in probabilistic terms.

3. OBJECTIVE PROBABILITY FORECASTS

As noted in Chapter 8, an objective weather forecasting system is a system that produces one and only one forecast from a given set of data. Thus, an objective forecast does not depend directly on the judgment and/or experience of a particular meteorologist, although subjective judgment plays an important role in the development of objective forecasting systems. In Section 3.1 we briefly describe several methods of objective forecasting that can be used to obtain probabilistic forecasts, and in Section 3.2 we present some results of recent operational objective probability forecasting programs in the U.S.

3.1 Methods of Objective Probability Forecasting

Objective probabilistic weather forecasts have been prepared on an experimental and/or operational basis using a variety of statistical, numerical-statistical, and other procedures. Many of these procedures are described in some detail in Chapter 8. Thus, the discussion of methods of objective forecasting in this section will be relatively brief and will focus on the use of such methods to obtain probability forecasts. References are provided for those readers who seek more detailed descriptions of the procedures themselves or examples of their application.

Contingency tables represent perhaps the simplest procedure for obtaining estimates of the probability of occurrence of future weather events. In this procedure the ranges of values of the predictand and each predictor are divided into two or more mutually exclusive and collectively exhaustive categories or classes, and the data available are used to obtain a discrete joint frequency distribution over the cells in the table. Conditional probability estimates (i.e., forecasts) for the predictand classes can then be obtained from this joint distribution and the relevant marginal relative frequencies. Two-way tables involving the predictand and a single predictor represent the simplest form of this procedure and have been used most frequently, but n-way (n > 2) or multidimensional tables also have been developed in some studies. A problem that frequently arises in connection with the use of contingency tables is that the number of cases in some cells may be quite small, thereby leading to unstable probability estimates. This problem necessarily becomes more acute as the number of predictors (i.e., the dimensions of the table) is increased. To reduce the effects of small sample sizes, it may be necessary to limit the number of predictors, to combine classes of the predictand and/or predictors, or to smooth the probability estimates. The contingency table procedure is quite flexible and does not require the analyst to make any assumptions concerning the form of the joint distribution of the variables involved. The work of Wahl et al. (1952) contains examples of the use of contingency tables to obtain objective probability forecasts. In a related vein, Gringorten (1966, 1971), Martin (1972), and others have developed simple conditional probability models that can be used in a manner similar to contingency tables to produce probabilistic forecasts on the basis of climatological data.

Scatter diagrams, or graphical regression, can also be used to obtain probability forecasts of events of interest. This procedure has been described in some detail in Chapter 8 and that description will not be repeated here. It is sufficient to indicate that the procedure can accommodate several predictors and provides the meteorologist or forecaster with a considerable degree of flexibility in analyzing relationships between variables (although care must be taken not to "over-analyze" the data). As noted in Chapter 8, it is a particularly suitable procedure for small data samples and generally has been based on a subjective analysis process, although efforts have been made to automate the procedure (e.g., see Freeman, 1961). Scatter diagrams were studied extensively by Brier (1946), who used them to obtain probability forecasts of precipitation amounts. Other examples of the application of this procedure to derive probability forecasts of various events include papers by Dickey (1949), Thompson (1950), and Schmidt (1951).

A variety of statistical procedures (or models) can be used to produce objective probability forecasts, and several of these procedures were described in some detail in Chapter 8. The model employed most frequently in objective statistical or numerical-statistical weather forecasting is multiple linear regression. Multiple regression is used to obtain probability forecasts by treating the predictand as a set of two or more Boolean (or dummy) variables. This variant of standard regression was developed and popularized in the meteorological community by Miller (1964), and it is usually referred to as Regression Estimation of Event Probabilities (REEP). As noted in Chapter 8, the REEP procedure has the important property that it minimizes the Brier score, a mean square error measure of the accuracy of probabilistic forecasts (see Chapter 10). The probabilities generated by the REEP model for a set of mutually exclusive and collectively exhaustive events must sum to one, but they are not constrained to the closed interval [0,1] (this latter problem is not serious and can be readily rectified). In meteorological applications, the REEP procedure generally is used in conjuction with a forward stepwise technique for predictor selection, and the predictors themselves frequently are transformed into Boolean variables. Extensive use of the REEP model is made in the current NWS operational objective weather forecasting program (see Section 4.2; see also Chapter 8).

Multiple discriminant analysis (MDA) is another statistical model that has been used to produce probabilistic forecasts of a variety of weather events. As indicated in Chapter 8, this procedure is particularly appropriate for nominal (or non-numerical) predictands, and it is also able to incorporate certain nonlinear relationships between the predictors and the predictand. A nonparametric approach to discriminant analysis, based on the work of Fix and Hodges (1951), was used by Miller (1962) to obtain probability forecasts for ceiling height and visibility events. This work also involved the development and application of a forward stepwise technique for predictor selection. Extensive use was made of the MDA procedure in a variety of objective probabilistic weather forecasting studies at the Travelers Research Center in Hartford, CT in the 1960's (e.g., see Enger et al., 1964). Subsequent comparisons of the MDA and REEP models revealed that MDA performed somewhat better than REEP for some predictands, but the former imposed a greater computational burden than the latter.

Other objective procedures that have been used to generate probabilistic weather forecasts on an experimental and/or operational basis include the logit model (Brelsford and Jones, 1967; Jones, 1968), adaptive logic (Glahn, 1964), and analogues (Kruizinga and Murphy, 1983). The logit model is currently used operationally to produce probability forecasts of precipitation type (see Section 3.2).

The objective procedures described in this section (and in Chapter 8) generally can be used in conjunction with any of three approaches to statistical or numerical-statistical weather forecasting. These approaches are the classical method, the perfect prog procedure, and the model output statistics (MOS) technique, and they are described and compared in detail in Chapter 8. Probabilistic forecasts can be generated using any of these approaches in combination with a suitable statistical procedure. Comparative evaluation of objective forecasts for a variety of weather elements indicates that forecasts based on the MOS approach usually outperform forecasts based on the other two approaches (see Chapter 8).

Another objective method of obtaining estimates of the uncertainty inherent in weather forecasts is stochastic-dynamic prediction (Epstein, 1969). This method involves the use of dynamical equations to describe the evolution of the variances and covariances (and possibly higher moments of the probability distributions) of the variables of interest - as well as the use of an equation for the evolution of the mean - and uncertainty is introduced by means of the initial variances and covariances. To date, only relatively simple dynamical models have been employed in conjunction with the stochastic-dynamic approach. The use of this approach together with the more sophisticated models currently employed in operational numerical weather prediction would impose a severe computational burden on the forecasting system, and this fact alone precludes the practical application of the stochastic-dynamic method at this time.

An extensive operational program of objective statistical weather forecasting based primarily on the MOS approach now exists in the NWS. As noted in Chapter 8, approximately 90,000 regression and logit equations are used daily to produce nearly 600,000 forecasts. A substantial fraction of these forecasts are formulated in probabilistic terms and many of these probabilities are provided to the NWS forecasters as guidance and to subscribers to the NWS communications system.

3.2   Some Recent Operational Results

In this section we briefly describe the current NWS operational objective probability forecasting program in the U.S. and present some recent results of this program. At the end of the section, objective probabilistic weather forecasting in other countries will be mentioned. In evaluating the results presented here, emphasis will be placed on the reliability and skill of the forecasts. For a detailed discussion of methods of evaluating probability forecasts, see Chapter 10.

As indicated in Section 3.1, the NWS objective probability forecasting program is based largely on the MOS approach, in which statistical techniques such as multiple regression analysis are employed to develop prediction equations relating numerical model output (and other data) and the weather variables of interest. This approach is used to produce probabilistic weather forecasts on an operational basis for a variety of weather variables, and these forecasts are provided to NWS forecasters as guidance. A list of the predictands for these probabilistic forecasts is presented in Table 1. Information regarding predictand element, number of categories or thresholds, category/threshold definitions, forecasts periods (where appropriate), and lead times is included in this table. The contents of the table demonstrate the wide range of predictands for which objective probabilistic guidance forecasts are now prepared by the NWS. In this section we present and briefly describe some results of the MOS probability forecasting program for five predictands; namely, precipitation occurrence, precipitation type, cloud amount, ceiling height, and visibility.

Objective probabilistic forecasts of precipitation occurrence have been prepared on an operational basis using a regression model in conjunction with the MOS approach since 1972 (Lowry and Glahn, 1976), and these forecasts generally are referred to as objective probability of precipitation (PoP) forecasts. Some recent results of the NWS objective PoP program are presented in Figure 1 and Table 2. Figure 1 contains reliability diagrams for these forecasts for three periods with lead times of 12-24 hours, 24-36 hours, and 36-48 hours, respectively. The data represent the 1980 warm season (April-September) and the 1980-1981 cool season (October-March) and consist of the composite results for 87 stations in the conterminous U.S. In a reliability diagram, the relative frequency of occurrency of the event of interest is plotted against the forecast probability for specific probability values (in this case, the thirteen permissible probability values are 0.00, 0.02, 0.05, 0.10, 0.20, ..., 1.00). The diagonal 45° line represents perfect reliability, in the sense that relative frequency is exactly equal to forecast probability (see Chapter 10 for further discussion of reliability). The reliability curves in Figure 1 indicate that the correspondence between forecast probabilities and observed relative frequencies is quite good for both seasons and all three periods. The insets in the reliability diagrams depict the relative frequency of use of the probability values. The frequency of use distributions for the three periods are similar, with the extreme probabilities (i.e., 0.00 and 1.00) being used more frequently in the cool season than in the warm season.

Average Brier scores and skill scores for these samples of objective PoP forecasts are presented in Table 2. The average Brier score is the mean square error of the forecasts and the skill score is the percent improvement in the average Brier score for the forecasts over the average Brier score for forecasts based solely on climatological probabilities (see Chapter 10 for formal definitions of these evaluation measures). The scores in Table 2 indicate that, as expected, the accuracy (Brier score) and skill (skill score) of the forecasts decrease as the lead time increases. Moreover, the overall

Table 1. Predictands for which objective probabilistic forecasts are prepared on an operational basis by the U.S. National Weather Service.

Predictand element[a]	Number of categories (C)/ thresholds (T)	Category/ threshold definitions	Forecast periods (hours)[b]	Lead times (hours)[c]
Precipitation occurrence	1(T)	≥ 0.01 inches	6,12	6-12,12-18,...,54-60 12-24,24-36,...,48-60
Precipitation amount	4(T)	≥ 0.25; ≥ 0.50; ≥ 1.00; ≥ 2.00 inches	6,12	6-12,12-18,...,36-42 12-24,24-36,36-48
Type of liquid precipitation	3(C)	drizzle; rain; showers		18,30,42
Precipitation type	3(C)	snow; freezing rain; rain		6,12,...,60
Snow amount	3(T)	≥ 2.0; ≥ 4.0; ≥ 6.0 inches	12	12-24
Cloud amount	4(C)	clear; scattered; broken; overcast		6,12,...,60
Ceiling height	6(C)	< 200; 200-400; 500-900; 1000-2900; 3000-7500; > 7500 feet		6,12,...,48
Visibility	6(C)	< 1/2; 1/2-7/8; 1-2 3/4; 3-4; 5-6; > 6 miles		6,12,...,48
Obstructions to vision	4(C)	none; haze/smoke; blowing phenomena; fog/ground fog		6,12,...,48
Thunderstorms	1(T)	VIP[d] level ≥ 3	4,12,24	2-6,12-24,24-36, 36-48,12-36
Severe local storms (watches)	1(T)	tornadoes, hail, and/or strong winds	4	2-6
Severe local storms (outlooks)	1(T)	tornadoes, hail, and/or strong winds	12,24	12-24,24-36, 36-48,12-36
Convective wind gusts	1(T)	thunderstorm wind gusts ≥ 25 knots	8	8-16,20-28
Dew (ground condensation) intensity	3(C)	light; moderate; heavy		36,60,84

[a]Forecasts of the type of liquid precipitation, precipitation type, and severe local storms (outlooks) are expressed in the form of conditional probabilities. Forecasts of all other predictand elements are expressed in the form of unconditional probabilities.

[b]Predictand elements with unspecified forecast periods (e.g., cloud amount) are valid at specific times (see entry under Lead times).

[c]Lead times are defined with respect to the latest numerical model output used as input in preparing the forecasts (this time is taken as the zero point).

[d]VIP = Radar Video Integrator and Processor.

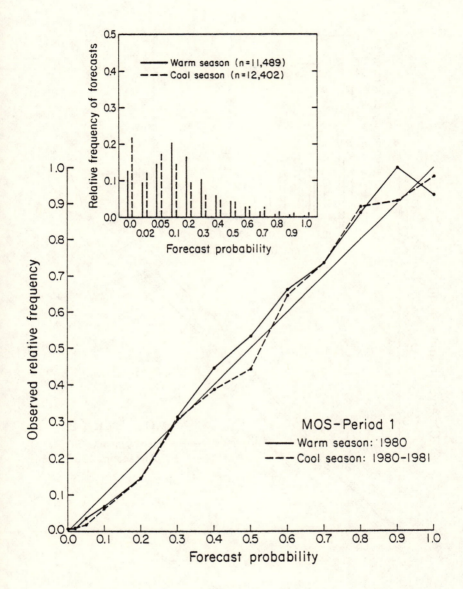

Figure 1a.   Reliability diagram for first-period (12-24 hours) objective PoP forecasts in 1980 warm season and 1980-1981 cool
season for approximately 85 stations in conterminous U.S.
Inset indicates relative frequency of use of probability
values in respective seasons.

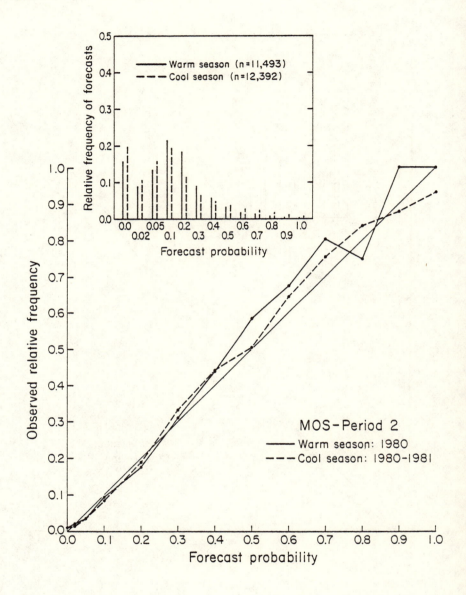

Figure 1b. Reliability diagram for second-period (24-36 hours) objective PoP forecasts in 1980 warm season and 1980-1981 cool season for approximately 85 stations in conterminous U.S. Inset indicates relative frequency of use of probability values in respective seasons.

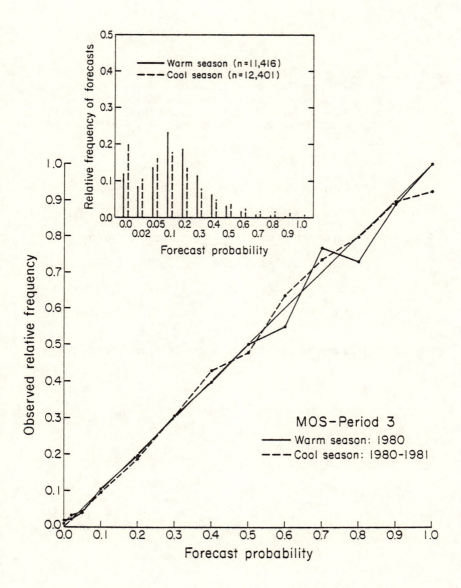

Figure 1c. Reliability diagram for third-period (36-48 hours) objective PoP forecasts in 1980 warm season and 1980-1981 cool season for approximately 85 stations in conterminous U.S. Inset indicates relative frequency of use of probability values in respective seasons.

Table 2. Average Brier scores and skill scores based on Brier score of MOS probability of precipitation forecasts for three 12-hour periods (initial time 0000 GMT) in 1980 warm season and 1980-1981 cool season for approximately 85 stations in the conterminous U.S. (Maglaras et al., 1981; Schwartz et al., 1981).

	Warm season: 1980			Cool season: 1980-1981		
Period	Number of forecasts	Average Brier score[a]	Skill score (%)	Number of forecasts	Average Brier score[a]	Skill score (%)
1	11,489	0.0993	26.5	12,402	0.0772	46.8
2	11,493	0.1023	21.8	12,392	0.0912	34.2
3	11,416	0.1144	14.7	12,401	0.0959	33.2

[a]Average Brier score reported here is one-half of original Brier score.

accuracy and skill of the objective PoP forecasts are considerably better in the cool season than in the warm season (presumably due primarily to the fact that the cool season, unlike the warm season, is dominated by large-scale weather systems that are quite well predicted by numerical models).

Conditional objective probabilistic forecasts of precipitation type in three categories (snow or sleet, freezing rain, rain) are prepared operationally by the NWS using a logit model in conjunction with the MOS approach (Bocchieri, 1979a). The forecasts are conditional in the sense that the probabilities indicate the likelihoods of occurrence of the three types of precipitation given that measurable precipitation occurs (i.e., the forecasts are not "in force" if precipitation does not occur). Average Brier scores and skill scores for a dependent sample of MOS precipitation type forecasts are presented in Table 3 for four lead times. These data represent the "winter" season (September-April) and consist of the composite results for 15 stations in the conterminous U.S. (see Table 3 for additional details). The scores indicate that these forecasts are quite skillful at all lead times. It should be noted that skill scores for an independent (e.g., operational) sample of objective forecasts generally would be somewhat lower than the scores reported in Table 3. For a more detailed description of this program and some additional results, including reliability diagrams, see Bocchieri (1979a).

A regression model is also used in conjunction with the MOS approach to produce objective probability forecasts of cloud amount, ceiling height, and visibility (Globokar, 1974; Carter and Glahn, 1976). These forecasts are expressed in the form of discrete probability distributions for four (cloud amount) and six (ceiling height, visibility) categories of the relevant variables (see Table 1). Some results of this operational program from the 1980-1981 cool season (October-March) for nine stations in the southeastern U.S. are presented in Table 4. In addition to average Brier scores and the associated skill scores, average ranked probability scores and the corresponding skill scores are also given for these forecasts. The latter is a quadratic scoring rule that is particularly appropriate for the evaluation of probability forecasts of multiple-category ordinal variables (see Chapter 10 for a description of the ranked probability score and its properties). These scores indicate that the 6-hour and 12-hour forecasts for the three elements all possess positive skill and that, as expected, skill decreases as lead time increases. Since the ranked probability score (unlike the Brier score) is sensitive to distance, the differences between the values of the two skill scores suggest that the MOS probability forecasts of these elements are quite successful in assigning high probabilities to the observed category or to categories close to the observed category (this attribute of the forecasts is not necessarily reflected in the Brier score).

Table 3. Average Brier scores and skill scores based on Brier score of conditional probability forecasts of precipitation type for four lead times (initial time 0000 GMT) in "winter" season (September-April) for 15 stations in the conterminous U.S.[a] Results based on developmental sample for nine years (1972-1981) for 12-hour and 24-hour forecasts and for five years (1976-1981) for 36-hour and 48-hour forecasts.

Lead time (hours)	Number of forecasts	Average Brier score	Skill score[b] (%)
12	3,533	0.1193	71.5
24	3,028	0.1329	67.7
36	2,159	0.1668	60.6
48	1,795	0.1796	56.8

[a]Stations include Albany, NY; New York, NY; Norfolk, VA; Charleston, WV; Atlanta, GA; Flint, MI; St. Louis, MO; Little Rock, AK; Minneapolis, MN; Grand Island, NE; Williston, ND; Amarillo, TX; Spokane, WA; Salt Lake City, UT; Winslow, AR.

[b]Skill score is based on climatological probabilities for developmental sample for winter season at each station.

Table 4. Average Brier and ranked probability scores and
corresponding skill scores for six-hour forecasts of
selected elements for 1980-1981 cool season at 233 stations
in the conterminous U.S.

		Brier score		Ranked probability score	
Element	Number of forecasts	Average score	Skill score[a] (%)	Average score	Skill score[a] (%)
Cloud amount	36,387	0.4087	41.5	0.1006	85.7
Ceiling height	36,391	0.2972	42.2	0.0445	91.0
Visibility	36,434	0.1828	32.7	0.0342	89.2
Obstructions to vision	36,436	0.1082	50.4	0.0453	84.4

[a]Skill scores based on sample climatological probabilites for all
stations combined for 1980-1981 cool season.

In addition to probability forecasts of precipitation occurrence, precipitation type, cloud amount, ceiling height, and visibility, the MOS approach currently is used to produce objective probabilistic forecasts of a variety of other elements on an operational basis (see Table 1). To conserve space, results of these other programs are not presented here. Instead, we provide a list of some references that the reader can study to find specific results of the element(s) of interest: precipitation amount (Bermowitz and Zurndorfer, 1979); snow amount (Bocchieri, 1979b); severe local storms (watches) (Charba, 1979); severe local storms (outlooks) (Reap and Foster, 1979); and dew (Jensenius and Carter, 1979). It should be noted that the probability forecasts produced by the MOS system for many of these variables frequently are converted into categorical forecasts either for evaluation purposes or when they are disseminated via the NWS communications system. Thus, it is sometimes difficult to find results in the published literature concerning the quality of these objective forecasts in their original probabilistic form. General references containing results of the MOS program, as well as comparisons of objective and subjective forecasts, include Zurndorfer et al. (1979), Charba and Klein (1980), Glahn (1980), and Chapter 8 in this book.

In recent years, objective probabilistic weather forecasting has also been undertaken on an operational and/or experimental basis in several other countries. For example, operational precipitation probability forecasts have been made in France since 1979 using a perfect prog approach (Javelle et al., 1980); extensive tests of an analogue/perfect prog approach to formulating probability forecasts of precipitation occurrence and precipitation amount have been conducted in Canada (Wilson and Yacowar, 1980); and a procedure based on decision trees has been used on an operational basis for thunderstorm probability forecasting in the German Democratic Republic since 1979 (Balzer, 1980). In addition, many other "small" studies of objective probability forecasting have been conducted on an experimental basis (e.g., Mason, 1977).

4. SUBJECTIVE PROBABILITY FORECASTS

Subjective probabilistic weather forecasts have been formulated on an experimental and/or operational basis for many years (see Murphy and Winkler, 1984). By subjective probability forecasts, we mean probability forecasts whose formulation depends ultimately on the real-time judgments of one or more human forecasters. Of course, objective guidance information, including objective probabilistic weather forecasts, may play an important role in the process of preparing these subjective forecasts. In any case, according to this definition, the probability of precipitation (PoP) forecasts currently disseminated to the general public in the U.S. are subjective forecasts.

Historically, subjective probability forecasting in meteorology has been conducted largely in an intuitive manner. That is, formal methods of assessing and/or aggregating probabilities have seldom been used. Nevertheless, we believe that such procedures are quite useful and that they should be employed more widely in the context of weather forecasting. Thus, a brief discussion of these assessment and aggregation procedures appears in Section 4.1. Some results of recent operational and experimental subjective probability forecasting programs are presented in Section 4.2.

4.1 Assessment and Aggregation of Subjective Probabilities

In formulating weather forecasts, forecasters generally assimilate information in a largely intuitive manner from a variety of sources. Here, we assume that this assessment and assimilation process leads to the formulation of internal judgments that represent the forecasters' true beliefs concerning the likelihoods (i.e., probabilities) of occurrence of the relevant events. Thus, weather forecasts stated or recorded by forecasters represent "externalizations" of these judgments, and the former may not always correspond to the latter. In this regard, since judgments are "expressed" in probabilistic terms, only probability forecasts can correspond exactly to the judgments. Close correspondence between judgments and forecasts obviously is quite desirable and this consideration plays an important role in both probability assessment and evaluation (e.g., see Winkler and Murphy, 1968; see also Chapter 10 in this book). Here, we discuss some methods of probability assessment that can help to ensure good correspondence between judgments and forecasts and then briefly consider the problem of the aggregation of probabilistic assessments.

As noted earlier, weather forecasters generally have made their probability assessments (i.e., forecasts) in an intuitive manner. That is, they have seldom used any of the formal procedures that are available to assist individuals in making subjective probability forecasts. Descriptions of some of these procedures can be found in the literature of statistics, operations research, and management science (e.g., Winkler, 1967; Staël von Holstein, 1970; Hampton et al., 1973; Spetzler and Staël von Holstein, 1975). Probability assessment procedures can be classified in different ways; here, we follow the scheme outlined in Spetzler and Staël von Holstein (1975). Most of these procedures are based on questions (posed by an interviewer or the assessor himself/herself) for which the answers can be represented as points on a cumulative distribution function. One dimension of this classification scheme relates to whether the assessor (or forecaster) is asked to assign probabilities (with the values of the variable fixed) or to assign values (with the probabilities fixed). To illustrate this difference, the fixed-width and variable-width credible interval temperature forecasts formulated in several recent experiments (e.g., Murphy and Winkler, 1974; Winkler and Murphy, 1979) represent examples of assigning probabilities (with values fixed) and of assigning values (with probabilities fixed), respectively. The second dimension of the classification scheme relates to whether the forecaster's response mode is direct or indirect. A direct response mode requires numbers as answers (to the questions), whereas an indirect response mode requires the assessor to choose between one or more bets or alternatives. In the case of the latter the bets are adjusted until the assessor is indifferent in choosing between them, and this indifference can be translated into a probability or value assignment. For example, if a forecaster is indifferent between a payoff of $3.00 for certain and a payoff of $10.00 if measurable precipitation occurs today, then his/her probability of the event "precipitation" on this occasion is 0.3. In the indirect response mode, it is possible to use either internal or external (reference) events. The probability wheel (Spetzler and Stael von Holstein, 1975) represents an example of a device that is useful in specifying external reference events. For further details concerning this two-dimensional classification scheme and for descriptions of probability assessment procedures corresponding to each combination of assignment method and response mode, see Spetzler and Stael von Holstein (1975). We briefly describe a fixed-value, indirect response mode procedure that has been used successfully in meteorological and other contexts in the next paragraph.

A formal procedure that is useful for assessing probabilities of (essentially) continuous variables is the interval technique or "method of successive subdivisions" (Raiffa, 1968). In this technique, the forecaster is required to make a series of equal-odds indifference judgments, each of which divides an interval of values of the variable into two equally likely subintervals. The first judgment leads to the identification of the median (i.e., the value that the forecaster feels is equally likely to be exceeded or not exceeded), and subsequent judgments specify the 0.25 and 0.75 fractiles (lower and upper quartiles), the 0.125 and 0.875 fractiles, etc. Then, 50%, 75%, etc., central credible intervals can be obtained from the values (of the variable) corresponding to these fractiles. Consistency checks can be incorporated into this procedure to ensure that the assessments actually reflect a forecaster's true judgments (e.g., values of the variable falling inside the 75% credible interval should be three times as likely as values falling outside the interval). Dialogues describing the application of this procedure can be found in Raiffa (1968, pp. 161-168), Peterson et al. (1972), and Air Weather Service (1978). The method of successive subdivisions has been applied successfully to variable-width (fixed-probability) credible interval temperature forecasting in several experiments (Peterson et al., 1972; Murphy and Winkler, 1974; Winkler and Murphy, 1979), and some results of these programs are presented in Section 4.2.

In the process of formulating a probability forecast, forecasters assimilate or aggregate information from a variety of sources. That is, they are continually revising their probabilistic judgments concerning the events of interest on the basis of new or additional

information. Weather forecasters generally perform this aggregation
process in an informal, intuitive manner, and the question arises as
to whether this process is being conducted in an efficient and effective manner. In this regard, formal models exist for the optimal
aggregation of information in probabilistic prediction (e.g., Winkler
and Murphy, 1973). In brief, these models involve the revision of
probabilities in the light of new information within the framework of
Bayes' theorem, and they can treat both conditionally independent and
dependent information sources (however, the latter may involve complex
probability assessments). Although relatively little experience has
been gained in the use of such models in real-world (as opposed to
laboratory) settings, they offer some promise as techniques for improving the process of information aggregation in subjective (probabilistic) weather forecasting. In a related vein, simple ad hoc procedures, such as hierarchical models (e.g., see Murphy and Winkler,
1971) or decision trees (e.g., Belville and Johnson, 1982), that can
be used to decompose the assessment and assimilation processes may be
of some value in this context.

If subjective probability forecasts of the same event are made by
more than one forecaster, then it is possible to combine these forecasts to form a "consensus" forecast. Considerable evidence now
exists in the context of probabilistic weather forecasting to indicate
that consensus forecasts formulated simply by averaging the individual
probabilities generally will perform as well as most if not all of the
forecasters (e.g., Sanders, 1963, 1973; Bosart, 1975). Winkler et
al. (1977) compared various procedures for combining individual subjective PoP forecasts, including schemes that ignore the forecasters'
past performance and schemes that consider their past performance.
These consensus procedures all performed about equally well, and they
all performed better than almost all of the forecasters. Thus, when
several forecasters are available, it may be possible to improve
probability forecasts by combining the forecasts of individual forecasters (of course, one or more of these "individuals" could represent
objective forecasting procedures).

It should be noted that subjective probability assessments
frequently suffer from certain biases or systematic errors (Tversky
and Kahneman, 1974). These biases may be due to the inherent
limitations of humans to assimilate information and quantify judgments
and/or to their use of heuristics (i.e., simple rules or procedures)
to simplify these difficult and complex tasks. Studies of subjective
probability assessments made in other contexts generally reveal
substantial overconfidence on the part of the forecasters (e.g.,
Fischhoff and MacGregor, 1982; Lichtenstein et al., 1982; Wallsten and
Budescu, 1983). Although weather forecasters frequently represent an
exception to this general result, overforecasting has been noted in
several experimental probability forecasting programs (e.g., Daan and
Murphy, 1982; Murphy et al., 1982; Murphy and Winkler, 1982). This
overforecasting usually has been attributed to the forecasters' lack
of experience in probability forecasting, to the failure to provide

the forecasters with feedback regarding their performance, and/or to a value-induced bias associated with the impacts of the events and forecasts of interest on the users. A recent experimental study of subjective probabilistic weather forecasting in a "laboratory" setting (Allen, 1982) investigated several possible explanations for overforecasting and other biases. In any case, it is certainly desirable to give forecasters some training in probability assessment prior to the start of any experimental or operational probability forecasting program and to provide them with feedback concerning their performance on a regular basis. Recently, some efforts have been made to provide weather forecasters with background information and training materials in the context of probability forecasting (e.g., Air Weather Service, 1978; Hughes, 1980).

4.2 Some Recent Operational and Experimental Results

Several operational and experimental subjective probability forecasting programs are briefly described in this section. First, recent results of the NWS operational precipitation probability forecasting program are presented and references to similar operational programs in other countries are given. Then, some results of probabilistic temperature forecasting experiments and of a recent probability forecasting experiment in The Netherlands are summarized. Finally, several other experimental subjective probability forecasting programs are mentioned.

As noted in Section 1, subjective PoP forecasts have been prepared operationally on a nationwide basis in the U.S. since 1965. Some recent results of this program are presented in Figure 2 and Table 5. These results relate to the same lead times, seasons, and stations as the results for the objective PoP forecasts described in Section 3.2. Figure 2 contains reliability diagrams for the subjective PoP forecasts for the three lead times. The diagrams indicate that these forecasts are quite reliable in both seasons (warm/cool) for all three periods. Comparison of the reliability diagrams for the SUB and MOS forecasts (cf. Figures 1 and 2) reveals only small differences in reliability. In this regard, it should be noted that the reliability of the SUB forecasts is not due solely to the reliability of the MOS forecasts, since the subjective PoP forecasts formulated by NWS forecasters prior to the availability of the objective MOS forecasts were also very reliable (e.g., Roberts et al., 1969). The frequency of use distributions (insets) are similar for the three periods, although a slight tendency exists for the extreme probabilities to be used less frequently as lead time increases. Comparison of these distributions with the corresponding distributions for the objective PoP forecasts indicates that extreme probabilities are assigned to this event (i.e., the occurrence of measurable precipitation) more often by the forecasters than by the MOS system.

Average Brier scores and skill scores for the subjective PoP

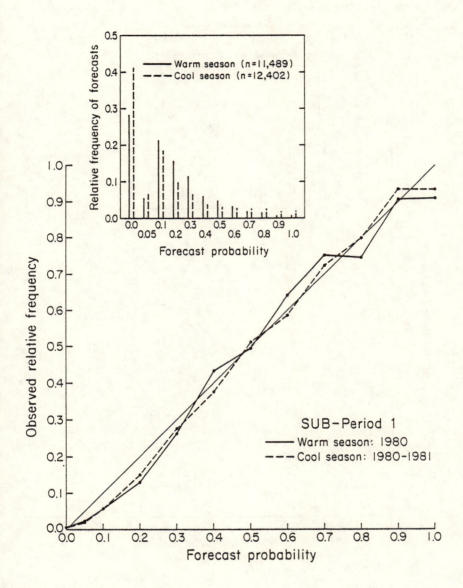

Figure 2a. Reliability diagram for first-period (12-24 hours) subjective PoP forecasts in 1980 warm season and 1980-1981 cool
season for approximately 85 stations in conterminous U.S.
Inset indicates relative frequency of use of probability
values in respective seasons.

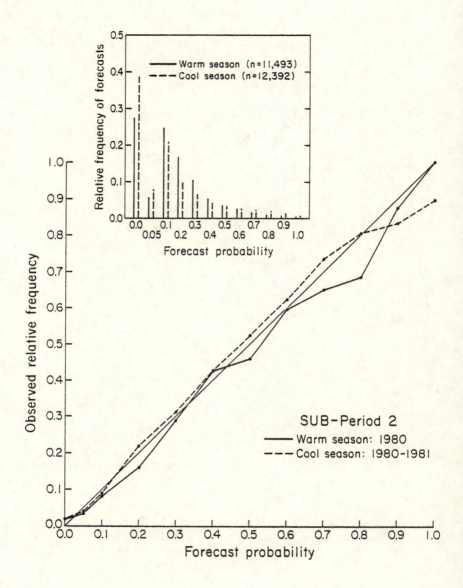

Figure 2b. Reliability diagram for second-period (24-36 hours) subjective PoP forecasts in 1980 warm season and 1980-1981 cool season for approximately 85 stations in conterminous U.S. Inset indicates relative frequency of use of probability values in respective seasons.

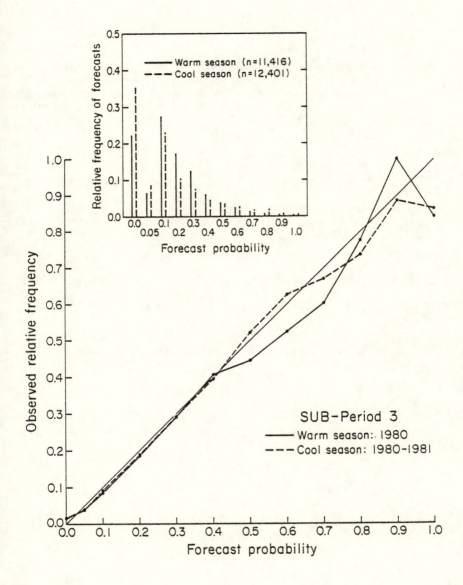

Figure 2c. Reliability diagram for third-period (36-48 hours) subjective PoP forecasts in 1980 warm season and 1980-1981 cool season for approximately 85 stations in conterminous U.S. Inset indicates relative frequency of use of probability values in respective seasons.

Table 5. Average Brier scores and skill scores based on Brier score of SUB probability of precipitation forecasts for three 12-hour periods (initial time 0000 GMT) in 1980 warm season and 1980-1981 cool season for approximately 85 stations in the conterminous U.S. (Maglaras et al., 1981; Schwartz et al., 1981).

	Warm season: 1980			Cool season: 1980-1981		
Period	Number of forecasts	Average Brier score[a]	Skill score (%)	Number of forecasts	Average Brier score[a]	Skill score (%)
1	11,489	0.0966	28.9	12,402	0.0733	49.5
2	11,493	0.1041	20.2	12,392	0.0926	33.4
3	11,416	0.1135	15.3	12,401	0.0946	34.2

[a]Average Brier score reported here is one-half of original Brier score.

forecasts are presented in Table 5. These results are similar to the results for the objective PoP forecasts (see Table 2); that is, skill generally decreases as a function of lead time and skill scores are higher in the cool season than in the warm season. Comparison of the skill scores for the subjective and objective PoP forecasts reveals that the subjective forecasts are more skillful for the first period, but that differences in skill for the second and third periods are quite small. It appears that (overall) the forecasters are able to improve appreciably upon the MOS guidance forecasts only in the first period.

Operational programs similar to the PoP program in the U.S. now exist in several countries. For example, meteorologists at the Royal Netherlands Meteorological Institute (KNMI) have made dry-period probability forecasts (the event of interest is the complement of the event of concern in PoP forecasts) for De Bilt on an operational basis since 1971 (Daan and Murphy, 1982). Moreover, the quality of these forecasts is comparable to the quality of the NWS precipitation probability forecasts. The KNMI dry-period forecasts are now available to the general public in The Netherlands via a teletext system. In Canada, the Atmospheric Environment Service (AES) initiated a nation-wide precipitation probability forecasting program in July 1982 (Grimes, 1982). It is of interest to note that this program was undertaken (in part) in response to the general public's requests for PoP forecasts and that the initial results of this AES activity are quite encouraging. In addition to these operational programs, extensive experimental/operational trials involving PoP forecasts have been conducted in a number of countries, including Australia (Mason, 1976) and Finland (Saarikivi, 1980).

In Section 4.1, the method of successive subdivisions was described as a means of obtaining variable-width credible interval forecasts for continuous variables such as maximum and/or minimum temperatures. This approach has been used by NWS forecasters in several experiments (e.g., Peterson et al., 1972; Murphy and Winkler, 1974; Winkler and Murphy, 1979), and some results from two of these experiments are presented in Table 6. The data in this table relate to forecasts of maximum and minimum temperatures made at Denver, Colorado and Milwaukee, Wisconsin during the periods 1972-73 and 1974-75, respectively (see references for additional details). The reliability of these subjective credible interval forecasts can be evaluated by comparing the observed relative frequencies of temperatures below, in, and above the intervals with the probabilities of the intervals (the latter are 25.0%, 50.0%, and 25.0% for the 50% intervals and 12.5%, 75.0%, and 12.5% for the 75% intervals). Comparison of the relative frequencies and probabilities indicates that these credible interval forecasts are very reliable. Climatological credible interval forecasts were used as a standard of reference, and comparison of the reliability of the climatological forecasts and the subjective forecasts reveals that the latter are more reliable than the former. Moreover, the precision of the credible interval forecasts is also of consider-

Table 6. Relative frequency of occurrence of observed temperatures below interval (BI), in interval (II), and above interval (AI) for (a) subjective variable-width credible interval forecasts formulated in Denver and Milwaukee experiments (Murphy and Winkler, 1974; Winkler and Murphy, 1979) and (b) climatological forecasts corresponding to subjective forecasts.

Experiment	Number of forecasts	Relative frequency of observed temperatures (%)						Average width (standard deviation of width) (°F)	
		50% intervals			75% intervals			50% intervals	75% intervals
		BI	II	AI	BI	II	AI		
(a) Subjective forecasts									
Denver	132	25.8	45.5	28.8	10.6	73.5	15.9	6.2(1.3)	11.7(2.2)
Milwaukee	432	18.1	53.9	28.0	8.1	79.4	12.5	5.9(2.0)	10.1(3.1)
(b) Climatological forecasts									
Denver	132	31.1	44.7	24.2	18.9	65.2	15.9	14.8(4.2)	24.2(5.7)
Milwaukee	432	17.8	56.9	25.2	8.1	81.7	10.2	14.5(3.9)	23.7(4.9)

able interest; precision is measured here by the (average) width of
the forecasts. Examination of the results in Table 6 indicates that
the average width of the subjective credible interval forecasts is
less than one-half the average width of the climatological forecasts.
Thus, the variable-width temperature forecasts formulated in the
Denver and Milwaukee experiments are both reliable and precise. It
should be noted that, unlike the case of the forecasters who make the
PoP forecasts, these forecasters had no prior experience making credible interval forecasts and they received no objective probabilistic
temperature forecasts as guidance.

Subjective probability forecasts of wind speed, visibility, and
precipitation events have been made on an experimental basis by KNMI
forecasters at Zierikzee since 1980 (Daan and Murphy, 1982; Murphy and
Daan, 1984). The forecasts specify the probabilities that certain
critical threshold values of the respective variables will be exceeded
or not exceeded in six-hour periods (see references for details).
Evaluation of the results of the first year of the experiment reveals
a considerable amount of overforecasting (Daan and Murphy, 1982); that
is, forecast probabilities tended to exceed observed relative frequencies for all events. The forecasters were provided with extensive
feedback at the beginning of the second year of the experiment, and
the results for the two periods have been compared in a recent paper
(Murphy and Daan, 1984). The reliability of the wind speed forecasts
for the first and second years of the Zierikzee experiment is depicted
in Figure 3 (reliability diagrams for the visibility and precipitation
forecasts are similar and are omitted to conserve space). Moderate
overforecasting occurred for the wind speed events for high probability values in year 1, but the feedback (and experience) appears to
have reduced the overforecasting appreciably in year 2. The inset
indicates that the frequency of use of the various probability values
is similar in the two years, although probabilities in the highest
range (0.90-1.00) were used less frequently in the second year.

Further evaluation of the subjective probabilistic wind speed,
visibility, and precipitation forecasts formulated in the first and
second years of the Zierikzee experiment is based upon the ranked
probability score (RPS), and the results are presented in Table 7.
The RPS itself is partitioned into three terms - variance of observations (UNC), reliability (REL), and resolution (RES) (with RPS = UNC +
REL - RES) - and a skill score is computed based on this scoring rule
(see Chapter 10 for definitions of these evaluation measures). The
results indicate substantial improvement in reliability and skill from
year 1 to year 2, whereas differences in resolution between the first
and second years of the experiment are quite small. The increases in
reliability and skill can be attributed primarily to the feedback
provided to the forecasters in October 1981 (see Murphy and Daan,
1984). With regard to the differences in skill between the wind speed
and visibility forecasts and the precipitation forecasts, it should be
noted that the former are of much greater concern than the latter to
the forecasters at Zierikzee (see Daan and Murphy, 1982). Finally,

364

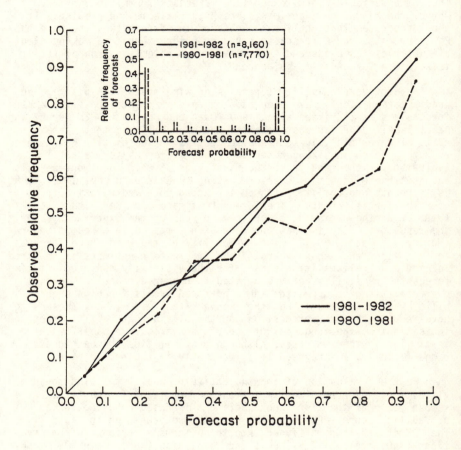

Figure 3.   Reliability diagram for subjective probabilistic wind
            speed forecasts formulated in first and second years of
            Zierikzee experiment. Inset indicates relative fre-
            quency of use of probability values in respective
            periods.

Table 7. Partition of average ranked probability score and skill score based on ranked probability score for subjective probability forecasts of wind speed, visibility, and precipitation forecasts formulated in first and second years of Zierikzee experiment (Daan and Murphy, 1982; Murphy and Daan, 1984). Figures for first year are given in parentheses.

Element	Number of thresholds	Number of probabilities	Variance of observations	Reliability	Resolution	Skill score (%)
			Partition of ranked probability score			
Wind speed	3	8160(7770)	0.148(0.145)	0.005(0.016)	0.043(0.036)	25.4(13.9)
Visibility	3	8160(7770)	0.138(0.141)	0.007(0.018)	0.038(0.035)	22.4(12.4)
Precipitation	2	5440(5140)	0.122(0.143)	0.017(0.056)	0.018(0.021)	0.5(-24.7)

the forecasters at Zierikzee had no experience in probability forecasting prior to the start of the experiment, and they did not receive
any objective probabilistic or categorical forecasts as guidance
during the first two years of this program.

In addition to these experiments, several other experimental subjective probability forecasting programs have been conducted in the
U.S. and elsewhere. For a review of the early history and developments in subjective probability forecasting in meteorology, see Murphy
and Winkler (1984). More recently (i.e., since 1960), such programs
in the U.S. have included probabilistic forecasting "games" conducted
in conjunction with synoptic laboratory courses at several universities (e.g., Sanders, 1963, 1973, 1979; Bosart, 1975) and studies in
which NWS forecasters have formulated probability forecasts on an
experimental or quasi-operational basis. With regard to the latter,
point and area precipitation probability forecasts have been prepared
in several experiments (Winkler and Murphy, 1976; Murphy and Winkler,
1977; Murphy, 1978); probabilistic quantitative precipitation forecasts have been formulated for several locations in Texas (Murphy et
al., 1982); forecasts of the probability that minimum temperatures
will fall below 28°F have been prepared for fruit growers in New
Mexico (Gregg, 1977); and probabilistic tornado forecasts have been
made by forecasters at the NWS National Severe Storms Forecast Center
(Murphy and Winkler, 1982). Recent probability forecasting experiments in other countries include the work of Mason (1976) in Australia
and Saarikivi (1980) in Finland.

In this chapter we have been concerned primarily with short-
range forecasts (i.e., forecasts of day-to-day weather events). However, monthly and seasonal forecasts of temperature and precipitation
have been formulated subjectively (with objective inputs) by the NWS
on an operational basis for more than two decades, and these forecasts
are now provided to subscribers in a new probabilistic format (U.S.
Department of Commerce, 1982). Specifically, the forecasts consist of
probabilities for three categories of either average temperature or
total precipitation in the respective periods (i.e., month or
season). The categories are defined in such a manner that the historical climatological probabilites associated with these ranges of
temperature or precipitation values are equal to 0.30, 0.40, and 0.30,
respectively. In the context of forecasting, the probability of the
central category ("near normal" in the case of temperature and "moderate" in the case of precipitation) is held constant at 0.40 on all
occasions. Thus, each temperature or precipitation forecast can be
expressed in the form of a single probability value. Evaluation of
the NWS monthly and seasonal forecasts in their previous format
reveals modest skill for temperature and only marginal skill for precipitation. The number of forecasts expressed in the new probabilistic format is still too small to support a systematic evaluation
of their reliability and skill.

5. COMMUNICATION OF UNCERTAINTY IN WEATHER FORECASTS

In Sections 3 and 4 we described objective and subjective methods of probability forecasting and presented some results of operational and experimental probability forecasting programs. These results demonstrate that reliable and skillful probability forecasts can be formulated by both objective and subjective procedures. Nevertheless, with the exception of the precipitation probability forecasting program in the U.S. (and similar programs in a few other countries), probability forecasts generally are not provided to the general public and specific users on a regular basis. In view of the many successful operational and experimental probability forecasting programs and the fact that the value of probability forecasts generally exceeds the value of traditional categorical forecasts (see Section 2), the failure of the meteorological community to make greater use of probabilities in public (and other) weather forecasts is particularly surprising. Current resistance to the introduction of probabilities into weather forecasts appears to be based largely on issues related to the mode of expression of uncertainty in forecasts and the understanding of probability forecasts, and these issues are treated briefly in this section. The section concludes with a short discussion of the relative value of probabilistic and categorical forecasts.

Uncertainty can be described qualitatively in terms of words or quantitatively in terms of numbers. Traditionally, uncertainty in weather forecasts has been expressed in terms of verbal modifiers such as "chance" or "likely." Are these verbal expressions of uncertainty interpreted by the general public in an appropriate and consistent manner? This question has been addressed in several studies related to terminology in weather forecasts (e.g., Bickert, 1967; Abrams, 1971; Rogell, 1972; McBoyle, 1974) and in laboratory experiments conducted by behavioral psychologists (e.g., Lichtenstein and Newman, 1967; Budescu and Wallsten, 1981; Beyth-Marom, 1982). These investigators all reached the same general conclusion - a large amount of variability exists when individuals are asked to assign numerical values to such verbal expressions and the amount of overlap among terms is substantial. To eliminate this overlap, it would be necessary to limit the forecaster's vocabulary to a very small set of "distinct" words (e.g., always, often, sometimes, never), thereby severely restricting the amount of information concerning uncertainty that could be transmitted to users. Budescu and Wallsten (1981) conclude that "... communication from one person to another regarding degrees of uncertainty would be degraded by the use of probabilistic phrases rather than numbers." Unlike verbal expressions of uncertainty, probabilities can provide a precise and unambiguous description of the uncertainty in weather forecasts.

A related issue involves the interpretation by the general public of probabilistic weather forecasts such as the PoP forecasts. In this regard, selected individuals have been asked in several studies to interpret a PoP forecast such as "the probability of precipitation today is 30%" (e.g., Bickert, 1967; Scoggins and Vaughn, 1971; Rogell, 1972). The results of these studies reveal that many individuals misinterpret PoP forecasts. However, it is important to distinguish between misinterpretation of the event (e.g., precipitation in an area instead of precipitation at a point) and misinterpretation of the probability associated with the event. In a recent study, Murphy et al. (1980) discovered that misinterpretation of PoP forecasts consists almost entirely of event misinterpretation rather than probability misinterpretation (the questions in the earlier studies were concerned exclusively with event misinterpretation). Moreover, this latter study revealed that a comparable amount of event misinterpretation exists for both categorical and probabilistic forecasts of precipitation occurrence. Thus, the misinterpretations of PoP forecasts cannot be attributed to the presence of probabilities in these forecasts. In fact, in view of the ambiguities associated with the interpretation of verbal expressions of uncertainty, the introduction of probabilities into public forecasts of precipitation occurrence actually may have led to a reduction in the overall level of misinterpretation of such forecasts.

The reactions of members of the general public to the precipitation probability forecasting program and their preferences regarding the mode of expression of uncertainty in forecasts are also of interest here. The PoP program initially encountered some resistance from both the forecasters and the public, but it is now generally agreed that these probabilities are an important and integral part of public weather forecasts in the U.S. (e.g., Bickert, 1967; American Telephone and Telegraph Company, 1971). Moreover, a recent nationwide statistical survey of 1300 members of the general public reveals that 70 percent of the participants prefer numerical probabilities to verbal modifiers as descriptors of uncertainty in forecasts of precipitation occurrence (M.S.I. Services, Inc., 1981). In addition, requests by the public in Canada for precipitation probability forcasts played a major role in the recent decision by the Atmospheric Environment Service to initiate a nationwide PoP program in July 1982 (Grimes, 1982).

In Section 2 we indicated that the value of probabilistic forecasts generally exceeds the value of categorical forecasts (e.g., see Thompson, 1962; Murphy, 1977; Krzysztofowicz, 1983). It should be noted that the studies on which this conclusion is based usually assume that the users of the forecasts process information and make decisions in a rational and optimal manner. Such an assumption may not be consistent with the information processing and decision-making procedures employed by most members of the general public or even by many specific users. However, Krzysztofowicz (1983) recently has compared the value of probabilistic and categorical forecasts for both optimal and suboptimal users and has concluded that the differences in value between the two types of forecasts are likely to be greater for suboptimal users than for optimal users. This result provides additional support for the use of probabilities in public weather forecasts.

ACKNOWLEDGMENTS

The preparation of this chapter was supported in part by the
National Science Foundation (Atmospheric Sciences Division) under
grant ATM-8004680. The author would like to thank the following staff
members at the NWS Techniques Development Laboratory (Silver Spring,
MD) for providing data and/or information on which the indicated
tables and figures are based: G. M. Carter (Table 1); J. R. Bocchieri
(Figures 1 and 2 and Table 3); and K. F. Hebenstreit (Table 4).

REFERENCES

Abrams, E. 1971: Problems in the communication of routine weather
information to the public. University Park, Pennsylvania State
University, Department of Meteorology, M.S. Thesis, 121 pp.

Air Weather Service, 1978: Probability forecasting: a guide for
forecasters and staff weather officers. Scott Air Force Base,
IL, U.S. Air Force, AWS Pamphlet 105-51, seven chapters and
attachments (paged separately).

Allen, G., 1982: Probability judgment in weather forecasting.
Preprints, Ninth Conference on Weather Forecasting and Analysis
(Seattle). Boston, MA, American Meteorological Society, pp. 1-6.

American Telephone and Telegraph Company, 1971: Weather announcement
study. New York, NY, American Telephone and Telegraph Company,
Market and Service Plans Department, Report, 28 pp.

Balzer, K., 1980: Automated short-range area probability forecasts of
thunderstorms, using the decision-tree method and observed temp
part A data. Preprints, Symposium on Probabilistic and
Statistical Methods in Weather Forecasting (Nice). Geneva,
Switzerland, World Meteorological Organization, pp. 453-458.

Belville, J. D., and G. A. Johnson, 1982: The role of decision trees
in weather forecasting. Preprints, Ninth Conference on Weather
Forecasting and Analysis (Seattle). Boston, MA, American
Meteorological Society, pp. 7-11.

Bermowitz, R. J., and E. A. Zurndorfer, 1979: Automated guidance for
predicting quantitative precipitation. Monthly Weather Review,
107, 122-128.

Beyth-Marom, R., 1982: How probable is probable? A numerical
translation of verbal probability expressions. Journal of
Forecasting, 2, 257-269.

Bickert, C. von E., 1967: A study of the understanding and use of probability of precipitation forecasts in two major cities. Denver, CO, Unversity of Denver, Denver Research Institute, Report (Review Draft), 21 pp. plus four appendices.

Bocchieri, J. R., 1979a: A new operational system for forecasting precipitation type. Monthly Weather Review, 107, 637-649.

Bocchieri, J. R., 1979b: The use of LFM output for automated prediction of heavy snow. Preprints, Fourth Conference on Numerical Weather Prediction (Silver Spring). Boston, MA, American Meteorological Society, pp. 77-81.

Bosart, L. F., 1975: SUNYA experimental results in forecasting daily temperature and precipitation. Monthly Weather Review, 103, 1013-1020.

Brelsford, W. M., and R. H. Jones, 1967: Estimating probabilities. Monthly Weather Review, 95, 570-576.

Brier, G. W., 1946: A study of quantitative precipitation forecasting in the TVA basin. Washington, D.C., U.S. Weather Bureau, Research Paper No. 26, 40 pp.

Budescu, D. V., and T. S. Wallsten, 1981: Intra and intersubject consistency in comparing probability phrases. Chapel Hill, University of North Carolina, Psychometric Laboratory, Research Memorandum No. 49, 25 pp.

Carter, G. M., and H. R. Glahn, 1976: Objective prediction of cloud amount based on Model Output Statistics. Monthly Weather Review, 104, 1565-1572.

Charba, J. P., 1979: Two to six hour severe local storm probabilities: an operational forecasting system. Monthly Weather Review, 107, 268-282.

Charba, J. P., and W. H. Klein, 1980: Skill in precipitation forecasting in the National Weather Service. Bulletin of the American Meteorological Society, 61, 1546-1555.

Cooke, W. E., 1906a: Forecasts and verifications in Western Australia. Monthly Weather Review, 34, 23-24.

Cooke, W. E., 1906b: Weighting forecasts. Monthly Weather Review, 34, 274-275.

Daan, H., and A. H. Murphy, 1982: Subjective probability forecasting in The Netherlands: some operational and experimental results. Meteorologische Rundschau, 35, 99-112.

Dickey, W. W., 1949: Estimating the probability of a large fall in temperature at Washington, D.C. Monthly Weather Review, 77, 67-78.

Enger, I., J. A. Russo, Jr., and E. L. Sorenson, 1964: A statistical approach to 2-7 hr prediction of ceiling and visibility, volumes I and II. Hartford, CT, Travelers Research Center, Inc., Contract No. Cwb-10704, 48 pp. and 195 pp., respectively.

Epstein, E. S., 1969: Stochastic dynamic prediction. Tellus, 21, 739-759.

Fischhoff, B., and D. MacGregor, 1982: Subjective confidence in forecasts. Journal of Forecasting, 1, 155-172.

Fix, C., and J. L. Hodges, Jr., 1951: Discriminatory analysis, non-parametric discrimination: consistency properties. Randolph Field, TX, USAF, School of Aviation Medicine, Report No. 4, 19 pp.

Freeman, M. H., 1961: A graphical method of objective forecasting derived by statistical techniques. Quarterly Journal of the Royal Meteorological Society, 87, 393-400.

Glahn, H. R., 1964: An application of adaptive logic to meteorological prediction. Journal of Applied Meteorology, 3, 718-725.

Glahn, H. R., 1980: Methods and accuracy of statistical weather forecasting in the United States. Preprints, Symposium on Probabilistic and Statistical Methods in Weather Forecasting (Nice). Geneva, Switzerland, World Meteorological Organization, pp. 387-395.

Globokar, F. T., 1974: Computerized ceiling and visibility forecasts. Preprints, Fifth Conference on Weather Forecasting and Analysis (St. Louis). Boston, MA, American Meteorological Society, pp. 228-233.

Gregg, G. T., 1977: Probability forecasts of a temperature event. National Weather Digest, 2, No. 2, 33-34.

Grimes, D., 1982: A preliminary verification of the Pacific Weather Centre's probability of precipitation forecasts. Vancouver, B.C., Canada, Atmospheric Environment Service, Pacific Region Technical Notes 82-013, 8 pp.

Gringorten, I. I., 1966: A stochastic model of the frequency and duration of weather events. Journal of Applied Meteorology, 5, 606-624.

Gringorten, I. I., 1971: Modelling conditional probability. Journal of Applied Meteorology, 10, 646-657.

Hampton, J. M., P. G. Moore, and H. Thomas, 1973: Subjective probability and its measurement. Journal of the Royal Statistical Society, Series A, 136, 21-42.

Hughes, L. A., 1980: Probability forecasting - reasons, procedures, problems. Silver Spring, MD, U.S. Department of Commerce, NOAA, National Weather Service, Technical Memorandum NWS FCST 24, 84 pp.

Javelle, J. P., J. M. Veysseire, A. M. Calvayrac, F. Duvernet, and G. Der Megreditchian, 1980: Operational forecasting of local weather parameters through statistical interpretation of NWP using "perfect prog" methods. Preprints, Symposium on Probabilistic and Statistical Methods in Weather Forecasting (Nice). Geneva, Switzerland, World Meteorological Organization, pp. 423-429.

Jensenius, J. S., and G. M. Carter, 1979: Automated forecasts of agricultural weather elements. Preprints, Fourteenth Conference on Agriculture and Forest Meteorology (Minneapolis). Boston, MA, American Meteorological Society, pp. 42-44.

Jones, R. H., 1968: A nonlinear model for estimating probabilities of k events. Monthly Weather Review, 96, 383-384.

Kruizinga, S., and A. H. Murphy, 1983: Use of an analogue procedure to formulate objective probabilistic temperature forecasts in The Netherlands. Monthly Weather Review, 111, 2244-2254.

Krzysztofowicz, R., 1983: Why should a forecaster and a decision maker use Bayes' theorem? Water Resources Research, 19, 327-336.

Lichtenstein, S., and J. R. Newman, 1967: Empirical scaling of common verbal phrases associated with numerical probabilities. Psychonomic Sciences, 9, 563-564.

Lichtenstein, S., B. Fischhoff, and L. D. Phillips, 1982: Calibration of probabilities: the state of the art to 1980. Judgment under Uncertainty: Heuristics and Biases (D. Kahneman, P. Slovic, and A. Tversky, Editors). London, Cambridge University Press, pp. 306-334.

Lindley, D. V., 1965: Introduction to Probability and Statistics from a Bayesian Viewpoint: Part I. Probability. London, Cambridge University Press, 259 pp.

Lowry, D. A., and H. R. Glahn, 1976: An operational model for forecasting probability of precipitation - PEATMOS PoP. Monthly Weather Review, 104, 221-232.

Maglaras, G. J., J. P. Dallavalle, K. F. Hebenstreit, G. H. Hollenbaugh, B. E. Schwartz, and D. J. Vercelli, 1981: Comparative verification of guidance and local aviation/public weather forecasts - No. 10 (April 1980 - September 1980). Silver Spring, MD, U.S. Department of Commerce, NOAA, National Weather Service, TDL Office Note 81-7, 61 pp.

Martin, D. E., 1972: Climatic presentations for short-range forecasting based on event occurrence and reoccurrence profiles. Journal of Applied Meteorology, 11, 1212-1223.

Mason, I. B., 1976: Weather forecasts as subjective probability statements: a report on an experiment. Canberra, Australia, Bureau of Meteorology, Regional Forecast Centre, unpublished manuscript, 36 pp.

Mason, I. B., 1977: Objective probabilities for fog at Canberra Airport using Bayesian aggregation and principal components. Australian Meteorological Magazine, 25, 121-134.

McBoyle, G. R., 1974: Public response to weather terminology in the Kitchener-Waterloo area. Climatological Bulletin (McGill University), No. 15, pp. 11-29.

Miller, R. G., 1962: Statistical prediction by discriminant analysis. Meteorological Monographs, 4, No. 25, 54 pp.

Miller, R. G., 1964: Regression estimation of event probabilities. Hartford, CT, Travelers Research Center, Contract Cwb-10704, Technical Report No. 1, 153 pp.

M.S.I. Services, Inc., 1981: Public requirements for weather information and attitudes concerning weather service. Washington, D.C., M.S.I. Services, Inc., Technical Report, 46 pp. plus three appendices.

Murphy, A. H., 1977: The value of climatological, categorical and probabilistic forecasts in the cost-loss ratio situation. Monthly Weather Review, 105, 803-816.

Murphy, A. H., 1978: On the evaluation of point precipitation probability forecasts in terms of areal coverage. Monthly Weather Review, 106, 1680-1686.

Murphy, A. H., and H. Daan, 1984: Impacts of feedback and experience on the quality of subjective probability forecasts: comparison of results from the first and second years of the Zierikzee experiment. Monthly Weather Review, 112, 413-423.

Murphy, A. H., W.-R. Hsu, and R. L. Winkler, 1982: Subjective probabilistic quantitative precipitation forecasts: some experimental results. Preprints, Ninth Conference on Weather Forecasting and Analysis (Seattle). Boston, MA, American Meteorological Society, pp. 94-100.

Murphy, A. H., S. Lichtenstein, B. Fischhoff, and R. L. Winkler, 1980: Misinterpretations of precipitation probability forecasts. Bulletin of the American Meteorological Society, 61, 695-701.

Murphy, A. H., and R. L. Winkler, 1971: Forecasters and probability forecasts: some current problems. Bulletin of the American Meteorological Society, 52, 239-247.

Murphy, A. H., and R. L. Winkler, 1974: Credible interval temperature forecasting: some experimental results. Monthly Weather Review, 102, 784-794.

Murphy, A. H., and R. L. Winkler, 1977: Experimental point and area precipitation probability forecasts for a forecast area with significant local effects. Atmosphere, 15, 61-78.

Murphy, A. H., and R. L. Winkler, 1982: Subjective probabilistic tornado forecasting: some experimental results. Monthly Weather Review, 110, 1288-1297.

Murphy, A. H., and R. L. Winkler, 1984: Probability forecasting in meteorology. Journal of the American Statistical Association, 79, in press.

Peterson, C. R., K. J. Snapper, and A. H. Murphy, 1972: Credible interval temperature forecasts. Bulletin of the American Meteorological Society, 53, 966-970.

Raiffa, H., 1968: Decision Analysis: Introductory Lectures on Choices under Uncertainty. Reading, MA, Addison-Wesley, 309 pp.

Reap, R. M., and D. S. Foster, 1979: Automated 12-36 hour probability forecasts of thunderstorms and severe local storms. Journal of Applied Meteorology, 18, 1304-1315.

Roberts, C. F., J. M. Porter, and G. F. Cobb, 1969: Report on Weather Bureau forecast performance 1967-68 and comparison with previous years. Silver Spring, MD, U.S. Department of Commerce, ESSA, Weather Bureau, Technical Memorandum WBTM FCST 11, 44 pp.

Rogell, R. H., 1972: Weather terminology and the general public. Weatherwise, 25, 126-132.

Saarikivi, P., 1980: The effect of the temporal variability of precipitation on the quality of probability forecasts. Preprints, Symposium on Probabilistic and Statistical Methods in Weather Forecasting (Nice). Geneva, Switzerland, World Meteorological Organization, pp. 229-234.

Sanders, F., 1963: On subjective probability forecasting. Journal of Applied Meteorology, 2, 191-201.

Sanders, F., 1973: Skill in forecasting daily temperature and precipitation: some experimental results. Bulletin of the American Meteorological Society, 54, 1171-1179.

Sanders, F., 1979: Trends in skill of daily forecasts of temperature and precipitation, 1966-78. Bulletin of the American Meteorological Society, 60, 763-769.

Schmidt, R. C., 1951: A method of forecasting occurrence of winter precipitation two days in advance. Monthly Weather Review, 79, 81-95.

Schwartz, B. E., J. R. Bocchieri, G. M. Carter, J. P. Dallavalle, G. H. Hollenbaugh, G. J. Maglaras, and D. J. Vercelli, 1981: Comparative verification of guidance and local aviation/public weather forecasts - No. 11 (October 1980 - March 1981). Silver Spring, MD, U.S. Department of Commerce, NOAA, National Weather Service, TDL Office Note 81-10, 77 pp.

Scoggins, J. R., and W. W. Vaughn, 1971: How some nonmeteorological professionals view meteorology and weather forecasting. Bulletin of the American Meteorological Society, 52, 974-979.

Spetzler, C. S., and C.-A. S. Staël von Holstein, 1975: Probability encoding in decision analysis. Management Science, 22, 340-358.

Staël von Holstein, C.-A. S., 1970: Assessment and Evaluation of Subjective Probability Distributions. Stockholm, Sweden, Stockholm School of Economics, Economic Research Institute, 225 pp.

Thompson, J. C., 1950: A numerical method for forecasting rainfall in the Los Angeles area. Monthly Weather Review, 78, 113-124.

Thompson, J. C., 1962: Economic gains from scientific advances and operational improvements in meteorological prediction. Journal of Applied Meteorology, 1, 13-17.

Tversky, A., and D. Kahneman, 1974: Judgment under uncertainty: heuristics and biases. Science, 185, 1124-1131.

U.S. Department of Commerce, 1982: Monthly and seasonal weather outlook. Washington, D.C., NOAA, National Weather Service, Climate Analysis Center, Volume 36, Number 14, 8 pp. (Published twice a month and available from U.S. Government Printing Office.)

Wahl, E. W., and collaborators, 1952: The construction and application of contingency tables in weather forecasting. Cambridge, MA, U.S. Air Force Cambridge Research Laboratories, Surveys in Geophysics No. 19, 37 pp.

Wallsten, T. S., and D. V. Budescu, 1983: Encoding subjective probabilities: a psychological and psychometric review. Management Science, 29, 151-173.

Wilson, L. J., and N. Yacowar, 1980: Statistical weather element forecasting in the Canadian Weather Service. Preprints, Symposium on Probabilistic and Statistical Methods in Weather Forecasting (Nice). Geneva, Switzerland, World Meteorological Organization, pp. 401-406.

Winkler, R. L., 1967: The assessment of prior distributions in Bayesian analysis. Journal of the American Statistical Association, 62, 776-800.

Winkler, R. L., 1972: Introduction to Bayesian Inference and Decision. New York, Holt, Rinehart and Winston, 563 pp.

Winkler, R. L., and A. H. Murphy, 1968: "Good" probability assessors. Journal of Applied Meteorology, 7, 751-758.

Winkler, R. L., and A. H. Murphy, 1973: Information aggregation in probabilistic prediction. IEEE Transactions on Systems, Man, and Cybernetics, SME-3, 154-160.

Winkler, R. L., and A. H. Murphy, 1976: Point and area precipitation probability forecasts: some experimental results. Monthly Weather Review, 104, 86-95.

Winkler, R. L., and A. H. Murphy, 1979: The use of probabilities in forecasts of maximum and minimum temperatures. Meteorological Magazine, 108, 317-329.

Winkler, R. L., A. H. Murphy, and R. W. Katz, 1977: The consensus of subjective probability forecasts: are two, three, ... heads better than one? Preprints, Fifth Conference on Probability and Statistics in Atmospheric Sciences (Las Vegas). Boston, MA, American Meteorological Society, pp. 57-62.

Zurndorfer, E. A., J. R. Bocchieri, G. M. Carter, J. P. Dallavalle, D. B. Gilhousen, K. F. Hebenstreit, and D. J. Vercelli, 1979: Trends in verification scores for guidance and local aviation/ public weather forecasts. Monthly Weather Review, 107, 799-811.

10
Forecast Evaluation

Allan H. Murphy and Harald Daan

1. INTRODUCTION

Forecast evaluation can be described as "the process and practice of determining the quality and value of forecasts." As such, it consists of a body of concepts and procedures for comparing forecasts and observations and for assessing the value or utility of forecasts to users. The practice of evaluating weather forecasts is almost as old as routine weather forecasting itself, with the first papers and reports on evaluation methodology and on the results of evaluation studies appearing nearly a century ago. Moreover, the history and development of forecast evaluation in meteorology have been marked by several colorful incidents and by considerable controversy. Unfortunately, a comprehensive review of the literature on the evaluation of weather forecasts is clearly beyond the scope of this chapter. The reader is referred to publications by Muller (1944), Bleeker (1946), Brier and Allen (1951), Gringorten (1951), Johnson (1957), Meglis (1960), Murphy and Allen (1970), Dobryshman (1972), National Center for Atmospheric Research (1976), World Meteorological Organization (1980), and Daan (1984) for reviews, collections of papers, and bibliographies concerning various aspects of this topic.

The purposes of this chapter are to discuss the basic principles and concepts on which current evaluation procedures and practices are based and to describe some measures and techniques that can be used to assess the quality and value of different types of forecasts. Emphasis here will be placed on methods of evaluating forecasts rather than on the practice of forecast evaluation or on the results of evaluation programs. The nature and purposes of forecast evaluation are discussed in Section 2. This discussion leads to the identification of two types of evaluation; namely, inferential (or empirical) evaluation and decision-theoretic (or operational) evaluation. In essence, the former is concerned with the quality of forecasts, whereas the latter is concerned with the value of forecasts. Inferential evaluation is the primary focus of this chapter. Various basic aspects of the forecasts themselves are described in Section 3, including the nature of forecast variables (i.e., predictands), the types of forecasts, and the principal attributes of forecasts. The two basic types of forecasts considered here are categorical forecasts and probabilistic forecasts. Desirable properties of evaluation measures are identified and defined in Section 4. Evaluation measures and other procedures for assessing the quality of categorical and probabilistic forecasts are described in Sections 5 and 6, respectively. Section 7 consists of a brief discussion of current issues in forecast evaluation and identifies some sources of results of evaluation programs in meteorology. In addition, this section contains lists of key references to studies involving decision-theoretic evaluation in meteorology and to the forecast evaluation literature in other fields.

2. NATURE AND PURPOSES OF EVALUATION

Forecast evaluation is undertaken for a variety of different purposes. The choice of a particular purpose or objective necessarily plays an important role in determining which measures and procedures should be used in the evaluation process. As Brier and Allen (1951) and others have indicated, the controversies associated with forecast evaluation have arisen in large measure because of the failure to take proper account of both the purposes of evaluation and the implications of these purposes for evaluation procedures and practices. For discussions of the purposes of evaluation, see Brier and Allen (1951), Johnson (1957), Murphy and Epstein (1967a), and Daan (1984).

At the most general level, two purposes for evaluation can be identified: (a) to determine the quality of forecasts and (b) to assess the value of forecasts. Each of these overall purposes defines a specific type of evaluation. The process and practice of determining the quality of forecasts (i.e., comparing forecasts and observations) will be referred to here as inferential evaluation. In the meteorological literature, this aspect of forecast evaluation frequently is called forecast verification. Assessment of the value or utility of forecasts, on the other hand, will be referred to in this chapter as decision-theoretic evaluation. Studies in this area generally focus on a particular type of evaluation, although some investigations have considered both inferential and decision-theoretic evaluation as well as the relationships between these two types of evaluation measures. In this regard, although we usually consider these two types of evaluation to be quite distinct, it is perhaps more realistic to think of inferential and decision-theoretic evaluation as the end points of an evaluation continuum (Murphy, 1976). Thus, measures designed for inferential (decision-theoretic) evaluation may also serve in part as measures of value (quality).

Within the context of inferential evaluation, it is possible to identify several specific purposes. These purposes include: (a) to determine the state of the art of weather forecasting and trends in the quality of forecasts over time; (b) to establish a system to monitor, on a continuing basis, the quality of forecasts (quality control); (c) to compare forecasts produced by different forecasters or by different forecasting procedures; (d) to provide feedback to forecasters concerning their individual and/or collective performance; (e) to assess the errors in forecasts to develop a basis for improving forecasting techniques; and (f) to provide users with information concerning the quality of the forecasts (e.g., the distribution of forecast errors). Many evaluation programs involve several of these purposes or objectives. As indicated above, the choice of a particular purpose, or set of purposes, has important implications for decisions related to the development of evaluation methodology as well as for the process and practice of evaluation.

Decision-theoretic evaluation also can serve several specific purposes. These purposes include: (a) to obtain information concerning the preferred form and mode of expression of forecasts (the forecast structure); (b) to describe the optimal procedure for using the forecasts; and (c) to derive estimates of the value or utility of the forecasts. Studies or programs designed to meet these objectives generally will require the formulation of descriptive and/or prescriptive models of the relevant decision-making situations, and a variety of such models have been used in an evaluation context in meteorology (see Section 7).

Whatever the purposes of an evaluation program, it is important to recognize that forecast evaluation in practice is a comparative process. That is, we are usually concerned with the relative, rather than the absolute, quality and/or value of the forecasts. Relative performance, in terms of quality or value, is determined by comparing the forecasts of interest with forecasts based on a reference procedure. A variety of such procedures can be defined, with climatology and persistence representing two of the most frequently used standards of reference in meteorology. Other procedures can quite naturally play the role of standards of reference in certain situations. For example, it may be appropriate to compare forecasts formulated locally with centrally prepared guidance or to compare subjective forecasts with objective forecasts. In any case, measures of relative performance play an important role in forecast evaluation, and they will be given special emphasis in this chapter.

3. PREDICTANDS, FORECASTS, AND ATTRIBUTES

Evaluation procedures also depend on the nature of the variable or element for which the forecasts are formulated and on the type of forecasts. The nature of these forecast elements, or predictands, and the principal types of forecasts are described briefly in this section. In addition, we identify and define desirable properties, or attributes, of the different types of forecasts.

3.1 Nature of Predictands

For the purposes of this discussion, it will be sufficient to differentiate between two types of predictands, nominal predictands and ordinal predictands. A nominal predictand consists of a set of N (> 2) mutually exclusive and collectively exhaustive events or states, and this set of events may be ordered or unordered. However, a predictand will be treated here as a nominal predictand when the ordering of the events, if they are ordered, is not considered to be relevant from the point of view of evaluation. On the other hand, a predictand will be treated as an ordinal predictand when it is represented either by the values of a variable defined on an ordinal scale or by ordered events when this ordering is considered to be essential for evaluation purposes. Thus, ordinal predictands may be either continuous or discrete variables, whereas nominal predictands are necessarily discrete variables. Since forecasts and observations are recorded on a finite scale, all ordinal predictands are in reality discrete variables. However, it is convenient, and not inappropriate, to treat some ordinal predictands as continuous variables. It should be noted that discrete variables (or sets of events) may be considered to be nominal predictands for some purposes and ordinal predictands for other purposes. Finally, the situation in which a predictand consists of only two events (i.e., N = 2) is frequently of special interest, and this situation is treated separately in some cases in Sections 5 and 6.

The distinction between nominal and ordinal predictands is made because the concept of distance (between events or values of a variable) plays an important role in forecast evaluation. In the case of ordinal predictands, whether discrete or continuous, the forecast error may be expressed (for example) by the difference, or distance, between the forecast and observed events or values of the variable. For nominal predictands, on the other hand, the distance concept is not meaningful and forecast errors must be described in different terms. Further discussion of distance as an attribute of forecasts appears in Section 3.3, and distance measures are described in Sections 5 and 6.

3.2 Types of Forecasts

Forecasts may be considered to be statements based on judgments regarding the likelihood of occurrence of specific events or particular values of predictands. These judgments may arise from the intuitive assimilation of information by human forecasters or they may represent the output of objective numerical and/or statistical forecasting procedures. In any case, we will assume here that such judgments always underlie the forecasts of concern and that the judgments consist of discrete or continuous probability distributions over the events or values of the variable. The existence of the judgments will be of particular interest, from the point of view of evaluation measures, when it is assumed that the forecasts are based on subjective forecasting procedures (see Section 4.1).

Two general types of forecasts can be identified, categorical forecasts and probability forecasts. Categorical forecasts consist of statements indicating that certain events or values, or sets of events or values, will or will not occur (i.e., the forecasts are stated without qualification). In the case of nominal predictands, a single-event categorical forecast is a statement that a particular event will occur and, by implication, that the other N-1 events will not occur. When such predictands consist of only two possible events (i.e., when N = 2), the categorical forecasts sometimes are referred to as "yes/no forecasts." In multiple-event categorical forecasts for nominal predictands, two or more events may be included in the forecasts (i.e., the forecasts may indicate that a set of events will occur). A categorical forecast for an ordinal predictand specifies a particular event or value (or set of events or values) of the relevant variable.

Probability forecasts for nominal predictands consist of discrete probability distributions, in which each probability represents the (stated) likelihood of occurrence of the corresponding event. Since the sum of the probabilities assigned to the N events must be equal to one, the probabilities for only N-1 events need to be specified explicitly. For example, when N = 2, only one probability is required. It should be noted that the probabilities assigned to the set of N events on a particular occasion can be treated as a single forecast in a vector framework or as N probabilities in a scalar framework. When these alternative frameworks are used, we will refer to the forecasts as vector and scalar forecasts, respectively. In the case of an ordinal predictand, a probability forecast may be expressed as a discrete or continuous probability distribution or in terms of the corresponding cumulative distributions. When the cumulative distributions are considered, we will refer to the forecasts as cumulative forecasts.

The form of the forecasts actually issued or disseminated to users depends on the nature of their requirements for information and on the structure of their payoff or loss functions, as well as on the judgments regarding the likelihood of occurrence of the events or values of the relevant variable. Users who can accommodate probability statements should be given forecasts that explicitly express the likelihood of occurrence of the events or ranges of values in the form of discrete or continuous probability distributions. Clearly, it would be desirable to have such probabilities correspond exactly to the underlying judgments, since the latter represent the "true" opinions of the forecasting system (see Section 4.1). The situation is somewhat more complicated when the users require forecasts expressed in categorical terms. If the structure of the users' payoff functions is known, then it should be possible to translate the (probabilistic) judgments into categorical forecasts specifically tailored for the users. On the other hand, when the structure of the payoff functions is unknown, "representative" events or values must be chosen as categorical forecasts. For example, in the case of nominal predictands, the event associated with the largest judgmental probability might be forecast categorically. For ordinal predictands, the mean or median of the judgmental probability distribution could be chosen as the categorical forecast. In many situations, however, the selection of an appropriate categorical forecast is a much more difficult problem (e.g., in situations involving rare or extreme events or in situations in which the climatological distribution of the variable is highly skewed or bimodal).

3.3 Attributes of Forecasts

When the purposes of an evaluation program have been specified, it is then usually possible to identify certain desirable properties, or attributes, of forecasts with respect to these purposes. Several important attributes are defined in this section, and interpretations of these attributes are provided for nominal and ordinal predictands and for categorical and probabilistic forecasts. In defining attributes for categorical forecasts, it is sometimes convenient to distinguish between cases in which a single value or event is forecast and cases in which intervals of values or multiple events are forecast. The evaluation measures to be described in subsequent sections generally are explicit functions of one or more of these basic attributes.

3.3.1 Accuracy.

Accuracy is defined as the average degree of correspondence between individual forecasts and observations. In the case of categorical forecasts for nominal predictands, a forecast is perfectly accurate if the event (or set of events) that is observed also was forecast. Specifically, suppose that the (row) vector $\underline{c} = (c_1, ..., c_N)$ denotes a single-event categorical forecast for such a predictand, where $c_n = 1$ if the nth event is forecast and $c_n = 0$ otherwise ($n = 1, ..., N$). Further, let the (row) vector $\underline{d} = (d_1, ..., d_N)$ denote the corresponding observation, where $d_n = 1$ if the nth event occurs and $d_n = 0$ otherwise. It is also convenient to let the subscripts i and j denote the forecast and observed events, respectively. The observation vector $\underline{d}$, then, consists of the components $d_j = 1$ and $d_n = 0$ for all $n \neq j$. In terms of this notation, the forecast $\underline{c}$ is perfectly accurate if the difference between $\underline{c}$ and $\underline{d}$ is zero (i.e., if $c_j = 1$ and $c_i = 0$ for all $i \neq j$). For a multiple-event forecast, two or more components of the forecast vector $\underline{c}$ may be set equal to one. In such cases, the forecast is perfectly accurate if the non-zero component of the observation vector $\underline{d}$ corresponds to one of the forecast events. With respect to nominal predictands, accurate and inaccurate forecasts generally are referred to as "hits" and "misses," respectively. The percentage of hits in a sample of forecasts (i.e., the "hit rate") frequently represents an important ingredient in the process of formulating evaluation measures for categorical forecasts of nominal predictands.

The accuracy of categorical forecasts for ordinal predictands can be defined in terms of the average difference (or distance) between the forecast and observed events or values of the variable. Let y denote the categorical forecast for a continuous (ordinal) predictand and let z denote the corresponding observed value of the variable. In this context, accuracy can be expressed in terms of the difference between y and z for continuous variables and in terms of the difference between i and j for discrete variables. These differences are zero for perfectly accurate forecasts.

In the case of probability forecasts, accuracy relates to the average degree of correspondence between individual discrete or continuous probability distributions (or cumulative distributions) and the singular distributions represented by the observation vectors. Specifically, suppose that a discrete probability forecast is denoted by a vector $\underline{r} = (r_1, \ldots, r_N)$ $(r_n \geq 0, \sum_n r_n = 1; n = 1, \ldots, N)$. Then for a nominal predictand, the accuracy of the forecast $\underline{r}$ is defined in terms of the difference between the vectors $\underline{r}$ and $\overline{\underline{d}}$. When this difference is zero (i.e., when $r_j = 1$), the forecast is perfectly accurate.

The accuracy of probability forecasts for ordinal predictands is defined in terms of the difference between cumulative distributions, since the order of the events or values is relevant for evaluation purposes. For continuous distributions, let $r(y)$ $[r(y) \geq 0, \int_y r(y)dy = 1]$ denote the forecast probability distribution defined on the variable y and let $R(y)$ denote the corresponding cumulative distribution. Then the difference of concern is $R(y) - D(y)$, where $D(y) = 0$ if $y < z$ and $D(y) = 1$ if $y \geq z$. In the case of discrete distributions, this difference can be expressed as $\underline{R} - \underline{D}$, where $\underline{R} = (R_1, \ldots, R_N)$ and $\underline{D} = (D_1, \ldots, D_N)$ in which $R_n = \sum_m r_m$ and $D_n = \sum_m d_m$ $(n = 1, \ldots, N; m = 1, \ldots, n)$. As in the case of nominal predictands, accuracy is perfect when this difference in cumulative distributions is zero [i.e., when $R(y) = D(y)$ in the case of continuous predictands and when $\underline{R} = \underline{D}$ in the case of discrete predictands].

3.3.2 Reliability (bias). Reliability is defined as the degree of correspondence between average forecasts and average observations over a sample of forecasts. Reliability and bias are complementary terms. It is convenient to distinguish between reliability in-the-large and reliability in-the-small. The latter concept applies only to probability forecasts. In the case of categorical forecasts for nominal predictands, a sample of K forecasts is perfectly reliable (or completely unbiased) in-the-large if the average forecast vector $\bar{c}$ = $(\bar{c}_1, \ldots, \bar{c}_N)$, where $\bar{c}_n = (1/K) \sum\limits_{k=1}^{K} c_{nk}$, corresponds exactly to the average observation vector $\bar{d} = (\bar{d}_1, \ldots, \bar{d}_N)$, where $\bar{d}_n = (1/K) \sum\limits_{k=1}^{K} d_{nk}$ $(n = 1, \ldots, N)$. Similarly, the reliability (or bias) in-the-large of the forecasts for the nth event can be described by comparing the scalar components $\bar{c}_n$ and $\bar{d}_n$. This comparison can be accomplished in a variety of ways, including the use of differences or ratios (see Section 5.2.2).

The reliability (or bias) in-the-large of categorical forecasts for ordinal predictands can be defined in terms of the difference between the average forecast and observed values of the variable. Let $\bar{y}$ and $\bar{z}$ denote the average forecast and observed values, respectively, for a continuous variable, where $\bar{y} = (1/K) \sum\limits_{k=1}^{K} y_k$ and $\bar{z} = (1/K) \sum\limits_{k=1}^{K} z_k$. In this context, reliability can be expressed in terms of the difference $\bar{y} - \bar{z}$ or the ratio $\bar{y}/\bar{z}$ for continuous variables. For discrete ordinal predictands, reliability can be defined in terms of the difference between, or ratio of, the average forecast value $\bar{i}$ and the average observed value $\bar{j}$, where $\bar{i} = (1/K) \sum\limits_{k=1}^{K} i_k$ and $\bar{j} = (1/K) \sum\limits_{k=1}^{K} j_k$, respectively. These differences are zero, and the ratios are one, for perfectly reliable (or completely unbiased) forecasts.

In the case of probability forecasts, reliability or bias relates to the degree of correspondence between average probabilities and observed relative frequencies over a sample of forecasts. Let $\bar{r}$ = $(\bar{r}_1, \ldots, \bar{r}_N)$ denote the overall average forecast vector for a sample of K forecasts, where $\bar{r}_n = (1/K) \sum\limits_{k=1}^{K} r_{nk}$. Then for a nominal predictand, the reliability in-the-large of the forecasts is described by the difference between $\bar{r}$ and $\bar{d}$. When this difference is zero, the probability forecasts are perfectly reliable (or completely unbiased) in-the-large. This overall indicator of reliability also has been referred to as global reliability.

Frequently, only a finite number of probability values are used in expressing the forecasts (e.g., 0.0, 0.1, ..., 1.0). Moreover, even when an essentially infinite number of values are used, it may be desirable to combine probabilities into a discrete set of classes (e.g., 0.0-0.1, 0.1-0.2, ..., 0.9-1.0). In such situations, the reliability of the subsamples of forecasts corresponding to a particular probability value or range of probability values may be of interest. This aspect of reliability is referred to as reliability in-the-small. Let the vector $\underline{r}^t = (r_1^t, \ldots, r_N^t)$ denote the set of probabilities corresponding to the t-th distinct vector forecast ($t = 1, \ldots, T$). Moreover, let the vector $\underline{d}^t = (d_1^t, \ldots, d_N^t)$ denote the relative frequencies of occurrence of the events in the subsample of K^t occasions on which the forecast $\underline{r}^t$ was used ($\Sigma K^t = K$). Then for a nominal predictand, the reliability of the forecasts in the t-th subsample is described in terms of the difference between $\underline{r}^t$ and $\underline{d}^t$. When this difference is zero, the subsample of forecasts is perfectly reliable (or completely unbiased). If a sample of forecasts is perfectly reliable in-the-small over all T distinct sets of probability values, then it is also completely reliable in-the-large (the converse is not necessarily true).

The reliability (or bias) of probability forecasts for discrete ordinal predictands is defined in terms of the difference between cumulative probabilities and cumulative relative frequencies. Let $\overline{R} = (\overline{R}_1, \ldots, \overline{R}_N)$ and $\overline{D} = (\overline{D}_1, \ldots, \overline{D}_N)$ denote the average cumulative forecast and cumulative observation, respectively, where $\overline{R}_n = (1/K) \sum_{k=1}^{K} R_{nk}$ and $\overline{D}_n = (1/K) \sum_{k=1}^{K} D_{nk}$. Then, the overall reliability is described in terms of the difference between $\overline{R}$ and $\overline{D}$. When this difference is zero, the forecasts are perfectly reliable in-the-large. Reliability in-the-small for ordinal predictands can be defined in a manner analogous to the definition for nominal predictands, except that attention is focused on differences between cumulative probabilities and cumulative relative frequencies. Let $\underline{R}^t = (R_1^t, \ldots, R_N^t)$ denote the set of probabilities corresponding to the t-th distinct cumulative vector forecast and let $\overline{D}^t$ denote the cumulative relative frequencies of occurrence of the events on the K^t occasions on which the forecast $\underline{R}^t$ was used. [It should be noted that a one-to-one correspondence exists between the forecast vector $\underline{r}^t$ (relative frequency vector $\overline{d}^t$) and the cumulative forecast vector R^t (cumulative relative frequency vector $\overline{D}^t$)]. Then for a (discrete) ordinal predictand, the reliability (or bias) of the forecasts in the t-th subsample is described by the difference between R^t and $\overline{D}^t$. When this difference is zero, the subsample of forecasts is perfectly reliable (or completely unbiased). If the sample is perfectly reliable in-the-small over all such subsamples, then it is also perfectly reliable in-the-large.

3.3.3 <u>Sharpness</u>. Sharpness represents the degree to which a forecast approaches a categorical forecast of a particular value in the case of a continuous variable or of a specific event in the case of a discrete variable (see Bross, 1953). Since a one-to-one correspondence exists between the set of N possible single-event categorical forecasts and the set of N possible observations, it will be convenient here to use the latter instead of the former in formulating definitions of this attribute. However, it should be kept in mind that sharpness relates only to characteristics of the forecasts; specifically, this attribute is not concerned with the correspondence between forecasts and observations.

For categorical forecasts, the concept of sharpness is meaningful only for nominal predictands, since order is not relevant in this context. A categorical forecast for a single event is, by definition, perfectly sharp. However, categorical forecasts may specify that a set of two or more events will occur. In such cases, the degree of sharpness may be characterized simply by the number of events included in the forecast.

For probabilistic forecasts, sharpness relates to the degree of correspondence between the individual forecasts of concern and categorical forecasts (or observations). Consider a probability forecast r for a nominal predictand, and let d^* denote the observation that "most closely" corresponds to the forecast r. The concept of closeness in this context will be discussed in Section 6.2.3 when measures of sharpness for probability forecasts are described. Sharpness for a nominal predictand is defined in terms of the difference between the vectors r and d^*. When this difference is zero (i.e., when $r_n = 1$ for some n), the probability forecast is perfectly sharp.

In the case of probability forecasts for a continuous ordinal predictand, sharpness is defined in terms of the difference between the cumulative forecast distribution $R(y)$ and the closest cumulative observation $D^*(y)$. Sharpness for a discrete ordinal predictand is defined in terms of the difference between the cumulative forecast R and the closest cumulative observation D^*. For both continuous and discrete ordinal predictands, sharpness is perfect only when the forecast distribution represents a categorical forecast for just one value or event.

3.3.4 <u>Resolution</u>. Resolution refers to the concept first described by Köppen (1884) when he stated that "different forecasts should be followed by different observations." Alternatively, resolution represents the ability of the forecaster (or forecast system) to sort the forecasts in a sample into subsamples whose corresponding frequencies of occurrence of the events are maximally different. The interpretation of resolution is essentially identical for both categorical and probabilistic forecasts. For categorical forecasts, resolution is defined only for nominal predictands. Let d^t denote the relative frequencies of occurrence of the N events in a sample of forecasts when the t-th distinct categorical forecast is made (t = 1, ..., T). Then, resolution in this context is concerned with the differences between the relative frequency vectors for different values of t (i.e., for different categorical forecasts). If these differences are large, then the resolution is relatively good; if these differences are small, then the resolution is relatively poor.

In the case of probability forecasts for nominal predictands, resolution relates to the correspondence between relative frequencies of occurrence of events in subsamples of forecasts associated with distinct probability forecasts and the overall observed relative frequencies for the entire sample of forecasts. Using previous notation, these relative frequencies are represented by $\bar{d}^t$ (t = 1, ..., T) and $\bar{d}$. Thus, resolution is described by the differences between $\bar{d}^t$ and $\bar{d}$. If these differences are large, then resolution is relatively good; if they are small, then resolution is relatively poor. This definition can be extended to the case of probability forecasts for discrete ordinal predictands, in which case the differences of concern involve the cumulative relative frequency vectors $\bar{D}^t$ and $\bar{D}$.

4. SOME DESIRABLE PROPERTIES OF EVALUATION MEASURES

The results of evaluation studies can be summarized in terms of contingency tables or frequency distributions, and such summaries contain useful information regarding forecast quality for many purposes (e.g., see Section 5.1). However, it is generally desirable to compute one or more evaluation measures to characterize various aspects of performance. Such measures should possess certain properties to minimize any adverse effects of the evaluation process on the forecasters (or forecasting systems) and to facilitate comparisons of performance over different samples of forecasts. Several desirable properties of evaluation measures are described in this section.

First, it is necessary to distinguish between an evaluation measure and a scoring rule. An evaluation measure is a mathematical function of the forecasts and/or the observations. This function assigns numerical values or scores to the forecasts, and a monotonic relationship is assumed to exist between these scores and the goodness of the forecasts as characterized by one or more specific attributes. Evaluation measures may or may not be defined for individual forecasts or observations. Scoring rules are a special class of evaluation measures. Specifically, a scoring rule is a mathematical function of both the forecasts and the observations. Moreover, scoring rules are defined for individual forecasts and observations. Thus, all scoring rules are evaluation measures, but the converse is not true.

4.1 Proper and Consistent Measures

It is a general principle of evaluation that the measures used to evaluate forecasts should not influence forecasters (or forecasting systems) to make forecast statements that differ from their judgments. Specifically, the measures should encourage forecasters to make their forecasts correspond exactly with their judgments by rewarding such behavior with the best possible (expected) scores. Since judgments are expressed in probabilistic terms (see Section 3.2), complete correspondence between forecasts and judgments is possible only when forecasts are expressed in probabilistic terms. Evaluation measures that encourage complete correspondence between judgments and forecasts are referred to as "proper" scoring rules and they are defined only for probability forecasts. With regard to the latter, the fact that categorical forecasts necessarily differ from judgments implies that the concept of proper measures is not meaningful in the context of categorical forecasting.

Consider a situation in which probability forecasts are made for a discrete variable consisting of N events in the presence of a scoring rule S with a positive orientation. An evaluation measure is said to have a positive (negative) orientation if a larger (smaller) score is better (see Section 4.2). Let the vector $\underline{p} = (p_1, \ldots, p_N)$ ($p_n \geq 0$, $\sum_n p_n = 1$; $n = 1, \ldots, N$) denote the forecaster's judgment concerning the likelihood of occurrence of these events. Further, let $S_j(\underline{r})$ denote the score assigned (by S) to the forecast $\underline{r}$ when the jth event occurs and let $ES(\underline{r},\underline{p})$ denote the forecaster's subjective expected score, where

$$ES(\underline{r},\underline{p}) = \sum_{j=1}^{N} p_j S_j(\underline{r}).$$ (1)

The scoring rule S is said to be strictly proper if $ES(\underline{p},\underline{p}) > ES(\underline{r},\underline{p})$ for all $r \neq p$ and proper if $ES(\underline{p},\underline{p}) \geq ES(\underline{r},\underline{p})$ for all r. That is, S is strictly proper if the forecaster can maximize his expected score only by making the forecast $\underline{r}$ correspond exactly with his judgment $\underline{p}$. If hedging can be said to occur when $\underline{r} \neq \underline{p}$, then strictly proper scoring rules discourage hedging. A proper scoring rule, on the other hand, yields the same maximum expected score for at least some forecasts for which $r \neq p$ as it does for the forecast $\underline{r} = \underline{p}$. Proper scoring rules do not encourage hedging, but they do not completely discourage it either. A measure that is neither strictly proper nor proper is improper. Such measures encourage hedging in that a forecaster can only maximize his expected score by making some forecast for which $r \neq p$. If the measure S has a negative orientation, then the inequalities between expected scores must be reversed, since the forecaster wants to minimize his (subjective) expected score in this case. Analogous definitions of strictly proper, proper, and improper measures can be formulated for situations in which the underlying variable is continuous and the judgments and forecasts are expressed in the form of continuous probability distributions.

Although the concept of proper scoring rules is not meaningful for categorical forecasts, it is possible to identify a less stringent condition that measures used to evaluate such forecasts should satisfy. This condition relates to the manner in which forecasters' probabilistic judgments are translated into categorical forecasts. It will be assumed here that the forecasters receive a "directive" concerning the procedure to be followed in the translation process and that it is desirable to choose an evaluation measure that is consistent with this process. An example may help to illustrate this concept. Consider a continuous ordinal predictand, and suppose that the directive states "forecast the expected (or mean) value of the variable." In this situation, the mean square error measure would be an appropriate scoring rule, since it is minimized by forecasting the mean of the (judgmental) probability distribution. Measures that correspond with a directive in this sense will be referred to as <u>consistent</u> scoring rules (for that directive).

Whenever possible, it is obviously desirable to use (strictly) proper scoring rules to evaluate probability forecasts and consistent measures to evaluate categorical forecasts. Thus, particular attention here will be focused on proper and consistent measures.

4.2 Scale of Goodness

A measure used to evaluate a sample of forecasts (and the corresponding observations) is assumed to yield a specific numerical value or score. Generally this value will be a finite real number. Since the measure should be an indicator of goodness with respect to a specific attribute of the forecasts, a monotonic relationship should exist between the score assigned by the measure and this aspect of goodness. The existence of such a relationship defines a scale of goodness and implies that the end points of this scale represent the "best" and "worst" possible forecasts. However, for evaluation measures in general, it frequently is not possible to uniquely identify the forecasts (or observations) associated with the end points of the scale. That is, more than one forecast (in some cases, many forecasts) may lead to the best and worst scores. However, if we are concerned only with scoring rules, then it generally is possible to identify uniquely at least the best possible forecast. For example, in the case of probability forecasts for a nominal predictand ($N > 2$), the best forecast is the forecast for which $r_j = 1$, and this forecast will (uniquely) yield the best score. On the other hand, the $N-1$ forecasts for which some $r_n = 1$ ($n = 1, \ldots, N; n \neq j$) may <u>all</u> yield the worst possible score.

Reference forecasting systems are frequently used to define additional points on the scale of goodness. Such systems play an important role in forecast evaluation as standards of comparison, and their use in formulating relative measures will be discussed in Section 4.3. It is sufficient here to note that the score associated with a reference system generally subdivides the range of possible scores into two intervals, an interval containing scores that are better than the score for the reference system and an interval containing scores that are worse than the score for the reference system. The interval between the score associated with the best possible forecast and the score associated with the reference system is usually referred to as the range of "skillful" forecasts. Since primary attention naturally is focused on the identification of situations in which forecasters and/or forecasting systems can improve upon the reference system, the range of scores from the best forecasts to the reference forecasts - the range of positive skill - is of particular importance here.

It would be desirable, in general, for the scores associated with the best forecasts and the reference forecasts to be fixed constants that could be determined prior to the start of an evaluation program. However, this condition cannot always be satisfied since these scores may be sample dependent. That is, they may depend on the observations in the sample. In such cases, it is sufficient to require that these scores approach fixed constants as the sample size increases and the sample relative frequencies approach the long-term climatological probabilities.

For some purposes it may be useful to establish a standard range and orientation of scores. As indicated in Section 4.1, an evaluation measure, and the associated scores, are said to have a positive (negative) orientation if larger (smaller) scores are better. Frequently, the range from zero to one is taken as such a standard, with one representing the best possible score and zero representing the score associated with the reference system (or the worst possible score). Of course, if the best score and the reference (or worst) score are known constants, then it is relatively easy to transform the original measure into a measure whose scores possess the desired range and orientation.

4.3 Absolute and Relative Measures

It is important to distinguish between absolute and relative measures of goodness. An absolute measure is a function of the forecasts and/or observations in the sample and is concerned only with the quality of these forecasts. A relative measure, on the other hand, is a function of these sample quantities and the forecasts and/or observations associated with the reference system. Specifically, relative measures are concerned with the quality of the forecasts in the sample relative to the quality of forecasts produced by the reference system. Since forecast evaluation is fundamentally a comparative process (see Section 2), relative measures are of primary concern. Of course, as is indicated below, relative measures are generally defined in terms of the differences between, and/or ratios of, absolute measures.

Let V denote an evaluation measure. Then V is, by definition, an absolute measure. Relative measures can be formulated in terms of the measure V by comparing the score for the forecasts (or forecast system) of concern, V_{FS}, with the score for the corresponding forecasts produced by the reference system, V_{RS}. This comparison could be accomplished simply by computing the difference between V_{FS} and V_{RS}. However, it is useful to compare this difference with the difference between the score for perfect forecasts and the score for the reference system according to the measure V. Thus, if V_{PF} denotes the score for perfect forecasts, then a natural relative measure would be defined by the ratio I_V, where

$$I_V = (V_{FS} - V_{RS})/(V_{PF} - V_{RS}). \qquad (2)$$

The measure I_V represents the improvement in the quality of the forecasts of concern over forecasts produced by the reference system relative to the maximum possible improvement specified by the difference in quality between perfect forecasts and reference forecasts. When V is taken as a measure of accuracy, then the index I_V represents a "skill score" according to the usual definition of this expression. It is important to note that the formulation of a relative measure amounts to a transformation of the scale of the original measure into a measure with the best possible score equal to one and the score for reference forecasts equal to zero. If the quality of the forecasts according to V is poorer than the quality of the reference forecasts, then the index I_V will yield a negative score.

Although the index I_V, as defined in (2), is an obvious and natural relative measure, it possesses certain deficiencies. These deficiencies relate to the fact that the denominator in the expression for I_V is not constant from sample to sample. As a result, although the basic measure V is strictly proper and additive (see Section 4.4), the index I_V may not possess these properties. An index I_V', which is strictly proper and additive (if V possesses these properties) and is similar in form to I_V, can be formulated by replacing V_{PF} and V_{RS} by their climatological expectancies, W_{PF} and W_{RS}, respectively. Then,

$$I_V' = (V_{FS} - W_{RS})/(W_{PF} - W_{RS}). \qquad (3)$$

This form of relative measure will be used frequently in this chapter.

4.4 Additivity of Scores

Many evaluation measures are defined for individual forecasts and/or observations. Obviously, it is not appropriate to judge the performance of a forecaster or forecast system on the basis of a single forecast. Nevertheless, it would be desirable for such measures to possess the property that the score for a sample of forecasts is equal to the average of the scores for the individual forecasts. This principle will be referred to as the principle of additivity. It should be noted that the relative measures I_v' and I_v in Section 4.3 possess and do not possess this property, respectively.

4.5 Equivalence of Events

This principle applies only in situations involving nominal predictands with two events (N = 2). In such situations evaluation measures should not be affected by an interchange of the events. For example, if the events of concern are "rain" and "no rain," then the score assigned by the measure should remain unchanged if the "rain" forecasts and observations are labeled "no rain" and the "no rain" forecasts and observations are labeled "rain." Most well-known evaluation measures for two-event predictands meet this requirement (see Section 5).

5. SOME INFERENTIAL MEASURES FOR CATEGORICAL FORECASTS

This section is concerned with methods of evaluating categorical (i.e., nonprobabilistic) forecasts. The basic results of studies involving the evaluation of categorical forecasts can generally be summarized in terms of classification or verification tables, and these tables are described briefly in Section 5.1. Some absolute measures of the quality of categorical forecasts - where quality is characterized by the attributes defined in Section 3.3 - are presented in Section 5.2. Relative measures, or skill scores, based on these absolute measures are discussed in Section 5.3.

5.1 Classification Tables

Classification or verification tables provide detailed information concerning the quality of categorical forecasts. In fact, these tables generally include the ingredients needed to compute most evaluation measures. Moreover, classification tables may be more useful to many users of the forecasts than the scores provided by such measures. For example, many of these tables contain the data required to determine the relative frequencies of observed weather events given a particular forecast event, thereby providing information concerning the overall uncertainty inherent in the forecasts.

For nominal or discrete ordinal predictands, verification tables are usually referred to as contingency tables. Such tables contain the joint distribution of forecasts and observed events and can be presented in the form of frequencies or relative frequencies. An example of a symmetric contingency table, in which the forecasts and observed events are defined in an identical manner, is presented in frequency form in Table 1. Let V denote this contingency table (or matrix). Then, using the notation introduced in Section 3, the matrix V can be expressed as follows:

$$\underline{V} = \sum_{k=1}^{K} \underline{c}_k' \underline{d}_k, \qquad (4)$$

where a prime denotes a transpose and K is the sample size. A relative frequency table V^* can be obtained simply by dividing the elements of V by K (i.e., $\overline{V}^* = V/K$). In creating a contingency table, it is not necessary that the forecast and observed events be defined in exactly the same manner. An asymmetric contingency table involving qualitative and quantitative definitions of amount-of-sunshine events is contained in Table 2. This matrix also can be formulated from the definitions of the forecast and observed events using (4). Perhaps the most common contingency table is the symmetric "two-by-two" table involving identical definitions of binary forecast and observed events. An example of such a table is presented in Table 3 for a sample of gale/no gale forecasts (a gale is defined here as a wind speed equaling or exceeding Beaufort force 7).

For continuous predictands, classification tables generally take the form of one-dimensional tables that describe the distribution of errors without specific reference to the forecast and observed values of the predictand. An example of such a table is presented in Table 4. This representation is most appropriate when the predictand has a symmetric, unimodal distribution and when the distribution of errors is independent of the forecast value. However, these conditions may not be satisfied for many continuous predictands. Moreover, the significance of an error of a given magnitude may differ greatly depending on the forecast value. In such cases, it may be desirable to transform the variable to approximate symmetry prior to the determination of a (one-dimensional) classification table or to display the results in the form of a contingency table such as Table 1.

5.2 Absolute Measures of Attributes

5.2.1 <u>Accuracy</u>. For nominal predictands, accuracy is measured by comparing an individual forecast, denoted by the row vector $\underline{c} = (c_1, ..., c_N)$, and the corresponding observation, denoted by the row vector $\underline{d} = (d_1, ..., d_N)$. A forecast is considered to be accurate if the same event is forecast and observed; otherwise, it is

Table 1. Symmetric classification table with four events for a
predictand consisting of the number of stations (out of ten)
in The Netherlands expected to receive measurable
precipitation. Forecasts issued at 1100 LST for the period
0700-1900 LST the following day during the period 1972-1979.

		Observed number of stations				
		0	1-2	3-7	8-10	Total
	0	468	62	34	6	570
	1-2	426	194	159	67	846
Forecast number of stations	3-7	234	267	436	426	1363
	8-10	14	3	40	85	142
	Total	1142	526	669	584	2921

Table 2. Asymmetric classification table for public forecasts of the amount of sunshine in three states at De Bilt, The Netherlands. Forecasts issued daily at 1100 LST for the following day during the period 1972-1979.

		Observed amount of sunshine (%)					
		0	1-20	21-40	41-60	61-80	81-100	Total
Forecast amount of sunshine	Clear or sunny	26	61	75	123	218	191	694
	Partly cloudy with sunny periods	132	238	211	184	131	23	919
	Cloudy or overcast	497	407	210	124	54	14	1306
	Total	655	706	496	431	403	228	2919

Table 3. Symmetric classification table with two events for gale/no gale forecasts at De Bilt, The Netherlands. Forecasts issued daily at 0500 LST for the 12-hour period 0700-1900 LST during the period 1972-1979.

| | | Observed event | | |
		Gale	No gale	Total
Forecast event	Gale	254	153	407
	No gale	143	2372	2515
	Total	397	2525	2922

Table 4. Classification table describing the distribution of errors in categorical forecasts of maximum temperature (°C) at De Bilt, The Netherlands. Forecasts issued daily at 1100 LST for the next day during the period 1972-1979.

Temperature error (°C)

	< -5	-4	-3	-2	-1	0	1	2	3	4	> 5	Total
Frequency	92	105	149	305	447	612	518	366	163	95	69	2921
Relative frequency	0.03	0.04	0.05	0.10	0.15	0.21	0.18	0.13	0.06	0.03	0.02	1.00

considered to be inaccurate. Thus, if the jth event is subsequently observed, then the forecast is accurate when $c_j = 1$ and inaccurate when $c_j = 0$. It should be noted that more than one event can be forecast; that is, two or more components of the forecast vector $\underline{c}$ can be set equal to one.

The most common measure of accuracy is the "hit rate" (or fraction correct) H. For a sample of K forecasts, H is defined as follows:

$$H = (1/K) \sum_{k=1}^{K} \underline{c}_k \underline{d}_k' = \overline{\underline{c}\,\underline{d}'}. \tag{5}$$

The range of H is the closed unit interval [0,1], and it has a positive orientation. As an example of the computation of H, consider the sample of forecasts and observations contained in Table 1. This situation is represented by a symmetric contingency table involving four events (N = 4). The four possible forecasts and observations are (1,0,0,0), (0,1,0,0), (0,0,1,0), and (0,0,0,1). Here, the measure H has the value 0.405 (= 1183/2921).

The computation of the hit rate H, as defined in (5), for an asymmetric contingency table involving some multiple-event forecasts can be illustrated using the hypothetical data in Table 5. In this case, we have five possible forecasts - namely, (1,0,0), (1,1,0), (0,1,0), (0,1,1), and (0,0,1). The value of H for this table is then 0.740 (= 740/1000). Entries in Table 5 contributing to the hit rate are underlined.

The hit-rate measure H may be considered to be an overall measure of accuracy, in the sense that it is concerned with the accuracy of the forecasts over all N events. It is also possible to define measures of partial accuracy with respect to individual events. Two such measures can be identified; namely, a measure of accuracy with respect to the number of observations of the event and a measure of accuracy with respect to the number of forecasts of the event. For the nth event (n = 1, ..., N) these measures are denoted by H_n^O and H_n^F, respectively, and they can be defined as follows:

$$H_n^O = \sum_{k=1}^{K} c_{nk} d_{nk} / \sum_{k=1}^{K} d_{nk} \tag{6}$$

and

$$H_n^F = \sum_{k=1}^{K} c_{nk} d_{nk} / \sum_{k=1}^{K} c_{nk}. \tag{7}$$

Table 5. Classification table for a hypothetical sample of single-
and multiple-event forecasts of temperature in three
classes; namely, below normal (B), normal (N), and above
normal (A). Five different forecasts were possible; namely,
B, B or N, N, N or A, and A.

| | | Observed class | | | |
		B	N	A	Total
	B	<u>162</u>	59	39	260
	B/N	<u>82</u>	<u>62</u>	6	150
Forecast class	N	53	<u>101</u>	36	190
	A/N	25	<u>54</u>	<u>81</u>	160
	A	8	34	<u>198</u>	240
	Total	330	310	360	1000

The range of H_n^0 and H_n^F is the closed unit interval $[0,1]$, and they both have a positive orientation. These measures sometimes are referred to as post agreement and prefigurance, respectively. However, it is more satisfactory to reformulate the basic measures of partial accuracy in terms of the probability of detection (POD) and the false alarm rate (FAR). In this framework,

$$POD_n = H_n^0 \tag{8}$$

and

$$FAR_n = 1 - H_n^F. \tag{9}$$

Note that the orientation of FAR_n is negative. Although these measures can be defined for any event $(n = 1, \ldots, N)$, they are generally used in conjunction with the verification of forecasts of rare or extreme events (e.g., storms, tornadoes). As an example of the computation of POD_n and FAR_n, consider the forecasts of gales in Table 3 $(n = 1)$. In this case, $POD_1 = 0.640$ $(= 254/397)$ and $FAR_1 = 0.376$ $(= 153/407)$. Another example of the use of these measures might involve the computation of POD_n and FAR_n for the below normal temperature event (B) in Table 5 $(n = 1)$. In this situation, $POD_1 = 0.739$ $(= 244/330)$ and $FAR_1 = 0.405$ $(= 166/410)$.

For ordinal predictands, accuracy can be measured in terms of the difference, or distance, between the forecast value y and observed value z (see Section 3.3.1). Common measures of accuracy, for a sample of K forecasts, include the mean absolute error (MAE) and the mean square error (MSE), where

$$MAE = (1/K) \sum_{k=1}^{K} |y_k - z_k| \tag{10}$$

and

$$MSE = (1/K) \sum_{k=1}^{K} (y_k - z_k)^2, \tag{11}$$

respectively. The MAE and MSE measures are both nonnegative and possess a negative orientation. These measures are consistent with directives that indicate that the forecaster (or forecast system) should forecast the median and mean, respectively, of the underlying judgmental distribution. The MSE measure, unlike the MAE measure, depends in part on the variance of the errors in the forecasts (see below). Frequently the root mean square error (RMSE) measure is used instead of the MSE measure, in order to obtain a scoring rule that is expressed in the same units as the original variable (and as the

MAE). Of course, RMSE = $(MSE)^{1/2}$. As an example of the application of these scoring rules, consider the sample of temperature forecasts in Table 4. In order to perform the relevant calculations, we assume that all errors in the forecasts greater than 4°C are exactly equal to 5°C. In this case, MAE = 1.66°C, MSE = 4.68 (°C)^2, and RMSE = 2.16°C.

It is of interest to note that the MSE measure can be partitioned into two terms that describe other attributes of the forecasts. Specifically,

$$MSE = (\overline{y} - \overline{z})^2 + (1/K) \sum_{k=1}^{K} [(y_k - z_k) - (\overline{y} - \overline{z})]^2. \qquad (12)$$

The first term on the right-hand side (RHS) of (12) is the square of the mean error (ME) of the forecasts. This term is a measure of overall forecast reliability and is discussed in Section 5.2.2. The second term on the RHS of (12) is the variance of the errors in the forecasts. Thus, the MSE measure is the sum of the variance of the forecast errors and the square of the mean forecast error.

The measures described in the previous two paragraphs are particularly suitable when the forecasts (and observations) relate to a variable with a symmetric, unimodal distribution. In the case of a variable with a skewed distribution, it may be desirable to transform the forecast values before they are evaluated. Logarithmic or power transformations can generally be used to obtain more nearly symmetric distributions for variables such as precipitation amount, ceiling height, and visibility. The MAE, MSE, and RMSE measures then can be computed in terms of the transformed values of the forecasts and observations.

Of course, the function used to transform the values of the forecasts need not be analytic in form. For example, a simple transformation has been proposed by Vernon (1953) for discrete ordinal predictands. First, the predictand is divided into events, or classes, in such a way that the overall frequency distribution of the events is of the desired form. Then accuracy is measured in terms of the distance between events, and Vernon (1953) defines measures of this attribute based on absolute and squared distances.

5.2.2 Reliability (bias). As noted in Section 3.3.2, reliability is defined as the degree of correspondence between average forecasts and observations over a sample of forecasts. For nominal predictands, reliability involves a comparison of the mean forecast (or forecast relative frequency distribution) $\overline{c}$ and the mean observation (or observed relative frequency distribution) $\overline{d}$. Single-event categorical forecasts would be completely unbiased for event n if $\overline{c}_n = \overline{d}_n$ (n = 1, ..., N) and completely unbiased for all N events if $\overline{c} = \overline{d}$. Thus, the difference between $\bar{c}$ and $\bar{d}$ would be a natural measure of bias. If we denote this vector measure by $\underline{b}$, then

$$\underline{b} = \bar{c} - \bar{d} \qquad (13)$$

(the components in this vector measure are defined as the differences between the corresponding components in $\bar{c}$ and $\bar{d}$). For completely un-biased forecasts, $\underline{b} = 0$. If an overall scalar measure of bias is desired, then we could use B, where

$$B = \underline{b}\,\underline{b}'. \qquad (14)$$

B is nonnegative and equals zero for completely unbiased forecasts. As an example of the computation of the bias measures, consider the sample of forecasts and observations in Table 1. For these data, $\underline{b}$ = (-0.196, 0.110, 0.238, -0.151) and B = 0.130 [$\bar{c}$ = (0.195, 0.290, 0.467, 0.049), $\bar{d}$ = (0.391, 0.180, 0.229, 0.200)].

In the previous paragraph it was assumed implicitly that the categorical forecasts are single-event forecasts (i.e., that one and only one event is forecast on each occasion). If multiple-event fore-casts are permitted, then it is no longer possible for $\bar{c}$ and $\bar{d}$ to be equal. One way to resolve this problem is to "normalize" $\bar{c}$ by divid-ing it by the number of events actually forecast. If we denote the normalized vector by $\underline{c}^*$, then $\underline{c}^* = c/(c\ c')$. Then, we can determine the bias of the forecasts by replacing $\bar{c}$ with $\bar{c}^*$ in (13) and (14). As an example of the computation of these bias measures in a multiple-event situation, consider the forecasts and observations in Table 5. For these data, $\underline{b}$ = (0.005, 0.035, -0.040) and B = 0.003 [$\bar{c}^*$ = (0.335, 0.345, 0.320), $\bar{d}$ = (0.330, 0.310, 0.360)].

For ordinal predictands, bias can be measured in terms of the difference between the mean forecast and the mean observation. Since this mean (or systematic) error is denoted by ME (see Section 5.2.1),

$$ME = \bar{y} - \bar{z}. \qquad (15)$$

For the temperature forecasts in Table 4, ME = 0.027°C. As in the case of measures of accuracy for ordinal predictands (see Section 5.2.1), it may be desirable to transform the predictand prior to the evaluation of the forecasts of interest. If $\overline{f(y)}$ and $\overline{f(z)}$ represent the mean values of the forecasts and observations after application of a transformation defined by the function f, then a measure of bias analogous to ME in (15) can be defined simply as the difference between $\overline{f(y)}$ and $\overline{f(z)}$.

5.2.3 <u>Sharpness</u>. As noted in Section 3.3.3, sharpness is an attribute that relates only to the forecasts. Moreover, this attribute is
similar to accuracy in the sense that it can be defined for individual
forecasts. Finally, sharpness has meaning only for nominal predictands, since categorical forecasts of ordinal predictands are
perfectly sharp by definition.

In the case of nominal predictands, sharpness relates to the
number of events that are forecast. If only one event is forecast
(single-event forecasts), then sharpness is always perfect. A natural
measure of the average sharpness of a sample of forecasts is the
measure S, where

$$S = (1/K) \sum_{k=1}^{K} c_k c'_k. \tag{16}$$

The range of S is the closed interval [1,N-1] and this measure has a
negative orientation. As an example of the computation of S, consider
the forecasts and observations in Table 5. For these data, S = 1.31.

When a nominal predictand is ordered, categorical forecasts may
be expressed as intervals of values. Such intervals generally consist
of one or more basic intervals or events. In this case, the sharpness
of a sample of multiple-event forecasts can be measured in terms of
the average width of the intervals.

5.2.4 <u>Resolution</u>. The attribute resolution relates to the extent to
which different forecasts are followed by different observations (see
Section 3.3.4). As is the case with the attribute reliability, resolution is defined only for samples of forecasts and observations.
Finally, resolution has no meaning for (continuous) ordinal
predictands.

For a nominal predictand, a measure of resolution necessarily
involves a comparison between relative frequencies of occurrence of
events for each distinct forecast and the overall relative frequencies
of occurrence of the events. Moreover, resolution may differ from
event to event. To illustrate the latter, consider the hypothetical
sample of forecasts and observations depicted in Table 6. Note that
event E_1 occurs in 50% of the cases regardless of the event that is
forecast. Thus, the forecasts exhibit no resolution for event E_1. On
the other hand, this sample of forecasts shows considerable resolution
for the events E_2 and E_3 (in a different vein, the accuracy of the E_3
forecasts is quite poor). It is important to note that, in the definition of resolution, the forecast event serves only as a "label" to
identify a specific subsample of observations.

Using the notation introduced in Section 3.3.4, we let $\bar{d}^t$
represent the relative frequencies of the N events when the $\bar{t}$-th

Table 6. Classification table for a hypothetical sample of
          categorical forecasts for a three-event nominal predictand.

		Observed event			
		E_1	E_2	E_3	Total
	E_1	25	0	25	50
Forecast event	E_2	10	10	0	20
	E_3	15	15	0	30
	Total	50	25	25	100

distinct categorical forecast is made and let $\bar{d}$ represent the overall relative frequencies of these events. Then a natural measure of resolution is R, where

$$R = (1/K) \sum_{t=1}^{T} K^t (\bar{d}^t - \bar{d})(\bar{d}^t - \bar{d})', \qquad (17)$$

in which K^t is the number of forecasts (or observations) in the t-th subsample, T is the number of subsamples (T = N for single-event forecasts), and K is the overall sample size. The range of R is the closed interval $[0, 2\bar{d}(1 - \bar{d})']$ and this measure has a positive orientation. To illustrate the computation of this measure of resolution, consider the sample of forecasts presented in Table 5. In this case, $\bar{d}^1$ = (0.623, 0.227, 0.150), $\bar{d}^2$ = (0.547, 0.413, 0.040), $\bar{d}^3$ = (0.279, 0.532, 0.190), $\bar{d}^4$ = (0.156, 0.338, 0.506), $\bar{d}^5$ = (0.033, 0.142, 0.825), and $\bar{d}$ = (0.330, 0.310, 0.360). Thus, R = 0.163.

5.3 Relative Measures of Attributes: Skill Scores

A relative measure is a measure of the quality of the forecasts of interest relative to the quality of reference forecasts, where quality is described in terms of the attributes defined in Section 3.3. Several different types of reference forecasts may be employed in defining relative measures, including perfect forecasts, climatological forecasts, and persistence forecasts. Relative measures are generally used to determine the performance of the forecasts of interest with respect to the performance of the reference forecasts, and they may or may not be defined directly in terms of absolute measures.

Although relative measures can be formulated for several attributes, relative measures of accuracy are of particular importance. The latter frequently are referred to as skill scores. Such measures generally are defined in such a way that perfect forecasts would obtain a score of one and that forecasts considered to be completely lacking in skill (e.g., forecasts based on climatology, persistence, or chance) would obtain a score of zero. Skill scores for categorical forecasts of nominal and ordinal predictands are described in Sections 5.3.1 and 5.3.2, respectively. A brief discussion of relative measures of other attributes is included in Section 5.3.3.

5.3.1 Skill Scores for Nominal Predictands. Most skill scores for nominal predictands are based in part on the hit rate H [see (5)], an absolute measure of accuracy for categorical forecasts. However, the full "derivation" of a skill score from an absolute measure frequently presents some problems. In this regard, the reference forecasts may not be uniquely defined in some cases; for example, when the reference system under consideration is climatology and the directive (vis-a-vis the reference forecasts) indicates that an event should be forecast when its probability of occurrence is greater than the climatological probability. To avoid this problem, the hit rate of the reference system can be defined as the climatological expectancy of the hit rate, assuming that the forecasts are made at random. As a result, the reference hit rate for climatological forecasts, H_r, is equal to $\underline{c}\ \underline{\delta}'$, where $\underline{\delta} = (\delta_1, \ldots, \delta_N)$ ($\delta_n \geq 0$, $\Sigma\delta_n = 1$; $n = 1$, $\ldots$, N) represents the (long-term) climatological probabilities.

Kuipers' performance index (Hanssen and Kuipers, 1965). Kuipers' performance index, I_K, is defined as follows:

$$I_K = (\overline{\underline{c}\ \underline{d}'} - \overline{\underline{c}}\ \underline{\delta}')/(1 - \underline{\delta}\ \underline{\delta}').\tag{18}$$

The index I_K has a form similar to that of the relative measure I_v' [see Section 4.3 and (3)]. In this regard, it yields a value of one for perfect forecasts and a value of zero for random forecasts (assuming, in both cases, that $\overline{d} = \delta$). The performance index possesses all the desirable properties described in Section 4. In particular, it is consistent with the directive that the nth event should be forecast if the judgmental probability p_n is greater than the climatological probability δ_n (this directive may lead to multiple-event forecasts). As an illustration of the computation of the index I_K, consider the sample of forecasts in Table 5. In this case $\overline{c\ d'} = 0.740$, $\overline{c}\ \delta' = 0.437$, and $\underline{\delta}\ \underline{\delta}' = 0.333$ [assuming $\underline{\delta} = (1/3, 1/3, 1/3)$]; thus, $I_K = 0.455$.

Gringorten's skill score (Gringorten, 1967). Gringorten's skill score, I_G, is defined as follows:

$$I_G = (1/N) \sum_{n=1}^{N} [(\overline{c_n d_n} - \overline{c}_n\delta_n - \delta_n\overline{d}_n + \delta_n^2)/\delta_n(1 - \delta_n)].\tag{19}$$

As in the case of most relative measures, the skill score I_G attains a value of one for perfect forecasts (assuming $\overline{d} = \underline{\delta}$) and a value of zero for random forecasts. The Gringorten skill score possesses all the desirable properties described in Section 4, and it is consistent with the same directive as Kuipers' performance index. The difference between the skill score I_G and the performance index I_K relates to the weighting of the various events. Specifically, the former assigns larger weights to relatively infrequent events and smaller weights to relatively frequent events than the latter. Computation of the skill score I_G for the sample of forecasts in Table 5 yields a value of 0.455, identical to the value of the performance index I_K. The two scores are equal in this case because the (long-term) climatological probabilities of the events are assumed to be equal [i.e., $\underline{\delta}$ = (1/3, 1/3, 1/3)].

Heidke's skill score (Heidke, 1926). Heidke's skill score, I_H, is defined as follows:

$$I_H = (\overline{c\ d}' - \overline{c}\ \delta')/(1 - \overline{c}\ \delta'). \tag{20}$$

As in the case of I_K and I_G, the measure I_H assumes a value of one for perfect forecasts and a value of zero for random forecasts (assuming $\overline{d} = \delta$). Although Heidke's skill score undoubtedly is the most widely used relative measure, it lacks some of the desirable properties described in Section 4. In this regard, I_H is consistent with a directive that depends on previous forecasts. Such a directive could lead to a situation in which two forecasters, starting from identical judgments regarding future weather conditions, would issue different forecasts in order to maximize their expected scores. Moreover, since the denominator of I_H is not constant (from sample to sample), this skill score is not additive.

It is instructive to compare Heidke's skill score, I_H, and Kuipers' performance index, I_K. The numerators in the two measures, which represent the difference between the hit rate of the forecasts ($\overline{c\ d}'$) and the climatological expectancy of that hit rate ($\overline{c}\ \delta'$), are identical [cf. (18) and (20)]. However, the denominator of I_H also contains the latter term (i.e., $\overline{c}\ \delta'$), in this case subtracted from the hit rate of perfect forecasts. On the other hand, the second term in the denominator of such a skill score should represent the "pure" climatological expectancy of the hit rate of perfect forecasts. It would then equal $\overline{d}\ \delta'$, which would approach $\delta\ \overline{\delta}'$ in the long term [the latter appears in the denominator of I_K; see (18)]. Thus, Kuipers' performance index could be viewed as the proper formulation of a skill score based on Heidke's original concept.

As an illustration of the computation of I_H, consider the sample of forecasts in Table 5. For these data, $\overline{c\ d}'$ = 0.740 and $\overline{c}\ \delta'$ = 0.437. Thus, I_H = 0.538.

Peirce's success index (Peirce, 1884). Peirce's success index was originally designed only for two-event ($N = 2$) situations. However, for $N\ (\geq 2)$-event situations, it can be defined as follows:

$$I_P = (\overline{c\ d}' - \overline{c}\ \overline{d}')/(1 - \overline{d}\ \overline{d}'). \tag{21}$$

The value of the measure I_P is one for perfect forecasts and zero

for random forecasts (in effect, assuming that $\bar{d} = \delta$). As is
the case with Heidke's skill score, Peirce's success index is not
additive. Moreover, I_P is not based on long-term climatological
probabilities (i.e., on δ); thus, it should not be used unless these
probabilities are unavailable. In this regard, it should be noted
that the measure I_P is identical to the performance index I_K when
the sample relative frequencies $\bar{d}$ are replaced by the climatological
relative frequencies δ. Finally, since the difference between $\bar{d}$ and δ
is quite small for the sample of forecasts and observations presented
in Table 5, the value of Peirce's index for these data - namely,
$I_P = 0.460$ - differs very little from the value of Kuiper's index
($I_K = 0.455$).

Appleman's score (Appleman, 1960). Appleman's score for
single-event categorical forecasts can be defined as follows:

$$I_A = (\overline{c\ d'} - \delta_{max})/(1 - \delta_{max}), \qquad (22)$$

where δ_{max} is the climatological probability of the most frequent
event. The value of the index I_A is one for perfect forecasts and
zero for forecasts in which the most frequent event is predicted on
each occasion (assuming $\bar{d} = \delta$). With regard to the latter, Appleman's
score is consistent with a directive that specifies that the most
frequent event should always be forecast. Although this directive may
appear reasonable at first glance, it implies that rare events will
seldom be forecast, even in those situations in which the likelihood
of their occurrence is considerably greater than the long-term climatological probability. Moreover, the index I_A permits only one
event to be forecast on each occasion. As a consequence of these
limitations, Appleman's score should not be considered to be a
"primary" skill score. However, it possesses all the desirable properties discussed in Section 4. The value of I_A for the data presented in Table 3 is 0.320 (it has been assumed that $\delta_{max} = 0.850$).

Other skill scores. Many other skill scores, or measures of
relative accuracy, have been developed over the last one hundred
years. Such measures include the threat score (Gilbert, 1884),
Doolittle's score (Doolittle, 1885), Wallen's correlation score
(Wallen, 1921), Clayton's "rating system" (Clayton, 1927), and more
recently Rousseau's skill score (Rousseau, 1980). We do not believe
that such measures can be considered to be as satisfactory, in terms
of their properties and performance, as the skill scores described
previously in this section. In this regard, a disadvantage shared by
all these scoring rules is that they correspond to a directive involving forecasts issued prior to the occasion of concern. That is,
in the presence of such scoring rules, two forecasters with exactly
the same judgmental probability distribution might issue different

forecasts because they possessed disparate "forecasting histories."
For further details concerning these other measures, we refer the
reader to the references cited above.

5.3.2 Skill Scores for Ordinal Predictands. For ordinal predictands,
it usually is possible to derive skill scores directly from absolute
measures of accuracy. Such skill scores generally are expressed in a
form similar to the relative measures I_V or I_V' in (2) and (3) (see
Section 4.3). For example, skill scores (or indices) based on the MAE
and MSE measures can be defined as follows:

$$I_{MAE} = 1 - (MAE_{FS}/MAE_{RS}) \tag{23}$$

and

$$I_{MSE} = 1 - (MSE_{FS}/MSE_{RS}), \tag{24}$$

respectively. The values of I_{MAE} and I_{MSE} are one for perfect
forecasts ($MAE_{FS} = MSE_{FS} = 0$) and zero for forecasts whose accuracy is identical to that of the reference system ($MAE_{FS} = MAE_{RS}$
and $MSE_{FS} = MSE_{RS}$). In order to ensure that the skill scores
I_{MAE} and I_{MSE} possess the property of additivity, the reference
system should be based on long-term climatology rather than sample
climatology.

As noted in Section 5.2.1, measures based on the mean absolute
and mean square error measures are consistent with directives pre-
scribing forecasts of the median and mean, respectively, of the judgmental probability distribution. If this distribution is approximately symmetric, then the two directives will lead to similar forecasts. However, for judgmental distributions that are asymmetric,
these directives may lead to quite different forecasts. In a related
vein, an advantage of using a measure or skill score based on the mean
absolute error is that the directive is not affected by monotonic
transformations of the scale of the predictand. On the other hand, if
the judgmental distributions are approximately symmetric, then a skill
score based on the mean square error possesses the advantage that the
denominator in such a score represents the climatological variance of
the predictand. The latter frequently is a well-known quantity.
Finally, a skill score based on the root mean square error (RMSE)
measure can be defined by replacing MSE_{FS} and MSE_{RS} in (24) with
$RMSE_{FS}$ and $RMSE_{RS}$, respectively.

Other types of skill scores can be based on correlations between
forecast and observed values of the predictand. Such skill scores
include the anomaly correlation and the tendency correlation. The
former refers to the correlation between forecast and observed departures from the climatological mean value, whereas the latter refers

to the correlation between departures from the previous observed value of the predictand. Correlations are defined only for samples of forecasts; thus, skill scores based on correlations are not additive. Moreover, the zero point for these correlation scores may not correspond with the reference system. For example, a constant forecast of the climatological mean value will yield a tendency correlation appreciably above zero (although distance measures also may be deficient in this respect, the effect is much larger in the case of correlation scores than in the case of distance measures). However, methods are available to define reference forecasts in such a way that reference systems based on persistence or climatology will yield a zero correlation (see Daan, 1980, 1984).

Another difficulty with scores based on correlation coefficients is that the directive generally depends on previous forecasts. This condition arises here because correlation scores neglect systematic errors in both the forecasts themselves and the magnitude of deviations from reference forecasts. Moreover, when a sample of forecasts contains such errors, the forecaster or forecast system is encouraged to maintain these errors in order to maximize the score on each new occasion.

5.3.3 Relative Measures for Other Attributes. Relative measures cannot be defined for the attribute reliability, since a reference system based on observed frequencies produces (almost) perfect forecasts with respect to this attribute. In particular, climatological forecasts are biased only due to differences between the long-term and sample frequencies, and persistence forecasts are biased only by the difference between the first and last observations in the sample.

For the attribute sharpness, a relative measure could be defined by replacing the number of forecast events by the climatological frequency of these events (i.e., by $\bar{c}_{\delta}'$). The latter quantity played an important role in many of the skill scores discussed in Section 5.3.1.

In the case of the attribute resolution, a relative measure may be derived from the absolute measure [see (17), Section 5.2.4] by dividing the latter by the maximum possible value of this measure. However, the usefulness of such a relative measure of resolution for categorical forecasts is debatable.

6. SOME INFERENTIAL MEASURES FOR PROBABILITY FORECASTS

In this section we describe some methods and measures for evaluating probability forecasts of nominal and ordinal predictands. Frequency distributions of forecasts and/or observations are considered in Section 6.1, including reliability tables and diagrams. Absolute measures of the attributes defined in Section 3.3 are presented in Section 6.2, and relationships among measures of these attributes

are also discussed. Finally, relative measures of these attributes -
that is, skill scores - are described in Section 6.3.

6.1 Distributions of Forecasts and Observations

As indicated in Section 5.1, distributions of forecasts and
observations - individually and jointly - contain useful information
for evaluation purposes, and these distributions can be presented in
the form of either tables or diagrams. The frequency distribution of
forecasts and the relationship between forecast probabilities and
observed relative frequencies are of particular interest. In describing and illustrating these distributions and relationships, it is
convenient to consider the two-event (N = 2) and N-event (N > 2)
situations separately.

As an example of distributions of forecasts and observations in
the two-event (N = 2) situation, we will consider a sample of subjective precipitation probability forecasts initially described in Chapter 9 (see Section 4). A tabular summary of these forecasts (and the
corresponding observations) by subsample is presented in Table 7,
where each subsample consists of all of the forecasts in the sample
with the same set of probabilities. Specifically, the t-th subsample
consists of K^t forecasts (and observations), and it is "represented"
here by the t-th distinct forecast vector $\underline{r}^t = (r_1^t, r_2^t)$ and the
corresponding relative frequency vector $\underline{\overline{d}}^t = (\overline{d}_1^t, \overline{d}_2^t)$ ($\Sigma K^t = K$;
t = 1, ..., T). The values of K^t, $\underline{r}^t$, $\underline{\overline{d}}^t$, K, and T for the
sample of precipitation probability forecasts are indicated in Table
7. Frequently, such a tabular summary is presented using only the
probability of precipitation r^t (= r_1^t = 1 - r_2^t) and the relative
frequency of precipitation $\overline{d}^t$ (= $\overline{d}_1^t$ = 1 - $\overline{d}_2^t$). In Table 7 we use
the vectors $\underline{r}^t$ and $\underline{\overline{d}}^t$ to emphasize the fact that precipitation
probability forecasts, in reality, are two-event forecasts.

The values of K^t (t = 1, ..., 12) in Table 7 indicate the frequency with which the different sets of probabilities (or forecasts)
were used in this sample. Here, more than 75% of the forecasts were
associated with probabilities of the event "precipitation" equal to or
less than 0.20. It should be noted that this distribution is an indicator of the sharpness of the sample of forecasts. Comparison of the
probability forecasts and the observed relative frequencies (i.e.,
$\underline{r}^t$ and $\underline{\overline{d}}^t$; t = 1, ..., 12) provides an indication of the reliability of the forecasts. In this case, the correspondence between
these probabilities and relative frequencies is quite close.

Table 7. Tabular summary by subsample of subjective precipitation probability forecasts formulated by U.S. National Weather Service (NWS) forecasters during the 1980-1981 cool season. See text for additional details.

Subsample number t	Probability forecast r^t	Number of forecasts K^t	Observed relative frequency $\bar{d}^t$
1	(0.00, 1.00)	5,100	(0.006, 0.994)
2	(0.05, 0.95)	832	(0.019, 0.981)
3	(0.10, 0.90)	2,273	(0.059, 0.941)
4	(0.20, 0.80)	1,223	(0.150, 0.850)
5	(0.30, 0.70)	764	(0.277, 0.723)
6	(0.40, 0.60)	454	(0.377, 0.623)
7	(0.50, 0.50)	376	(0.511, 0.489)
8	(0.60, 0.40)	341	(0.587, 0.413)
9	(0.70, 0.30)	303	(0.723, 0.277)
10	(0.80, 0.20)	273	(0.799, 0.201)
11	(0.90, 0.10)	211	(0.934, 0.066)
12 (= T)	(1.00, 0.00)	252	(0.933, 0.067)
Total/average	(0.177,0.823)(= $\bar{r}$)	12,402(= K)	(0.162, 0.838)(= $\bar{d}$)

The correspondence between forecast probabilities and observed relative frequencies, as well as the frequency distribution of the forecasts (i.e., the distribution of K^t), can be presented in graphical form, and these characteristics of the sample of precipitation probability forecasts are depicted in Figure 1. This figure contains a reliability diagram, in which observed relative frequency is plotted against forecast probability (i.e., $\bar{d}^t$ versus r^t when N = 2). The diagonal 45° line in this diagram represents perfect reliability, in the sense that forecast probability exactly equals observed relative frequency. The corresponding frequency distribution of forecasts also is depicted in Figure 1. Since the sampling variability inherent in a point on the reliability "curve" is inversely related to the corresponding subsample size, the frequency distribution of forecasts is an important adjunct to the respective reliability diagram.

The hypothetical sample of probability forecasts for a three-event predictand summarized in Table 8 will be used to illustrate some aspects of the distributions of forecasts and observations in N-event (N > 2) situations. This sample is stratified by subsample in the same manner as the sample of precipitation probability forecasts presented in Table 7. First, it should be noted that the number of distinct forecasts, T, in N > 2 situations generally greatly exceeds the number of distinct probability values, S say (unlike N = 2 situations in which T = S). Murphy (1972b) presents a formula for the computation of T as a function of S and N (under some assumptions concerning the "uniformity" of the set of probability values). As in the case of precipitation probability forecasts, the values of K^t (t = 1, ..., 15) in Table 8 represent the frequency distribution of the distinct three-event forecasts. In addition, comparison of r^t and $\bar{d}^t$ (t = 1, ..., 12) provides an indication of the reliability of this hypothetical sample of probability forecasts.

A graphical representation of the correspondence between r^t and $\bar{d}^t$ - and the distribution of K^t - for this sample of three-event forecasts is presented within the framework of an equilateral triangle in Figure 2. The points in this diagram are defined by a barycentric coordinate system, in which the forecast $r^t = (r_1^t, r_2^t, r_3^t)$ is located at a distance r_n^t from the side of the triangle opposite the vertex for which $d_n^t = 1$ (n = 1, 2, 3) (see Murphy, 1972b). The reliability of the subsamples of forecasts is indicated here by the lengths of the line segments joining the points r^t and $\bar{d}^t$ (t = 1, ..., 12). The reliability of the t-th subsample of forecasts is perfect if and only if $r^t = \bar{d}^t$ (i.e., if and only if the line segment is of zero length). In this three-event situation, the distribution of forecasts is indicated by entering the value of K^t next to the line segment joining the points r^t and $\bar{d}^t$. The geometrical description of forecasts and observations employed in Figure 2 can be

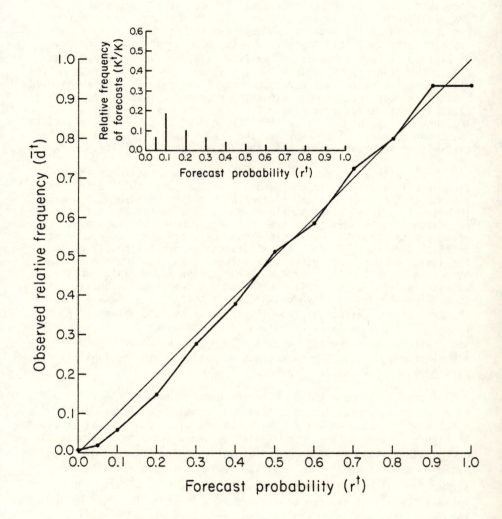

Figure 1. Reliability diagram for subjective precipitation probability forecasts formulated by NWS forecasters during the 1980-1981 cool season. Inset represents the relative frequency of use of various probability values. See Table 7 for a tabular summary of these data and see text for additional details.

Table 8. Tabular summary of hypothetical sample of probability forecasts for three-event predictand. Summary organized by subsample for each distinct forecast $\underline{r}^t$ (t = 1, ..., 18).

Subsample number t	Probability forecast $\underline{r}^t$	Number of forecasts K^t	Observed relative frequency $\underline{d}^t$
1	(0.9,0.1,0.0)	12	(0.750,0.167,0.083)
2	(0.7,0.2,0.1)	10	(0.600,0.300,0.100)
3	(0.6,0.3,0.1)	7	(0.429,0.429,0.143)
4	(0.6,0.1,0.3)	2	(0.500,0.000,0.500)
5	(0.5,0.4,0.1)	5	(0.400,0.400,0.200)
6	(0.4,0.5,0.1)	3	(0.333,0.333,0.333)
7	(0.4,0.6,0.0)	4	(0.250,0.750,0.000)
8	(0.3,0.6,0.1)	9	(0.111,0.778,0.111)
9	(0.3,0.5,0.2)	5	(0.400,0.600,0.000)
10	(0.2,0.7,0.1)	8	(0.250,0.625,0.125)
11	(0.2,0.5,0.3)	2	(0.000,0.000,1.000)
12	(0.2,0.3,0.5)	6	(0.167,0.167,0.667)
13	(0.2,0.1,0.7)	1	(0.000,0.000,1.000)
14	(0.1,0.8,0.1)	5	(0.200,0.600,0.200)
15	(0.1,0.6,0.3)	7	(0.000,0.714,0.286)
16	(0.0,0.8,0.2)	4	(0.250,0.500,0.250)
17	(0.0,0.6,0.4)	3	(0.000,0.667,0.333)
18 (= T)	(0.0,0.1,0.9)	7	(0.143,0.143,0.714)
Total/average	(0.373,0.417,0.210)(= $\bar{r}$)	100(= K)	(0.320,0.440,0.240)(= $\bar{d}$)

418

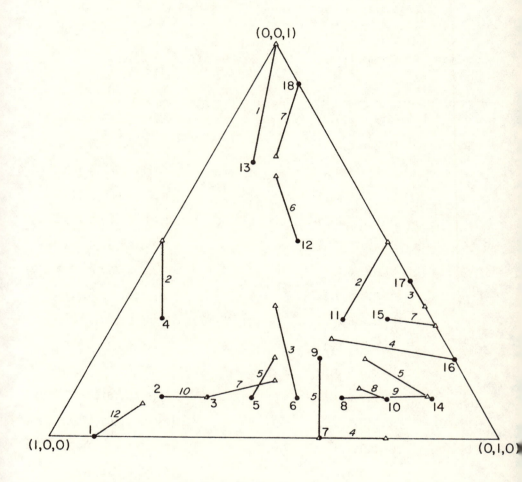

Figure 2. Reliability diagram for hypothetical sample of three-event probability forecasts. Line segments connect forecast probabilities (r^t, denoted by dots) and corresponding observed relative frequencies ($\overline{d}^t$, denoted by triangles) for the 18 subsamples. The subsample number t is entered next to each distinct forecast r^t (t = 1, ..., 18). The numbers adjacent to the line segments indicate the subsample sizes (K^t). See Table 8 for a tabular summary of these data and see text for additional details.

generalized to the N-event situation. The appropriate framework is a (N-1)-dimensional regular simplex (e.g., a regular tetrahedron when N = 4) with a barycentric coordinate system. It is of interest to note that this simplex is the unit line segment when N = 2. Thus, such a line segment provides an alternative framework to the "standard" reliability diagram (see Figure 1) for the representation of forecasts and observations in the two-event situation.

In the two previous paragraphs, we treated the three-event predictand as a nominal variable. If instead it is considered to be an ordinal variable, then the appropriate description of the distribution of data in Table 8 would involve <u>cumulative</u> forecasts and observations (see Section 3.3.1), and these da̲ta̲ are summarized in terms of such forecasts and observations in Table 9. It also is possible to represent cumulative forecasts and observations in geometrical terms, using a right isosceles triangle with cartesian coordinates R_1^t and R_2^t such that $R_1^t \leq R_2^t$ (see Murphy and Staël von Holstein, 1975).

In this section the forecasts and observations of interest have been treated as vectors [i.e., as sets of probabilities (in the case of the forecast vector), in which a unique correspondence is maintained between individual probabilities and specific events]. As an alternative to this vector approach, each component of the vector r_k (k = 1, ..., K) can be treated as a separate scalar probability forecast. In the scalar approach, the sample of K vector forecasts and observations is "transformed" into a sample of N x K scalar forecasts and observations. This sample can be divided into subsamples, each of which consists of all of the scalar forecasts (and observations) corresponding to a distinct probability value (see Murphy, 1972a,b). Then, the distributions of (scalar) forecasts and observations can be presented in tabular and/or graphical form (in this approach, the standard reliability diagram will suffice regardless of the number of events). We prefer the vector approach to the scalar approach because the latter fails to maintain the unique correspondence between forecast probabilities and events.

6.2 Absolute Measures of Attributes

6.2.1 <u>Accuracy</u>. In the case of a probability forecast for a nominal predictand, accuracy is measured by comparing the forecast, denoted by the row vector $\underline{r} = (r_1, ..., r_N)$ $(r_n \geq 0, \sum_n r_n = 1; n = 1, ..., N)$, and the corresponding observation, denoted by the row vector $\underline{d} = (d_1, ..., d_N)$ $(d_n = 1$ if the nth event occurs and $d_n = 0$ otherwise; n = 1, ..., N). Such a forecast is considered to be completely accurate if $r_j = 1$ (assuming $d_j = 1$; see Section 3.3.1) and to be completely inaccurate if $r_n = 1$ (n = 1, ..., N; n ≠ j). Probability forecasts for which $r_n < 1$ for all n can be said to be "partially accurate" (or "partially inaccurate").

Table 9. Tabular summary of hypothetical sample of data in Table 8 by cumulative forecasts and observations. Summary organized by subsample for each distinct cumulative forecast $\underline{R}^t$ (t = 1, ..., 18).

Subsample number t	Cumulative probability forecast $\underline{R}^t$	Number of forecasts K^t	Observed relative frequency $\underline{D}^t$
1	(0.9,1.0,1.0)	12	(0.750,0.917,1.000)
2	(0.7,0.9,1.0)	10	(0.600,0.900,1.000)
3	(0.6,9.0,1.0)	7	(0.429,0.857,1.000)
4	(0.6,0.7,1.0)	2	(0.500,0.500,1.000)
5	(0.5,0.9,1.0)	5	(0.400,0.800,1.000)
6	(0.4,0.9,1.0)	3	(0.333,0.667,1.000)
7	(0.4,1.0,1.0)	4	(0.250,1.000,1.000)
8	(0.3,0.9,1.0)	9	(0.111,0.889,1.000)
9	(0.3,0.8,1.0)	5	(0.400,1.000,1.000)
10	(0.2,0.9,1.0)	8	(0.250,0.875,1.000)
11	(0.2,0.7,1.0)	2	(0.000,0.000,1.000)
12	(0.2,0.5,1.0)	6	(0.167,0.333,1.000)
13	(0.2,0.3,1.0)	1	(0.000,0.000,1.000)
14	(0.1,0.9,1.0)	5	(0.200,0.800,1.000)
15	(0.1,0.7,1.0)	7	(0.000,0.714,1.000)
16	(0.0,0.8,1.0)	4	(0.250,0.750,1.000)
17	(0.0,0.6,1.0)	3	(0.000,0.667,1.000)
18 (= T)	(0.0,0.1,1.0)	7	(0.143,0.286,1.000)
Total/average	(0.373,0.790,1.000)(= $\underline{\bar{R}}$)	100(= K)	(0.320,0.760,1.000)(= $\underline{\bar{D}}$)

The most common measure of accuracy is the Brier, or probability, score (BS) (Brier, 1950). For a sample of K probability forecasts, the BS is defined as follows:

$$BS = (1/K) \sum_{k=1}^{K} (\underline{r}_k - \underline{d}_k)(\underline{r}_k - \underline{d}_k)'. \tag{25}$$

The range of BS is the closed interval [0,2] and it has a negative orientation. Moreover, BS is a strictly proper scoring rule (Murphy and Epstein, 1967b; Winkler and Murphy, 1968). Finally, it also should be noted that the BS is the mean square error of forecasts expressed in probabilistic terms. As an example of the computation of BS, consider the sample of subjective precipitation probability forecasts presented in Table 7. For this sample, BS = 0.146. Brier scores for other samples of operational objective and subjective precipitation probability forecasts are reported in Tables 2 and 5 in Chapter 9 [the Brier scores reported in these tables are one-half of the original BS in (25)]. As another example, the hypothetical sample of three-event forecasts in Table 8 yields a BS of 0.545.

It is of interest to note that the BS is a quadratic scoring rule and that this measure frequently is referred to as the "quadratic score" by nonmeteorologists. Moreover, the BS is usually written in a slightly different form in these other contexts. If we let QR denote this form of the quadratic scoring rule, then it often is expressed as follows:

$$QR = 1 - (1/2) \sum_{k=1}^{K} (\underline{r}_k - \underline{d}_k)(\underline{r}_k - \underline{d}_k)'. \tag{26}$$

Thus, the measure QR is a linear transformation of the BS and, as such, it is a strictly proper scoring rule (Winkler and Murphy, 1968). The range of QR is the closed unit interval [0,1], and it has a positive orientation.

Several other measures of the accuracy of probability forecasts have been formulated, including the logarithmic scoring rule (LR) and the spherical scoring rule (SR). For a sample of K forecasts, these measures can be defined as follows:

$$LR = (1/K) \sum_{k=1}^{K} \ln \underline{r}_k \, \underline{d}_k' \tag{27}$$

and

$$SR = (1/K) \sum_{k=1}^{K} \underline{r}_k \, \underline{d}_k' / (\underline{r}_k \, \underline{r}_k')^{1/2} \tag{28}$$

(e.g., see Winkler and Murphy, 1968). The range of LR is the half-open interval $[-\infty, 0]$ and the range of SR is the closed unit interval $[0,1]$. These measures have positive orientations, and they are both strictly proper scoring rules. Other properties of LR and SR are also of interest. Examination of (28) reveals that LR (unlike QR and SR) depends only on the probability assigned to the event that subsequently occurs. In this sense, the measure LR represents a "partial" measure of the attribute accuracy. Moreover, LR assumes a value of minus infinity when the probability assigned to the event that subsequently occurs is zero (i.e., when $r_j = 0$). This situation, which may present practical problems in the computation of LR for some samples of forecasts, can be alleviated by replacing the zero probabilities by a small positive probability (e.g., 0.005). Despite these apparent deficiencies, the measure LR possesses many desirable properties (e.g., see Winkler, 1969), and it is used frequently by non-meteorologists to evaluate probability forecasts. On the other hand, the spherical score SR has received relatively little attention or use in meteorology or in other fields.

In the case of probability forecasts for ordinal predictands, accuracy is defined in terms of the difference between cumulative forecasts and observations (see Section 3.3.1). A natural measure of the accuracy of such forecasts is the ranked probability score (RPS), originally formulated by Epstein (1969). For a sample of K probability forecasts, the RPS can be expressed as follows:

$$RPS = (1/K) \sum_{k=1}^{K} (\underline{R}_k - \underline{D}_k)(\underline{R}_k - \underline{D}_k)', \qquad (29)$$

where $\underline{R}_k = (R_{k1}, \ldots, R_{kN})$, in which $R_{kn} = \sum_{m=1}^{n} r_{km}$, and $\underline{D}_k = (D_{k1}, \ldots, D_{kN})$, in which $D_{kn} = \sum_{m=1}^{n} d_{km}$ (Murphy, 1971). The range of the RPS is the closed interval $[0, N-1]$ and it has a negative orientation. This measure is also a strictly proper scoring rule (Murphy, 1969). Moreover, since the vectors $\underline{R}_k$ and $\underline{D}_k$ ($k = 1, \ldots, K$) represent cumulative forecasts and observations, the RPS is the mean square error of such forecasts. Finally, it is of interest to note that the RPS represents the sum of the N-1 one-half Brier scores for two-event forecasts associated with the N-1 possible dichotomous representations of the predictand.

As examples of the computation of the RPS, consider the samples of probability forecasts in Tables 7 and 8. First, it should be noted that the RPS is equivalent (i.e., linearly related) to the BS in two-event (N=2) situations. Specifically, RPS = (1/2)BS when N = 2. Thus, the RPS for the sample of probability forecasts in Table 7 is

0.073. In the case of the three-event forecasts in Table 8, we have expressed these forecasts in terms of cumulative forecast probabilities and observed relative frequencies in Table 9. The RPS for these forecasts is 0.327.

Staël von Holstein and Murphy (1978) described a family of quadratic scoring rules that contains both weighted and unweighted measures of accuracy. For a sample of K probability forecasts, this general quadratic scoring rule GQR can be expressed as follows:

$$GQR = (1/K) \sum_{k=1}^{K} (\underline{r}_k - \underline{d}_k)\underline{C}(\underline{r}_k - \underline{d}_k)', \qquad (30)$$

in which $\underline{C}$ is a $N \times N$ matrix of weights. The values of GQR range from zero for perfect forecasts to a maximum value that depends upon the numerical values of the elements of the matrix $\underline{C}$, and it has a negative orientation. If the matrix C is positive definite, then GQR is strictly proper. The fact that GQR represents an entire family of quadratic scoring rules allows the evaluator considerable flexibility in tailoring the choice of $\underline{C}$ to the particular problem at hand. It is of interest to note that the measures BS and RPS are special cases of GQR (see Staël von Holstein and Murphy, 1978).

6.2.2 Reliability. The reliability in-the-large (or global bias) of probability forecasts for nominal predictands is concerned with the correspondence between the average forecast $\overline{r}$ and the average observation, or observed relative frequency vector, $\overline{d}$ (see Section 3.3.2). This correspondence could be assessed using a variety of measures. We propose a measure GB, where

$$GB = (\overline{r} - \overline{d})(\overline{r} - \overline{d})'. \qquad (31)$$

Note that GB = 0 for forecasts that are completely reliable in-the-large (i.e., $\overline{r}_n = \overline{d}_n$ for all n) and GB > 0 for all other forecasts, and it obviously has a negative orientation. An analogous measure of the reliability in-the-large of probability forecasts for ordinal predictands can be defined by replacing $\overline{r}$ and $\overline{d}$ in (31) with $\overline{R}$ and $\overline{D}$, respectively.

Recall that the reliability in-the-small of probability forecasts for nominal predictands involves the correspondence between the T distinct probability forecasts r^t and the respective observed relative frequencies $\overline{d}^t$ (t = 1, ..., T) (see Section 3.3.2). Although the correspondence between these vectors could be measured in a variety of different ways, a natural measure of reliability in this context is the quantity REL (Sanders, 1963; Murphy, 1972a,b), where

$$REL = (1/K) \sum_{t=1}^{T} K^t (\underline{r}^t - \overline{d}^t)(\underline{r}^t - \overline{d}^t)'. \qquad (32)$$

Note that REL = 0 for forecasts that are completely reliable in-the-small (i.e., $\underline{r}^t = \overline{\underline{d}}^t$ for all t) and REL > 0 for all other forecasts, and thus it has a negative orientation. It also should be noted that GB = 0 when REL = 0, whereas the converse relationship does not hold. The measure REL is simply the average squared euclidean distance between the respective forecast probability and observed relative frequency vectors (when N = 2, REL is the average squared distance between the points on the reliability curve and the corresponding points on the 45° line; see Figure 1). For additional motivation for the choice of REL in (32) as a measure of the reliability in-the-small of probability forecasts, refer to Section 6.2.5. An analogous measure of the reliability in-the-small of probability forecasts for ordinal predictands can be formulated by replacing $\underline{r}^t$ and $\overline{\underline{d}}^t$ in (32) with $\underline{R}^t$ and $\overline{\underline{D}}^t$, respectively.

6.2.3. Sharpness. The sharpness of probability forecasts for nominal predictands relates to the correspondence between the forecast $\underline{r}_k$ and the closest observation $\underline{d}_k^*$ (see Section 3.3.3). As in the case of reliability measures, many different measures of sharpness could be defined. We propose a measure SHP for the average sharpness of a sample of K probability forecasts $\underline{r}_k$ (k = 1, ..., K), where

$$SHP = (1/K) \sum_{k=1}^{K} (\underline{r}_k - \underline{d}_k^*)(\underline{r}_k - \underline{d}_k^*)', \tag{33}$$

in which $\underline{d}^*$ is the observation vector $\underline{d}_j$ (j = 1, ..., N) that minimizes the quantity $(\underline{r}_k - \underline{d}_j)(\underline{r}_k - \underline{d}_j)'$. Note that SHP = 0 for completely sharp forecasts ($r_{kn} = 1$ for some n for all k) and SHP > 0 for all other forecasts, and it has a negative orientation. It also should be noted that the use of the measure SHP implies that "closeness" is measured in terms of (squared) euclidean distance. An analogous measure of the average sharpness of probability forecasts for ordinal predictands can be formulated by replacing r_k and $\underline{d}_k^*$ in (33) with $\underline{R}_k$ and $\underline{D}_k^*$, respectively.

6.2.4 Resolution. The resolution of probability forecasts for nominal predictands relates to the correspondence between relative frequencies of occurrence of events in subsamples of observations associated with distinct forecasts, $\overline{\underline{d}}^t$ (t = 1, ..., T), and the overall observed relative frequencies for the entire sample of observations, $\overline{\underline{d}}$ (see Section 3.3.4). This correspondence could be measured in several different ways. Here, we propose a measure RES (Murphy, 1973), where

$$RES = (1/K) \sum_{t=1}^{T} K^t (\overline{\underline{d}}^t - \overline{\underline{d}})(\overline{\underline{d}}^t - \overline{\underline{d}})' \tag{34}$$

[the expression for RES in (34) is identical to the expression for R in (17); thus, the proposed measures for the resolution of categorical and probabilistic forecasts are identical]. Note that RES = 0 when $\overline{d}^t = \overline{d}$ for all t and RES > 0 when $\overline{d}^t \neq \overline{d}$ for some t (t = 1, ..., T), and thus it has a positive orientation (i.e., larger values of RES indicate greater resolution). The measure RES is simply the average squared euclidean distance between the subsample relative frequency vectors $\overline{d}^t$ (t = 1, ..., T) and the sample relative frequency vector $\overline{d}$ (when N = 2, RES is the average squared distance between the points on the reliability curve and the corresponding points on the line $\overline{d}^t = \overline{d}$; see Figure 1). For additional motivation for the choice of RES in (34) as a measure of resolution for probability forecasts of nominal predictands, refer to Section 6.2.5. An analogous measure of the resolution of probability forecasts for ordinal predictands can be formulated by replacing $\overline{d}^t$ and $\overline{d}$ in (34) with $\overline{D}^t$ and $\overline{D}$, respectively.

Another measure of resolution of forecasts for nominal predictands has been developed by Sanders (1963). If we denote this measure by RES', then

$$RES' = (1/K) \sum_{t=1}^{T} K^t \underline{d}^t (\underline{1} - \underline{d}^t)',$$ (35)

in which $\underline{1} = (1, ..., 1)$. Note that RES' = 0 when $\overline{d}_n^t = 1$ for some n for all t, and it has a negative orientation. A measure of the resolution of probability forecasts for ordinal variables analogous to RES' in (35) could be formulated by replacing $\underline{d}^t$ by $\underline{D}^t$.

6.2.5 Relationships among Measures. Relationships exist among some of the measures described in Sections 6.2.1-6.2.4, and such relationships are discussed briefly in this section. The relationships considered here are based on the fact that it is possible to partition quadratic scoring rules such as the BS and RPS into two or more terms, under the assumption that only a finite number of probability values are used in the forecasts (or that the range of probabilities is divided into a finite set of intervals represented by "typical" values). We shall focus on a particular class of partitions - the so-called vector partitions. Using the notation introduced in Section 3.3.2, the three-term vector partition of the BS can be expressed as follows:

$$BS = \overline{d} (\underline{1} - \overline{d})' + (1/K) \sum_{t=1}^{T} K^t (\underline{r}^t - \overline{d}^t)(\underline{r}^t - \overline{d}^t)'$$

$$- (1/K) \sum_{t=1}^{T} K^t (\overline{d}^t - \overline{d})(\overline{d}^t - \overline{d})'$$ (36)

(Murphy, 1973). The first term on the RHS of (36) has two interpretations: (1) the variance of the observations in the sample; and (2) the Brier score for a constant forecast of the sample climatological probabilities $\overline{d}$ (it should be noted that this term is not a function of the forecasts). We adopt the latter interpretation and denote this term by BS_{SC}. Then, noting that the second and third terms on the RHS of (36) are REL and RES, respectively [cf. (32) and (34)], BS in (36) can be rewritten as follows:

$$BS = BS_{SC} + REL - RES. \tag{37}$$

Thus, the Brier score for a sample of forecasts equals the Brier score for constant forecasts based on the sample climatology plus the reliability measure REL minus the resolution measure RES.

The two-term vector partition of the BS, originally formulated by Sanders (1963), can be expressed as follows:

$$BS = (1/K) \sum_{t=1}^{T} K^t (\underline{r}^t - \overline{\underline{d}}^t)(\underline{r}^t - \overline{\underline{d}}^t)' + (1/K) \sum_{t=1}^{T} K^t \overline{\underline{d}}^t (\underline{1} - \overline{\underline{d}}^t)'. \tag{38}$$

From (32) and (35), BS in (38) can be rewritten as follows:

$$BS = REL + RES'. \tag{39}$$

Thus, BS is also the sum of the reliability measure REL and the resolution measure RES'.

Note that (37) and (39) imply that

$$BS_{SC} = RES + RES'. \tag{40}$$

Thus, the sum of the two resolution terms equals the Brier score for forecasts based on the sample climatological probabilities.

Vector partitions of the RPS analogous to the two-term and three-term partitions of the BS can also be formulated (see Murphy, 1973). Moreover, it is possible to derive scalar partitions of the BS and the RPS (e.g., see Murphy, 1972a,b) - that is, partitions based on the treatment of individual probabilities as distinct forecasts (see Section 3.3.2). Unlike the vector partitions, the scalar partitions do not maintain a unique correspondence between the probabilities and the events (see Section 6.1).

6.3 Relative Measures of Attributes: Skill Scores

As indicated in Section 5.3, a relative measure is a measure of the quality of the forecasts of interest relative to the quality of reference forecasts. When the forecasts of interest are probability forecasts, the reference forecasts frequently are based on climatological probabilities (although standards of reference based on other types of probability forecasts may be more appropriate in some situations; see Section 2). The latter can be either the long-term climatological probabilities (δ) or the sample relative frequencies ($\overline{d}$). Use of the former generally is preferred, since the relative measure then will be strictly proper (assuming that the absolute measure on which it is based possesses this property) and additive (see Sections 4.3 and 4.4). However, the climatological probabilities are not always available; moreover, use of sample relative frequencies may be quite convenient and mathematically attractive in some cases (for an example of such a situation, see below).

Attention here will be focused on relative measures of accuracy (i.e., on skill scores). These skill scores generally are expressed in the form of a ratio similar to the relative measures I_V and I'_V in Section 4.3. Such skill scores can be formulated in terms of any of the absolute measures of accuracy described in Section 6.2.1. For example, a skill score for nominal variables based on the Brier score can be written as follows:

$$I_{BS} = (BS_C - BS)/BS_C = 1 - (BS/BS_C),\qquad(41)$$

in which

$$BS_C = \underline{\delta}\,(\underline{1} - \underline{\delta})' = 1 - \underline{\delta}\,\underline{\delta}'\qquad(42)$$

is the Brier score for constant forecasts based on the climatological probabilities. I_{BS} equals one for perfect forecasts ($BS = 0$), and it equals zero for climatological forecasts under the assumption that $\overline{d} = \underline{\delta}$ (in which case, $BS = BS_C$). Similar skill scores can be defined using the logarithmic score (LR) and the spherical score (SR). An analogous skill score for ordinal variables could be formulated in terms of the ranked probability score (RPS).

Another useful skill score for nominal variables based on the Brier score, but employing sample relative frequencies, can be developed using the three-term partition of BS in (37). Thus,

$$I^*_{BS} = (BS_{SC} - BS)/BS_{SC} = (RES - REL)/BS_{SC}.\qquad(43)$$

This "sample skill score" is positive (negative) when the magnitude of the resolution term is greater (less) than the magnitude of the reliability term. An analogous sample skill score for ordinal variables can be formulated by employing the three-term partition of the RPS.

7. SOME RELATED TOPICS

In the previous sections we have discussed some basic principles and concepts involved in the formulation and selection of evaluation measures (Sections 2-4) and described some suitable measures for evaluating categorical and probabilistic forecasts (Sections 5 and 6, respectively). As noted in Section 1, it was not the purpose of the chapter to provide a comprehensive review of evaluation methods or to discuss in detail either the practice of forecast evaluation or the results forthcoming from evaluation programs. Nevertheless, in this concluding section, we will briefly (a) consider some current issues and activities in forecast evaluation; (b) discuss some aspects of decision-theoretic evaluation (we had initially planned to devote an entire section of the chapter to this topic, but the section was omitted to conserve space); (c) identify some sources of information regarding the results of evaluation programs; and (d) describe some activities related to forecast evaluation in fields other than meteorology. Emphasis here will be placed on providing the interested reader with a representative set of references to relevant material in each of these areas.

The material in Sections 5 and 6 indicates that a variety of measures and procedures are currently available to evaluate categorical and probabilistic forecasts, and that many of these measures possess desirable properties. Despite the availability of such a relatively rich collection of procedures, a need still exists for evaluation measures that are more sensitive to differences in the characteristics of forecasts produced by alternative forecasting systems and that more accurately reflect the "impact" of such differences on selected groups of users of the forecasts. For example, better procedures are needed to evaluate forecasts of relatively rare but frequently damaging events such as heavy precipitation or tornadoes (the problem of evaluating forecasts of rare events would be simplified to a considerable degree if the forecasts were expressed in probabilistic rather than categorical terms). It would also be helpful to identify additional criteria that would allow forecasters (or evaluators) to choose appropriate scoring rules on a rational basis. At present, several measures frequently satisfy all of the existing criteria (see Section 4), and it is generally necessary to make relatively arbitrary decisions concerning the particular measure to use. Another issue of considerable importance is the formulation of suitable reference systems to serve as standards of comparison. The information available to potential users of weather forecasts (in the absence of the forecasts themselves) is generally not limited to simple persistence or climatology, and reference systems should account for this state of knowledge whenever possible. Finally, it clearly would be desirable to develop sound procedures to test differences in scores from a statistical point of view. Relatively little attention has been devoted to problems of inference in forecast evaluation in the meteorological literature; a noteworthy exception to this statement is the paper by Gruza and Radyukhin (1980). Other publications of interest with respect to these issues include works by Woodcock (1976), Mason (1982), and Daan (1984). Relatively recent collections of papers concerning forecast evaluation in meteorology appear in publications by the National Center for Atmospheric Research (1976) and the World Meteorological Organization (1980).

As noted in Section 1, this chapter has focused almost exclusively on inferential evaluation. However, decision-theoretic evaluation is also important, and meteorologists have undertaken a variety of studies related to the value of their forecasts to users. Much of the work in this area has been concerned with the development of measures of the monetary value of forecasts in simple, prototype decision-making problems such as the cost-loss ratio situation (see Chapter 13 for a description of this situation). Work of this type is exemplified by the papers of Bijvoet and Bleeker (1951), Thompson (1952), Thompson and Brier (1955), Murphy (1977), and Stuart (1982). In a related vein, several studies have identified relationships between familiar measures of the quality of forecasts and their monetary value in specific situations and/or for particular users. Investigations illustrating work of this type include papers by Murphy (1966, 1972c), Epstein (1969), and Daan (1982b). All of the above-mentioned studies have involved static (i.e., "one-shot") decision-making problems and have employed an ex post approach, in which the economic value is determined after the forecasts and associated observations become available. Recently, several studies of the ex ante value of forecasts have been undertaken in both static and dynamic (i.e., repetitive) decision-making problems (e.g., see Katz et al., 1982; Winkler et al., 1983). The ex ante approach is consistent with Bayesian or decision-analytic methods of modeling such problems and of assessing the value of information (see Chapter 13). Notwithstanding the usefulness of decision-analytic evaluation measures developed in conjunction with these studies, a need exists for additional measures of this type that are applicable to various classes of important decision-making problems. Moreover, decision-theoretic evaluation studies have focused almost exclusively on the attribute "monetary value," and this attribute may not accurately reflect the preferences of the users of weather forecasts in many situations. Utility is a more appropriate attribute than monetary value in this sense (see Chapter 13 for an introduction to the concept of utility), and measures based on this attribute should be employed in evaluation studies whenever possible. In this regard, Winkler and Murphy (1970) conducted a preliminary study of the impact of nonlinear utility functions on the Brier or quadratic score.

Operational weather forecasts are routinely evaluated in many countries. Unfortunately, relatively few papers summarizing the results of these evaluation programs have appeared in the published literature, although unpublished results are available in a few countries in the form of internal publications of the respective national weather services. In the United States, for example, Zurndorfer et al. (1979) summarized the results of objective and subjective weather forecasting programs, and Charba and Klein (1980) described the state of the art of precipitation forecasting in the National Weather Service (NWS) through the 1970's. Other very interesting and useful summaries of the quality of NWS forecasts appear in the form of occasional or regular technical memoranda (e.g., Polger, 1983) and office notes (e.g., Carter et al., 1983). In a different vein, several papers have reported the results of experimental but long-term forecasting activities conducted in conjunction with synoptic laboratory programs in U.S. universities (e.g., Sanders, 1973, 1979; Bosart, 1975). With regard to evaluation programs in other countries, some aspects of the quality of weather forecasts in The Netherlands have been reported recently by Daan (1982a) and Daan and Murphy (1982).

Interest in methods of evaluating forecasts is not restricted to the meteorological community. Forecasts are now prepared on a regular or quasi-regular basis in many fields (e.g., business, economics, management science, medicine), and forecasters and others in these fields are devoting increasing attention to the problems of developing and applying more sensitive and sophisticated methods of evaluating their forecasts. Thus, although very few in-depth treatments of evaluation methods had appeared in these fields prior to 1970, a rapidly growing body of literature on this topic exists today. Relatively early discussions of evaluation measures appeared in works by Goodman and Kruskal (1954, 1959), Thiel (1965, 1966), and Zarnowitz (1967) for categorical forecasts and in publications by de Finetti (1965), Winkler (1967), Raiffa (1969), and Stael von Holstein (1970) for probability forecasts (the latter three works contain references to early work on scoring rules for probability forecasts). The classic paper by Savage (1971) represented a milestone in the development of scoring rules, in that a general framework was presented for the formulation of "proper" scoring rules and it was demonstrated that all such scoring rules represent decision-theoretic evaluation measures with respect to some user of the forecasts. Decision-theoretic evaluation has also been investigated by Nelson and Winter (1964), Pearl (1978), and Krzysztofowicz (1983) among others. Recent treatments of the evaluation problem include the work of Matheson and Winkler (1976), Granger and Newbold (1977), Meeden (1979), Haim (1982), Yates (1982), and Winkler (1984). A particularly interesting and potentially useful set of papers appeared recently in the medical literature (Habbema et al., 1978; Hilden et al., 1978a,b; Habbema and Hilden, 1981; Habbema et al., 1981). Meteorologists genuinely concerned with developing new and improved evaluation procedures should acquaint themselves with the literature on this topic in these other fields.

ACKNOWLEDGMENT

The preparation of this chapter was supported in part by the National Science Foundation (Atmospheric Sciences Division) under grants ATM-8004680 and ATM-8209713.

REFERENCES

Appleman, H. S., 1960: A fallacy in the use of skill scores. Bulletin of the American Meteorological Society, 41, 64-67.

Bijvoet, H. C., and W. Bleeker, 1951: The value of weather forecasts. Weather, 6, 36-39.

Bleeker, W., 1946: The verification of weather forecasts. De Bilt, Royal Netherlands Meteorological Institute, Mededeelingen en Verhandelingen, Serie B, Deel 1, No. 2, 23 pp.

Bosart, L. F., 1975: SUNYA experimental results in forecasting daily temperature and precipitation. Monthly Weather Review, 78, 1-3.

Brier, G. W., 1950: Verification of forecasts expressed in terms of probability. Monthly Weather Review, 78, 1-3.

Brier, G. W., and R. A. Allen, 1951: Verification of weather forecasts. Compendium of Meteorology (T. F. Malone, Editor). Boston, MA, American Meteorological Society, pp. 841-848.

Bross, I. D. J., 1953: Design for Decision. New York, NY, MacMillan Co., 276 pp.

Carter, G. M., J. P. Dallavalle, G. W. Hollenbaugh. G. J. Maglaras, and G. E. Schwartz, 1983: Comparative verification of guidance and local aviation/public weather forecasts - No. 15 (October 1982 - March 1983). Silver Spring, MD, NOAA, National Weather Service, TDL Office Note 83-16, 76 pp.

Charba, J. P., and W. H. Klein, 1980: Skill in precipitation forecasting in the National Weather Service. Bulletin of the American Meteorological Society, 61, 1546-1555.

Clayton, H. H., 1927: A method of verifying weather forecasts. Bulletin of the American Meteorological Society, 8, 144-146.

Daan, H., 1980: Climatology and persistence as reference forecasts in verification scores. Preprints of WMO Symposium on Probabilistic and Statistical Methods in Weather Forecasting (Nice). Geneva, Switzerland, World Meteorological Organization, pp. 195-201.

Daan, H., 1982a: The practical use of weather forecast verification figures. Preprints, Ninth Conference on Weather Forecasting and Analysis (Seattle). Boston, MA, American Meteorological Society, pp. 166-168.

Daan, H., 1982b: Verification scores as measures of value for a "standard user." Preprints, Ninth Conference on Weather Forecasting and Analysis (Seattle). Boston, MA, American Meteorological Society, pp. 175-177.

Daan, H., 1984: Scoring rules in forecast verfication. De Bilt, The Netherlands, Royal Netherlands Meteorological Institute, manuscript, 60 pp.

Daan, H., and A. H. Murphy, 1982: Subjective probability forecasting in The Netherlands: some operational and experimental results. Meteorologische Rundschau, 35, 99-112.

Dobryshman, E. M., 1972: Review of forecast verification techniques. Geneva, Switzerland, World Meteorological Organization, Technical Note No. 120, 51 pp.

Doolittle, M. H., 1885: The verification of predictions. American Meteorological Journal, 2, 327-329.

Epstein, E. S., 1969: A scoring system for probabilities of ranked categories. Journal of Applied Meteorology, 8, 985-987.

de Finetti, B., 1965: Methods for discriminating levels of partial knowledge concerning a test item. The British Journal of Mathematical and Statistical Psychology, 18, 87-123.

Gilbert, G. K., 1884: Finley's tornado predictions. American Meteorological Journal, 1, 166-172.

Goodman, L. A., and W. H. Kruskal, 1954: Measures of association for cross classifications. Journal of the American Statistical Association, 49, 732-764.

Goodman, L. A., and W. H. Kruskal, 1959: Measures of association for cross classifications. II. Further discussion and references. Journal of the American Statistical Assoication, 54, 123-163.

Granger, C. W. J., and P. Newbold, 1977: Forecasting Economic Time Series. New York, NY, Academic Press, 333 pp.

Gringorten, I. I., 1951: The verification and scoring of weather forecasts. Journal of the American Statistical Association, 46, 279-296.

Gringorten, I. I., 1967: Verification to determine and measure
    forecasting skill. Journal of Applied Meteorology, 6, 742-747.

Gruza, G. V., and V. T. Radyukhin, 1980: Evaluation of probabilistic
    predictions and statistical inference. WMO Symposium on
    Probabilistic and Statistical Methods in Weather Forecasting
    (Nice). Geneva, Switzerland, World Meteorological Organization,
    pp. 211-218.

Habbema, J. D. F., and J. Hilden, 1981: The measurement of per-
    formance in probabilistic diagnosis. IV. Utility
    considerations in therapeutics and prognostics. Methods of
    Information in Medicine, 20, 80-96.

Habbema, J. D. F., J. Hilden, and B. Bjerregaard, 1978: The
    measurement of performance in probabilistic diagnosis. I. The
    problem, descriptive tools, and measures based on classification
    matrices. Methods of Information in Medicine, 17, 217-226.

Habbema, J. D. F., J. Hilden, and B. Bjerregaard, 1981: The
    measurement of performance in probabilistic diagnosis. V.
    General recommendations. Methods of Information in Medicine,
    20, 97-100.

Haim, E., 1982: Characterization and construction of proper scoring
    rules. Berkeley, University of California, Department of
    Industrial Engineering and Operations Research, Ph.D.
    Dissertation, 52 pp.

Hanssen, A. W., and W. J. A. Kuipers, 1965: On the relationship
    between the frequency of rain and various meteorological
    parameters. De Bilt, Royal Netherlands Meteorological
    Institute, Mededeelingen en Verhandelingen, No. 81, 65 pp.

Heidke, P., 1926: Berechnung des erfolges und der güte der
    windstarkevorhersagen im sturmwarnungsdienst. Geografiska
    Annaler, 8, 301-349.

Hilden, J., J. D. F. Habbema, and B. Bjerregaard, 1978a: The
    measurement of performance in probabilistic diagnosis. II.
    Trustworthiness of the exact values of the diagnostic
    probabilities. Methods of Information in Medicine, 17, 227-237.

Hilden, J., J. D. F. Habbema, and B. Bjerregaard, 1978b: The
    measurement of performance in probabilistic diagnosis. III.
    Methods based on continuous functions of the diagnostic
    probabilities. Methods of Information in Medicine, 17, 238-246.

Johnson, D. H., 1957: Forecast verification: a critical survey of
    the literature. London, England, Air Ministry, Meteorological
    Research Committee, M.R.P. 1056, S.C. 11/237, 40 pp.

434

Katz, R. W., A. H. Murphy, and R. L. Winkler, 1982: Assessing the value of frost forecasts to orchardists: a dynamic decision-making approach. Journal of Applied Meteorology, 21, 518-531.

Köppen, W., 1884: Eine rationelle methode zur prufung der wetter-prognosen. Meteorologische Zeitschrift, 1, 39-40 and 397-404.

Krzysztofowicz, R., 1983: Why should a forecaster and a decision maker use Bayes' theorem. Water Resources Research, 19, 327-336.

Mason, I. B., 1982: A model for assessment of weather forecasts. Australian Meteorological Magazine, 30, 291-303.

Matheson, J. E., and R. L. Winkler, 1976: Scoring rules for con-tinuous probability distributions. Management Science, 22, 1087-1096.

Meeden, G., 1979: Comparing two probability appraisers. Journal of the American Statistical Association, 74, 299-302.

Meglis, A. J., 1960: Annotated bibliography on forecast verification. Meteorological and Geoastrophysical Abstracts and Bibliography, 11, 1129-1174.

Muller, R. H., 1944: Verification of short-range weather forecasts (a survey of the literature). Bulletin of the American Meteorological Society, 25, 18-27, 47-53, and 88-95.

Murphy, A. H., 1966: A note on the utility of probabilistic predictions and the probability score in the cost-loss ratio decision situation. Journal of Applied Meteorology, 5, 534-537.

Murphy, A. H., 1969: On the "ranked probability score." Journal of Applied Meteorology, 8, 988-989.

Murphy, A. H., 1971: A note on the ranked probability score. Journal of Applied Meteorology, 10, 155-156.

Murphy, A. H., 1972a: Scalar and vector partitions of the probability score: Part I. Two-state situation. Journal of Applied Meteorology, 11, 273-282.

Murphy, A. H., 1972b: Scalar and vector partitions of the probability score: Part II. N-state situation. Journal of Applied Meteorology, 11, 1183-1192.

Murphy, A. H., 1972c: Ordinal relationships between measures of the "accuracy" and "value" of probability forecasts: preliminary results. Tellus, 24, 531-542.

Murphy, A. H., 1973: A new vector partition of the probability score. Journal of Applied Meteorology, 12, 595-600.

Murphy, A. H., 1976: Evaluation of probabilistic forecasts: some procedures and practices. Weather Forecasting and Weather Forecasts: Models, Systems, and Users - Volume 2 (A. H. Murphy and D. L. Williamson, Editors). Boulder, CO, National Center for Atmospheric Research, pp. 807-830.

Murphy, A. H., 1977: The value of climatological, categorical and probabilistic forecasts in the cost-loss ratio situation. Monthly Weather Review, 105, 803-816.

Murphy, A. H., and R. A. Allen, 1970: Probabilistic prediction in meteorology: a bibliographty. Silver Spring, MD, ESSA, Weather Bureau Technical Memorandum WBTM TDL 35, 60 pp.

Murphy, A. H., and E. S. Epstein, 1967a: Verification of probabilistic predictions: a brief review. Journal of Applied Meteorology, 6, 748-755.

Murphy, A. H., and E. S. Epstein, 1967b: A note on probability forecasts and "hedging." Journal of Applied Meteorology, 6, 1002-1004.

Murphy, A. H., and C.-A. S. Staël von Holstein, 1975: A geometrical framework for the ranked probability score. Monthly Weather Review, 103, 16-20.

National Center for Atmospheric Research, 1976: Weather Forecasting and Weather Forecasts: Models, Systems, and Users - Volume 2 (A. H. Murphy and D. L. Williamson, Editors). Boulder, CO, National Center for Atmospheric Research, pp. 699-899.

Nelson, R. R., and S. G. Winter, 1964: A case study of the economics of information and coordination: the weather forecasting system. Quarterly Journal of Economics, 78, 420-441.

Pearl, J., 1978: An economic basis for certain methods of evaluating probabilistic forecasts. International Journal of Man-Machine Studies, 10, 175-183.

Peirce, C. S., 1884: The numerical measure of the success of predictions. Science, 4, 453-454.

Polger, P. D., 1983: National Weather Service public forecast verification summary April 1978 to March 1982. Silver Spring, MD, NOAA, National Weather Service, Technical Memorandum NWS FCST 28, 112 pp.

Raiffa, H. 1969: Assessment of probabilities. Cambridge, MA, Harvard University, School of Business, unpublished paper, 23 pp.

Rousseau, D., 1980: A new skill score for the evaluation of yes/no forecasts. Symposium on Probabilistic and Statistical Methods in Weather Forecasting (Nice). Geneva, Switzerland, World Meteorological Oraganization, pp. 167-174.

Sanders, F., 1963: On subjective probability forecasting. Journal of Applied Meteorology, 2, 191-201.

Sanders, F., 1973: Skill in forecasting daily temperature and precipitation: some experimental results. Bulletin of the American Meteorological Society, 54, 1171-1179.

Sanders, F., 1979: Trends in skill of daily forecasts of temperature and precipitation, 1966-78. Bulletin of the American Meteorological Society, 60, 763-769.

Savage, L. J., 1971: Elicitation of personal probabilities and expectations. Journal of the American Statistical Association, 66, 783-801.

Staël von Holstein, C.-A. S., 1970: Assessment and Evaluation of Subjective Probability Distributions. Stockholm, Sweden, Stockholm School of Economics, Economic Research Institute, 225 pp.

Staël von Holstein, C.-A. S., and A. H. Murphy, 1978: The family of quadratic scoring rules. Monthly Weather Review, 106, 917-924.

Stuart, A., 1982: On the economic value of probability of precipitation forecasts in Canada. Journal of Applied Meteorology, 21, 495-498.

Thiel, H., 1965: Economic Forecasts and Policy. Amsterdam, The Netherlands, North-Holland Publishing Co., 567 pp.

Thiel, H., 1966: Applied Economic Forecasting. Amsterdam, The Netherlands, North-Holland Publishing Co., 474 pp.

Thompson, J. C., 1952: On the operational deficiencies in categorical weather forecasts. Bulletin of the American Meteorological Society, 33, 223-226.

Thompson, J. C., and G. W. Brier, 1955: The economic utility of weather forecasts. Monthly Weather Review, 83, 249-254.

Vernon, E. M., 1953: A new concept of skill score for rating quantitative forecasts. Monthly Weather Review, 81, 326-329.

Wallen, A., 1921: Sur le controle des annonces de tempetes. Geografiska Annaler, 3, 267-277.

Winkler, R. L., 1967: The quantification of judgment: some method-ological suggestions. Journal of the American Statistical Association, 62, 1105-1120.

Winkler, R. L., 1969: Scoring rules and the evaluation of probabil-ity assessors. Journal of the American Statistical Association, 64, 1073-1078.

Winkler, R. L., 1984: On "good probability appraisers." In Bayesian Inference and Decision Techniques with Applications: Essays in Honor of Bruno de Finetti (P. K. Goel and A. Zellner, Editors). Amsterdam, The Netherlands, North-Holland Publishing Co., in press.

Winkler, R. L., and A. H. Murphy, 1968: "Good" probability assessors. Journal of Applied Meteorology, 7, 751-758.

Winkler, R. L., and A. H. Murphy, 1970: Nonlinear utility and the probability score. Journal of Applied Meteorology, 9, 143-148.

Winkler, R. L., A. H. Murphy, and R. W. Katz, 1983: The value of climate information: a decision-analytic approach. Journal of Climatology, 3, 187-197.

Woodcock, F., 1976: The evaluation of yes/no forecasts for scientific and administrative purposes. Monthly Weather Review, 104, 1209-1214.

World Meteorological Organization, 1980: Symposium on Probabilistic and Statistical Methods in Weather Forecasting (Nice). Geneva, Switzerland, WMO, 498 pp. (see pp. 145-234).

Yates, J. F., 1982: External correspondence: decompositions of the mean probability score. Organizational Behavior and Human Performance, 30, 132-156.

Zarnowitz, V., 1967: An Appraisal of Short-Term Economic Forecasts. New York, NY, National Bureau of Economic Research, Occasional Paper 104, 144 pp.

Zurndorfer, E. A., J. R. Bocchieri, G. M. Carter, J. P. Dallavalle, D. B. Gilhousen, K. F. Hebenstreit, and D. J. Vercelli, 1979: Trends in comparative verification scores for guidance and local aviation/public weather forecasts. Monthly Weather Review, 107, 799-811.

11
Design and Evaluation of Weather Modification Experiments

Paul W. Mielke, Jr.

1. INTRODUCTION

Weather modification experiments are based on the comparison of treated (seeded) and nontreated (control) values which are sequentially dealt out according to an unknown meteorological scheme. For example, the meteorological scheme associated (a) with wintertime orographic experiments is often a sequence of specifically defined temporal periods (e.g., 3-hour, 6-hour, or 24-hour intervals) and (b) with summertime cumulus experiments is often a sequence of cumulus clouds which are both appropriate and available. As a consequence, weather modification experiments are more closely related to comparative surveys (i.e., sample surveys involving temporally changing populations) rather than to comparative experiments where any number of meaningful replications of a given experiment are easily implemented (perhaps even simultaneously). Because of meteorological variation over time, replications of a weather modification experiment are usually time consuming to implement and then questionable when obtained (regardless of the problems, replications of a weather modification experiment are essential).

Because of the many perplexing problems associated with weather modification experiments, a major effort has occurred and is continuing in the development of statistical procedures which are appropriate for evaluating data from such experiments (National Academy of Sciences, 1973; Brier, 1974; Brillinger et al., 1978; Braham, 1979). In fact, an entire volume (LeCam and Neyman, 1967) of the Proceedings of the Fifth Berkeley Symposium on Mathematical Statistics and Probability is solely devoted to research topics in weather modification. High speed computing equipment was not available during the earlier stages of weather modification experimentation. Thus, with the exception of a few simple nonparametric tests, analyses of previous weather modification experiments were usually based on distribution bound inference techniques; for example, likelihood ratio tests,

C(α) tests (National Academy of Sciences, 1973, pp. 216-222; Neyman and Scott, 1965), and classical tests based on the normal distribution such as Student's t test and analysis of variance. An excellent review of the distribution-bound inference techniques used in evaluating weather modification experiments is given by Moran (1970). Since the availability of high speed computing equipment is no longer a constraint, the presently recommended analyses for evaluating weather modification experiments are based on distribution-free inference techniques (e.g., rerandomization, nonparametric tests, and permutation tests). Recent discussion and motivation for using distribution-free inference techniques are given by Gabriel (1979) and Brillinger et al. (1978).

The goal of this chapter is to describe concepts which are applicable for current weather modification experiments. Section 2 describes and discusses commonly employed experimental designs used in weather modification experiments. In particular, specific details such as experimental units, randomization of treatments, double blind treatment and placebo allocations, and some consequences of partitioning before and after an experiment are included in Section 2. A discussion of selected analysis techniques for evaluating weather modification experiments is given in Section 3. While the emphasis of Section 3 is placed on distribution-free techniques, brief comments of issues regarding distribution-bound inference techniques are also included.

2. EXPERIMENTAL DESIGNS

If any experiment is expected to maximize the signal of a treatment induced difference (e.g., differences in precipitation amounts for seeded and non-seeded events) and also provide confidence in the results (i.e., that the differences are meaningful and not merely due to natural occurrences), then the design of the experiment must be carefully conceived. Many years of concentrated efforts have been spent to develop the best possible designs for experiments in fields such as agriculture and medicine. While specific problems associated with weather modification experiments may differ from these other fields, the design of a weather modification experiment requires the same type of intensive effort. Detecting precipitation amount differences between seeded and non-seeded events is usually a difficult task because of the high degree of natural variability associated with various meteorological phenomena.

Because of the diversity of purposes associated with different types of weather modification experiments (e.g., wintertime orographic experiments, summertime cumulus experiments, lightning suppression experiments, hail mitigation experiments), various considerations of this section necessarily have a greater bearing on certain types of experiments than others. Specific topics discussed in this section include experimental units, descriptions and some properties of commonly used experimental designs, randomization and double blind procedures, and partitioning of data in relation to experimental designs.

2.1 Experimental Units

The basic constituent of any experimental design is the experimental unit. The experimental unit is an object defined in space and time for a specific weather modification experiment which receives the treatment (e.g., seeding). The treatments are usually assigned according to a random schedule which preserves a reasonable balance between treated and non-treated experimental units at any given time. Depending on the goals of a specific weather modification experiment, the experimental unit can conceivably be defined in various ways. For example, the experimental units of an experiment concerned with effects of seeding on summertime cumulus clouds might consist of a sequence of specifically defined cumulus clouds, whereas the experimental units of a weather modification experiment concerned with effects of seeding on wintertime orographic clouds might consist of a sequence of specifically defined temporal periods having a given length (e.g., 3 hours, 6 hours, 24 hours, or even 6 months, depending on the purpose of the particular experiment). A further point worth noting is the type of information which may be collected during each experimental unit. The data will usually consist of many responses where the importance of each response is given a priority. Such responses include measurements which are anticipated to be altered by a treatment together with concomitant measurements which, hopefully, are independent of the treatment and might serve as predictors (covariates) for the measurements anticipated to be altered by the treatment. As a result of the multitude of responses associated with a weather modification experiment, the subsequent discussions should obviously provide the capability for dealing with multivariate data in an acceptable manner.

2.2 Specific Designs

Among the wide variety of weather modification experimental designs which are conceivable, only the most commonly considered designs will be discussed. The presently considered designs include (a) the target-only (simple target) design, (b) the target-control (continued-covariate) design, (c) the balanced cross-over design, and (d) the augmented (unbalanced) cross-over design.

The target-only design yields comparisons of measured responses from treated (usually seeded with dry ice or silver iodide) and non-treated experimental units. The measured responses might consist of observations such as specific collections of measurements from a particular class of cumulus clouds or surface precipitation measurements from a specified instrumented target area. Target-only designs are essential when satisfactory covariates (uncontaminated predictors) are not available. In particular, the subset of experimental units to be treated should be selected from the totality of experimental units in accordance with a well defined random selection procedure. Examples of weather modification experiments using the target-only design include the Florida single cumulus dynamic seeding experiments (Simpson et al., 1971), the Montana lightning mitigation experiment (Baughman et al., 1976), and the Climax wintertime orographic cloud seeding experiment (Mielke et al., 1971).

The target-control design also yields comparisons of response measurements from treated and non-treated experimental units. However, correlated and uncontaminated (by treatment) observations such as either control area measurements or physically defined predictor measurements (e.g., numerical model response predictions based on rawinsonde data) are used to improve these comparisons (e.g., with the aid of appropriate regression techniques). Once again a well defined random selection procedure should be used in selecting the subset of experimental units to be treated. Examples of weather modification experiments using the target-control design include the Snowy Mountains experiment (Smith et al., 1963), the Santa Barbara experiment (Elliott et al., 1971), the Project Whitetop experiment (Braham, 1979), and the Climax wintertime orographic cloud seeding experiment (Mielke et al., 1981c).

The cross-over design involves two areas (say area I and area II) and yields comparisons between event A (where area I is treated and area II is non-treated) and event B (where area I is non-treated and area II is treated). Providing that the measured responses of the two areas are positively correlated and that there is no contamination between the two areas, Moran (1959) showed that the asymptotically most powerful design for detecting an additive treatment effect (assuming the paired response measurements of the two areas are distributed as a bivariate normal distribution) is the balanced cross-over design (i.e., a cross-over design where events A and B occur equally often). A well defined selection procedure should again be used in selecting event A or event B. Examples of weather modification experiments using the balanced cross-over design include the New England experiment (Smith et al., 1965), the Rapid Project experiment (Dennis and Koscielski, 1969), and the Israeli rainfall stimulation experiment (Gabriel and Neumann, 1978).

The augmented cross-over design is an intermediate design between the target-control design and the balanced cross-over design. Assume the previously stated conditions for asymptotic optimality of the balanced cross-over design are satisfied (i.e., there is no contamination between the two areas and the paired response measurements of the two areas are positively correlated and distributed as a bivariate normal distribution). If historical data (a finite number of events in which no treatment was applied) are available prior to the initiation of the experiment and the treatment induces an additive effect, then the optimum design for a specified finite number of future events may be a design in the extended class of designs which vary from a target-control design, through augmented cross-over designs, and up to the balanced cross-over design (Wu et al., 1972). The optimum design depends on the amount of historical data, the number of events available for treatment, the relative difference in the response measurement variances associated with the two areas, and the correlation of the paired response measurements for the two areas. When historical data are available and a small number of events are available for treatment, then the optimum design is likely to specify a disproportionate (unequal) number of events A and B. The previously mentioned asymptotic optimality of the balanced cross-over design for the present situation (Moran, 1959) is also consistent with the results of Wu et al. (1972).

2.3 Randomization and Double Blind Procedures

The randomization procedure for sequentially selecting treated and non-treated experimental units (analogously, events A and B in the case of a cross-over design) should maintain a reasonably close sequential balance. This procedure should also be quite unpredictable relative to the ability of any individual being able to predict the next selection, even if the entire past history of previous selections is known. The following is an example showing the construction of a simple list of randomized selections which accomplishes both goals. Although the present example yields an equal proportion of treated and non-treated experimental units, a simple modification of this example would yield any desired proportion. The first of two steps involves the construction of a sequential list of two block sizes, for example, block sizes of four and six experimental units. This first step is based on an unrestricted random selection with specified probabilities for selecting either block size. The second step then involves a restricted random assignment of either treatment or non-treatment labels to experimental units within a block so that exactly half of the experimental units within a block receive a treatment label. For example, each of the twenty possible permutations involving the sequential occurrences of the three treated and three non-treated experimental units in a block of size six has an equal chance of being selected.

A double blind procedure provides an intermediate coding which keeps the selection decision of either a treatment or non-treatment from the individual applying the treatment or non-treatment. The random coding yields alternate uninformative names for treatment and non-treatment labels which are selected according to an unrestricted randomization procedure. Also, if the non-treatment has been selected, then a placebo should be substituted for the treatment. A placebo is simply a device which insures that the treatment and non-treatment appear identical in applications.

2.4 Partitioning

The common and useful practice of partitioning experimental units according to meteorologically defined conditions presents an interesting question which pertains to the experimental design. This question is whether partitioning of experimental units should be accomplished before (based on forecasted meteorological conditions) or after (when the most reliable information concerning actual meteorological conditions is available) a specific experimental unit is realized. Because of great uncertainties inherent in meteorological forecasts, the most reasonable choice at present is to partition the experimental units after they have been realized (cf. Gabriel, 1979). Though the desired balance between treated and non-treated experimental units may admittedly be sacrificed for any particular partition, this circumstance seems far less objectionable than having a multitude of erroneously classified experimental units simply due to inaccurate forecasts.

3. EVALUATION PROCEDURES

Statistical inference techniques provide measures of credibility for effects attributed to treatments in weather modification experiments. In particular, this measure of credibility is termed a P-value, which is the probability of obtaining a result due to chance alone that is as extreme or more extreme than an observed result. If the probability of a type I error is designated as α, then the null hypothesis (of no treatment effect) is rejected when the P-value is less than α. Also, the probability of a type II error is designated by β (power is $1-\beta$) and is dependent on the choice of a statistical inference technique, the value of α, the assumed distribution, the specific alternative in question, the total sample size, and the proportion of the total sample that is allocated to different treatments. Though not emphasized in the remainder of this section, these basic error concepts provide the basis for selections and comparisons among the various statistical inference techniques that are available.

This section is comprised of two distinct parts. The first part involves a very brief discussion of distribution-bound inference techniques. The techniques include many well known parametric tests that have been described in detail elsewhere. The second part provides a discussion of distribution-free inference techniques, including various types of permutation tests. The specific permutation tests emphasized in the second part are (a) rerandomization, (b) classes of nonparametric tests, and (c) some multivariate permutation tests.

3.1 Distribution-Bound Inference Techniques

As previously indicated, distribution-bound inference techniques include the commonly employed parametric procedures such as likelihood ratio tests, $C(\alpha)$ tests, and the various well known univariate and multivariate tests based on normality (e.g., Student's t test, analysis of variance, Hotelling's T^2, and multivariate analysis of variance). Because of the numerous distributionally-related assumptions and concerns regarding these types of techniques (adequacy of approximation and tractability of chosen distribution, dependence structure of observations, effects of transformations, use of predictors, etc.), associated inferences based on such techniques are usually plagued with severe criticisms and nagging doubts. A very disconcerting transformation induced effect on results of a distribution-bound inference technique based on a bivariate normal distribution is given by Mielke (1979a). While a large amount of literature exists on various topics pertaining to distribution-bound inference techniques, some major reference sources include LeCam and Neyman (1967), Moran (1970), National Academy of Sciences (1973), Brier (1974), and special issues of Communications in Statistics - Theory and Methods (Vol. A8, Nos. 10 and 11).

3.2 Distribution-Free Inference Techniques

In contrast with distribution-bound inference techniques, distribution-free inference techniques are all based on specific permutation distributions of objects in a realized collection (e.g., experimental units of a sample). If two or more objects are exchanged with one another, then the resulting permutation distribution of objects remains the same (invariant) under the null hypothesis. In particular, distribution-free inference techniques include the permutation tests associated with any well defined test statistic (e.g., all nonparametric tests such as the sign test, Wilcoxon-Mann-Whitney test, and Kruskal-Wallis test, together with well known permutation tests such as the Fisher-Pitman test).

Distribution-free inference techniques are based on either the exact null permutation distribution of a given test statistic (e.g., exact small sample tables for the Wilcoxon-Mann-Whitney test statistic) or an approximation of the null permutation distribution of the test statistic (e.g., normal distribution as an approximation for the Wilcoxon-Mann-Whitney test statistic). In general, approximations of the null distribution of test statistics are usually obtained from either an empirical distribution of test statistic values based on a large number of random samples (termed rerandomization) or an approximating distribution which matches a few (say two, three, or four) exact lower order moments of a null permutation distribution.

The remainder of this section considers a small selection of distribution-free inference techniques. Specifically included are brief descriptions and discussions of rerandomization, some classes of univariate nonparametric tests, and a broad group of multivariate permutation tests designated multi-response permutation procedures.

3.2.1 <u>Rerandomization</u>. Rerandomization is a comparison of an actual experimental outcome with a large sample of simulated outcomes based on re-randomizations of seeded and non-seeded designations to the same experimental units. The seeded and non-seeded designation frequencies of each simulation correspond to those of the actual experimental outcome. In many instances, the permutation distribution of a statistic in question might be very difficult to describe. This situation is a frequent occurrence even when the description is restricted to a few exact low order moments of the statistic such as the mean, variance, and skewness. In such instances, rerandomization provides an appropriate technique for coping with situations that might otherwise be intractable. While rerandomization is a valuable device for providing inferences for statistics having permutation distributions that might otherwise be intractable, an inherent difficulty of rerandomization is that it is necessarily at the mercy of all routine statistical errors (i.e., the luck of the draw can of course yield a peculiar collection of alternative outcomes). This difficulty is not encountered if either the actual permutation distribution is attainable or it is possible to provide an adequate approximation of the permutation distribution. In summary, the advantage of having the actual permutation distribution or at least having an adequate approximation of the permutation distribution is that the inherent sampling errors associated with rerandomization are avoided. An excellent discussion of various topics involving rerandomization (including confidence intervals and power comparisons) is given by Gabriel (1979) and Brillinger <u>et al</u>. (1978).

3.2.2 <u>Univariate Nonparametric Tests</u>. Envisioned treatment effects associated with weather modification experiments involve complex differential variations rather than merely a simple scale or location shift of a designated response such as precipitation amount. As a consequence, there is a need for inference techniques capable of suggesting the type of change (if any) being induced on the natural precipitation amount distribution. Some classes of nonparametric tests are discussed that attempt to address this issue.

Suppose that the target-only design is considered. Among a collection of N observations (such as precipitation amounts), assume that a randomly obtained subset of n observations is associated with treated (seeded) experimental units and that the remaining subset of N-n observations is associated with non-treated (control) experimental units. A suggested class of two-sample nonparametric test statistics for this case is given by

$$A_r = \sum_{j=1}^{N} j^r Z_j,$$

where $r > 0$ and Z_j is one or zero if the jth ordered observation from below of the N pooled observations is associated with a treated or non-treated experimental unit, respectively. This class of non-parametric tests was initially introduced by Tamura (1963). However, the treatment of ties and large sample characteristics for this class of tests have been described more recently by Mielke (1967, 1972). An intuitive interpretation of the results associated with this class of non-parametric tests is that small or large values of r place an emphasis on changes (increases or decreases) that occur in the lower or higher order statistics, respectively. A suggested inference procedure is to calculate A_r for r equal to 1/4, 1/2, 1, 2, and 4.

Next suppose that the target-control design is considered. Let x_i and y_i (i = 1, 2, ..., N) denote paired control and target observations associated with the ith of N experimental units. Potentially very important advantages to be gained by the appropriate use of control observations include: (a) increasing the sensitivity (perhaps by a substantial amount) in test procedures for detecting treatment induced changes; and (b) adjusting for the effects of uncontrolled natural variability (an unfortunate situation that appears to be quite common in weather modification experiments). Again assume that a randomly obtained subset of n paired control and target observations is associated with treated experimental units (where the treatment supposedly affects only the target observation) and the remaining subset of N-n paired control and target observations is associated with non-treated experimental units. If the model given by $y = \beta x$ is fit by least squares, then the residual value associated with the ith of the N experimental units is given by $e_i = y_i - bx_i$ where i = 1, 2, ..., N and

$$b = (\sum_{i=1}^{N} x_i y_i)/(\sum_{i=1}^{N} x_i^2).$$

In this instance, a suggested class of two-sample nonparametric test statistics is given by

$$C_r = \sum_{j=1}^{N} \phi_j Z_j,$$

where

$$\Phi_j = \begin{cases} -|j - \frac{N+1}{2}|^r & \text{if } j < \frac{N+1}{2}, \\[2mm] 0 & \text{if } j = \frac{N+1}{2}, \\[2mm] |j - \frac{N+1}{2}|^r & \text{if } j > \frac{N+1}{2}, \end{cases}$$

$r > -1$, and Z_j is one or zero if the jth ordered residual value from below of the N pooled residual values is associated with a treated or non-treated experimental unit, respectively. The treatment of ties and large sample characteristics for this class of nonparametric tests have also been described (Mielke, 1972, 1974; Mielke and Sen, 1981). An intuitive interpretation of the results associated with this class of nonparametric tests is that small or large values of r place an emphasis on changes (increases or decreases) that occur in the middle or extreme (lower and/or upper) order statistics, respectively. A suggested inference procedure is to calculate C_r for r equal to -1/2, 0, 1, 2 and 4. A specific application involving these statistics is contained in Mielke et al. (1981c). Incidentally, this class of non-parametric tests is also applicable in analyses of experiments based on the cross-over design (e.g., let $e_i = x_i - y_i$ where x_i and y_i are the paired observations associated with the two target areas of the cross-over design, n cases of event A and N-n cases of event B as described in Section 2).

3.2.3 <u>Multivariate Permutation Tests</u>. In recently designed random-ized weather modification experiments involving cumulus clouds, a number of measured responses are obtained from each experimental unit (an individual cloud). Examples of the cumulus cloud response measurements (or measured differences) include ice crystal con-centration, liquid water content, maximum vertical velocity, radar echo top, visual cloud top, cloud top temperature, and cloud depth. In order to circumvent the unjustified assumption that the joint dis-tribution of these various dependent response measurements (trans-formed or otherwise) is multivariate normal, alternative multivariate inference techniques are sought that beneficially replace multivari-ate analysis of variance, Hotelling's T^2, and other related inference techniques based on the multivariate normal distribution (cf. Morrison, 1967). For this purpose, a large number of multivariate nonparametric and permutation procedures have been developed (cf. Puri and Sen, 1971). An additional class of conceptually simple multi-variate response permutation procedures (designated MRPPs) appears to be a useful tool for detecting the multivariate treatment effects anticipated with the previously mentioned cumulus cloud weather modi-fication experiments. Although descriptions, examples, computational procedures, and theoretical aspects of MRPPs are given in detail else-where (Mielke et al., 1976, 1981a, 1981b; Mielke, 1979b; O'Reilly and Mielke, 1980), the following paragraphs provide a brief description of these procedures.

Let $\Omega = \{\omega_1, \omega_2, \ldots, \omega_N\}$ be a finite population of N objects, let $x'_I = (x_{1I}, x_{2I}, \ldots, x_{rI})$ denote r commensurate response measurements (including functions of response measurements or residuals adjusted by predictors) for object ω_I (I = 1, 2, ..., N), and let S_1, $S_2, \ldots, S_{g+1}$ be an exhaustive partitioning of Ω into g+1 disjoint subgroups. Also let $\Delta_{I,J}$ be a monotone increasing function of a normed distance between ω_I and ω_J that is zero when the normed distance is zero. The form of $\omega_{I,J}$ considered here is confined to

$$\Delta_{I,J} = [\sum_{k=1}^{r} (x_{kI} - x_{kJ})^2]^{v/2},$$

where v > 0 (if v = 1, then $\Delta_{I,J}$ is Euclidean distance). The statistic on which MRPPs are based is given by

$$\delta = \sum_{i=1}^{g} C_i \xi_i,$$

where

$$\xi_i = \binom{n_i}{2}^{-1} \sum_{I<J} \Delta_{I,J} I_{S_i}(\omega_I) I_{S_i}(\omega_J)$$

is the average between object distance for all objects in subgroup S_i (i = 1, 2, ..., g), $\sum_{I<J}$ is the sum over all I and J such that $1 \leq I < J \leq N$,

$$I_{S_i}(\omega_I) = \begin{cases} 1 \text{ if } \omega_I \text{ belongs to } S_i, \\ 0 \text{ otherwise,} \end{cases}$$

for I = 1, 2, ..., N, $n_i \geq 2$ is the number of (classified) objects in subgroup S_i (i = 1, 2, ..., g),

$$K = \sum_{i=1}^{g} n_i,$$

$n_{g+1} = N - K$ is the number of remaining (unclassified) objects in group S_{g+1}, $C_i > 0$ for i = 1, 2, ..., g, and

$$\sum_{i=1}^{g} C_i = 1.$$

The recommended choice of C_i is $C_i = n_i/K$ for $i = 1, 2, \ldots, g$ (cf. Mielke, 1979b). In the specific case of a cumulus cloud weather modification experiment, $g = 2$, S_1 is the collection of n_1 treated experimental units (i.e., cumulus clouds), S_2 is the collection of n_2 non-treated experimental units, and $S_3 = S_{g+1}$ is empty (i.e., $n_1 + n_2 = K = N$). The underlying permutation distribution of δ implies that each of the

$$M = N!/(\prod_{i=1}^{g+1} n_i!)$$

possible allocation combinations of the N objects to the g+1 subgroups will have the same chance of occurring. With this underlying distribution of δ assumed, let μ_δ, σ_δ^2, and γ_δ respectively denote the mean, variance, and skewness of δ. If δ_j denotes the jth value among the M possible values of δ, then

$$\mu_\delta = \frac{1}{M} \sum_{j=1}^{M} \delta_j,$$

$$\sigma_\delta^2 = \frac{1}{M} \sum_{j=1}^{M} \delta_j^2 - \mu_\delta^2,$$

$$\gamma_\delta = (\frac{1}{M} \sum_{j=1}^{M} \delta_j^3 - 3\mu_\delta\sigma_\delta^2 - \mu_\delta^3)/\sigma_\delta^3.$$

Efficient computational techniques for obtaining μ_δ, σ_δ^2, and γ_δ for a realized set of data when $C_i = n_i/K$ ($i = 1, 2, \ldots, g$) are described elsewhere (Mielke et al., 1976; Mielke, 1979b). An intuitive interpretation of the results of applying MRPPs is simply that small values of δ infer a concentration of response measurements within the g subgroups.

Since the calculation of the M possible values of δ is not computationally feasible even for relatively small values of N, an adequate approximation of the underlying permutation distribution of δ is indispensable. The standardized test statistic given by

$$T = \frac{\delta - \mu_\delta}{\sigma_\delta}$$

is approximately distributed as the Pearson type III distribution. In particular, the Pearson type III distribution approximation compensates for the fact that the underlying permutation distribution of δ is often very skewed in the negative direction (i.e., $\gamma_\delta < 0$). Substantial negative skewness is not uncommon even asymptotically; that is, as N tends to infinity (Mielke, 1979b). The P-value for a realized value of δ (say δ_0) is given by

$$P(\delta \le \delta_0) \simeq \int_{-\infty}^{T_0} f(u)du,$$

where $T_0 = (\delta_0 - \mu_\delta)/\sigma_\delta$ and $f(u)$ is the density function of the Pearson type III distribution specified by the skewness parameter $\gamma = \gamma_\delta \le -0.001$ (cf. Mielke et al., 1981a). A P-value approximation based on the normal distribution is reported if $\gamma_\delta > -0.001$. The P-value approximation based on the Pearson type III distribution is evaluated by Simpson's rule over the interval $(T_0 - 9, T_0)$. In addition, extensive tables for the Pearson type III distribution have been published (cf. Harter, 1969).

The following numerical examples of MRPPs involve synoptic comparisons based on two data sets. These examples include comparisons of 700 mb joint wind direction and velocity observations between two groups of 24-hour periods having observed precipitation amounts of at least 0.015 inches. One group (termed the "light" group) consists of 24-hour periods with observed precipitation amounts less than or equal to a specified data set's median precipitation amount. The other group (termed the "heavy" group) consists of 24-hour periods with observed precipitaiton amounts greater than the specified data set's median precipitation amount.

The first data set is from the 1960-70 Climax wintertime orographic cloud seeding experiment. This data set consists of 100 Fremont Pass Summit group average precipitation amounts for non-seeded 24-hour experimental units having recorded 500 mb temperatures greater than or equal to -20°C (Mielke et al., 1971). The second data set is from the 1964-70 Wolf Creek Pass wintertime orographic cloud seeding experiment. This data set consists of 101 three-sensor average precipitation amounts for non-seeded 24-hour periods having recorded 500 mb temperatures greater than or equal to -23°C (Grant and Elliot, 1974; Mielke et al., 1977).

Let the Ith pair of 700 mb wind direction and velocity observations from upper air soundings be designated by θ_I and v_I, respectively. The associated Ith pair of transformed wind pattern rectangular coordinates for these examples are denoted by x_{1I} and x_{2I}, where

$$x_{1I} = v_I\cos(450° - \theta_I),$$

$$x_{2I} = v_I\sin(450° - \theta_I).$$

As defined, x_{1I} and x_{2I} should respectively be interpreted as the Ith pair of east-west and north-south wind pattern coordinates. Figure 1 shows the wind pattern coordinates of the 51 24-hour experimental units in the light group (each coordinate indicated by a dot) and the 49 24-hour experimental units in the heavy group (each coordinate indicated by a cross) associated with the 1960-70 Climax experiment. Figure 2 shows the wind pattern coordinates of the 51 24-hour periods in the light group (each coordinate indicated by a dot) and the 50 24-hour periods in the heavy group (each coordinate indicated by a cross) associated with the 1964-70 Wolf Creek Pass experiment.

The form of $\Delta_{I,J}$ used in these examples is Euclidean distance (i.e., $v = 1$). The initial comparison made here examines the light group versus the heavy group of the Climax experiment. In this case, $N = 100$, $r = 2$, $g = 2$, $n_1 = 51$, $n_2 = 49$, $\xi_1 = 9.819$, $\xi_2 = 9.242$, $\delta = 9.536$, $\mu_\delta = 9.607$, $\sigma_\delta^2 = 0.002493$, and $\gamma_\delta = -1.725$. The standardized test statistic is $T = -1.427$ with an associated P-value of 0.0895. This same comparison for the Wolf Creek Pass experiment yielded $N = 101$, $r = 2$, $g = 2$, $n_1 = 51$, $n_2 = 50$, $\xi_1 = 7.250$, $\xi_2 = 9.489$, $\delta = 8.358$, $\mu_\delta = 9.208$, $\sigma_\delta^2 = 0.002698$, and $\gamma_\delta = -1.980$. The standardized test statistic is $T = -16.35$ with an associated P-value of 2.67×10^{-8}. These results are consistent with the graphical representation of the light and heavy group wind patterns in Figures 1 and 2. Differences in the light and heavy group wind patterns are not apparent in Figure 1 for the Climax experiment. However, the corresponding wind pattern differences for the Wolf Creek Pass experiment are visually apparent in Figure 2.

A further analysis involves examining the light group of the Climax experiment separately, with the heavy group treated as unclassified data. In this case $N = 100$, $r = 2$, $g = 1$, $n_1 = 51$, $\delta = 9.819$, $\mu_\delta = 9.607$, $\sigma_\delta^2 = 0.3912$, and $\gamma_\delta = -0.07581$. The standardized test statistic is $T = 0.3385$ with an associated P-value of 0.628. This same analysis for the Wolf Creek Pass experiment yielded $N = 101$,

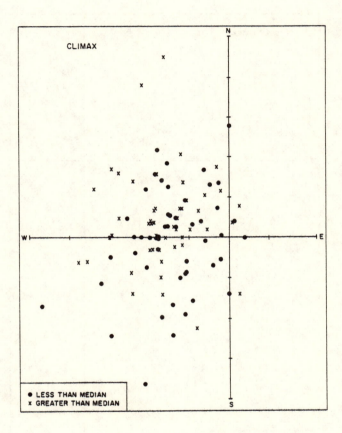

Figure 1. The 700-mb wind pattern coordinates for light (·) and heavy (x) groups of the Climax experiment.

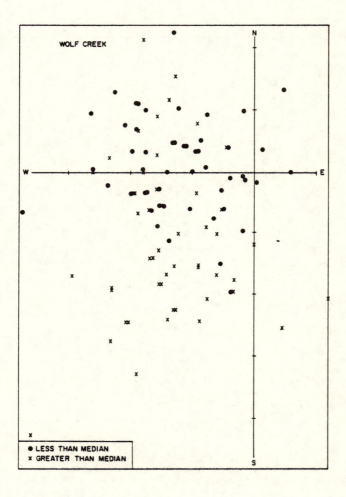

Figure 2. The 700-mb wind pattern coordinates for light (·) and heavy (x) groups of the Wolf Creek Pass experiment.

$r = 2$, $g = 1$, $n_1 = 51$, $\delta = 7.250$, $\mu_\delta = 9.208$, $\sigma_\delta^2 = 0.2645$, and $\gamma_\delta = -0.0786$. The standardized test statistic is $T = -3.807$ with an associated P-value of 1.32×10^{-4}. The last result indicates that the wind pattern of the light group for the Wolf Creek Pass experiment is more concentrated than anticipated by chance alone. This feature is also visually apparent in Figure 2.

Although the present examples have focused on a specific application of MRPPs for identifying associations among synoptic measurements, similar applications are conceptually appropriate for investigations involving any number of synoptic measurements ($r > 1$) and disjoint groups ($g > 1$). Additional numerical examples of MRPPs are given by Mielke et al. (1981a).

If a small P-value is attained in an analysis of some specific data, then a concentration of object measurements within some subgroups is suggested. A comparison of μ_δ (the average object distances for all pairs of objects in the population) with ξ_1, ξ_2, ..., ξ_g (the average object distances for all pairs of objects within each subgroup) provides a basis for interpreting the concentration. If all values of ξ_1, ξ_2, ..., ξ_g are smaller than μ_δ, then the concentration is primarily attributed to a separation (location shift) of measurements associated with the g subgroups. However, if not all of the ξ_i's are smaller than μ_δ, then the concentration is primarily attributed to the clumping of measurements for specific subgroups (those groups with values of ξ_i which are smaller than μ_δ) relative to the population of object measurements.

Being permutation-based procedures, MRPPs routinely avoid specific problems associated with the highly restrictive null hypothesis feature (i.e., level of significance being specified by an imaginatively conceived null distribution) that is assumed in distribution-bound inference techniques. In addition, the efficient discrimination characteristics of MRPPs are consistent with the fact that special cases of MRPPs coincide with well known parametric and nonparametric test procedures that provide optimum efficiency under prescribed conditions. For example, if $v = 2$, $r = 1$, $g \geq 2$, and S_{g+1} is empty, then the univariate version of MRPPs is asymptotically equivalent to the two-sample t and one-way analysis of variance class of tests. It is equivalent if $C_i = (n_i - 1)/(K - g)$ instead of $C_i = n_i/K$. In contrast with the two-sample t and one-way analysis of variance class of tests, this corresponding version of MRPPs is always appropriate for inferences (i.e., when the distributional assumptions associated with normality are false) since it is a permutation test. Also if $v = 2$, $r = 1$, $g \geq 2$, S_{g+1} is empty, and the univariate response measurements are replaced with ranks, then the resulting univariate nonparametric version of MRPPs is asymptotically equivalent [again equivalent if $C_i = (n_i - 1)/(K - g)$ instead of $C_i = n_i/K$] to the two-sample Wilcoxon-Mann-Whitney and Kruskal-Wallis class of tests. In contrast, if $v = 1$, $r = 1$, $g \geq 2$, S_{g+1} is empty, and the univariate response measurements are replaced by ranks, then the resulting univariate version of MRPPs is substantially superior in many instances to the previously mentioned univariate nonparametric version of MRPPs with $v = 2$ (Mielke et al., 1981b). Specific instances in which the version of MRPPs involving $v = 1$ is substantially superior to the version involving $v = 2$ include the detection of location shifts associated with either a double exponential or a U-shaped distribution. Even when location shifts associated with either a logistic or a normal distribution are considered, the version involving $v = 2$ represents at best a very minor improvement over the version involving $v = 1$ (the version involving $v = 2$ is known to have optimum properties or at least to be very efficient for these distributions).

A point worth mentioning is that none of the well known nonparametric tests coincide with the nonparametric versions of MRPPs indexed by $v > 0$, with the stated exception of $v = 2$. However, the space of distances associated with δ is defined by v (Mielke et al., 1981b). This space is non-metric when $v > 1$ and metric when $v \leq 1$. The space is a non-distorted metric space (Euclidean space) only when $v = 1$. Therefore, in addition to the previously mentioned advantages of $v = 1$ over $v = 2$, the choice of $v = 1$ also has an obvious logical appeal versus $v = 2$.

Statistics based on combinations of three or more objects at a time (MRPPs are based on only two objects at a time) are natural and intriguing extensions. However, such extensions are presently prohibitive for all but fairly small sample sizes because of the voluminous computational requirements. In conclusion, MRPPs are a general inference technique that requires very few assumptions and appears to possess a broad spectrum of potentially efficient inferential analyses for various types of experiments.

ACKNOWLEDGMENT

This work has been supported by the Division of Atmospheric Resources Research, Bureau of Reclamation, U.S. Department of the Interior, Contract No. 8-07-83-V0009.

REFERENCES

Baughman, R. G., D. M. Fuquay, and P. W. Mielke, 1976: Statistical analyses of a randomized lightning modification experiment. Journal of Applied Meteorology, 15, 790-794.

Braham, R. R., 1979: Field experimentation in weather modification. Journal of the American Statistical Association, 74, 57-104.

Brier, G. W., 1974: Design and evaluation of weather modification experiments. Weather and Climate Modification (W. N. Hess, editor). New York, Wiley, 206-225.

Brillinger, D. R., L. V. Jones, and J. W. Tukey, 1978: The Management of Weather Resources, Volume II: The Role of Statistics. Washington, D.C., U.S. Government Printing Office.

Dennis, A. S., and A. Koscielski, 1969: Results of a randomized cloud-seeding experiment in South Dakota. Journal of Applied Meteorology, 8, 556-565.

Elliott, R. D., P. St. Amand, and J. R. Thompson, 1971: Santa Barbara pyrotechnic cloud seeding test results 1967-70. Journal of Applied Meteorology, 10, 785-795.

Gabriel, K. R., 1979: Some statistical issues in weather modification. Communications in Statistics - Theory and Methods, A8, 975-1015.

Gabriel, K. R., and J. Neumann, 1978: A note of explanation of the 1961-67 Israeli rainfall stimulation experiment. Journal of Applied Meteorology, 17, 552-554.

Grant, L. O., and R. E. Elliott, 1974: The cloud seeding temperature window. Journal of Applied Meteorology, 13, 355-363.

Harter, H. L., 1969: A new table of percentage points of the Pearson type III distribution. Technometrics, 11, 177-187.

LeCam, L. M., and J. Neyman, Eds., 1967: Proceedings of the Fifth Berkeley Symposium on Mathematical Statistics and Probability, Volume 5. Berkeley, University of California Press.

Mielke, P. W., 1967: Note on some squared rank tests with existing ties. Technometrics, 9, 312-314.

Mielke, P. W., 1972: Asymptotic behavior of two-sample tests based on powers of ranks for detecting scale and location alternatives. Journal of the American Statistical Association, 67, 850-854.

458

Mielke, P. W., 1974: Squared rank test appropriate to weather modification cross-over design. Technometrics, 16, 13-16.

Mielke, P. W., 1979a: Some parametric, nonparametric and permutation inference procedures resulting from weather modification experiments. Communications in Statistics - Theory and Methods, A8, 1083-1096.

Mielke, P. W., 1979b: On asymptotic non-normality of null distributions of MRPP statistics. Communications in Statistics - Theory and Methods, A8, 1541-1550. Errata: A10, 1795; A11, 847.

Mielke, P. W., K. J. Berry, and G. W. Brier, 1981a: Application of multi-response permutation procedures for examining seasonal changes in monthly sea-level pressure patterns. Monthly Weather Review, 109, 120-126.

Mielke, P. W., K. J. Berry, P. J. Brockwell, and J. S. Williams, 1981b: A class of nonparametric tests based on multi-response permutation procedures. Biometrika, 68, 720-724.

Mielke, P. W., K. J. Berry, and E. S. Johnson, 1976: Multi-response permutation procedures for a priori classifications. Communications in Statistics - Theory and Methods, A5, 1409-1424.

Mielke, P. W., G. W. Brier, L. O. Grant, G. J. Mulvey, and P. N. Rosenzweig, 1981c: A statistical reanalysis of the replicated Climax I and II wintertime orographic cloud seeding experiments. Journal of Applied Meteorology, 20, 643-659.

Mielke, P. W., L. O. Grant, and C. F. Chappell, 1971: An independent replication of the Climax wintertime orographic cloud seeding experiment. Journal of Applied Meteorology, 10, 1198-1212. Corrigendum: 15, 801.

Mielke, P. W., and P. K. Sen, 1981: On asymptotic non-normal null distributions for locally most powerful rank test statistics. Communications in Statistics - Theory and Methods, A10, 1079-1096.

Mielke, P. W., J. S. Williams, and S. C. Wu, 1977: Covariance analysis technique based on bivariate log-normal distribution with weather modification applications. Journal of Applied Meteorology, 16, 183-187.

Moran, P. A. P., 1959: The power of a cross-over test for the artificial stimulation of rain. Australian Journal of Statistics, 1, 47-52.

Moran, P. A. P., 1970: The methodology of rain-making experiments. Review of the International Statistical Institute, 38, 105-119.

Morrison, D. F., 1967: Multivariate Statistical Methods. New York, McGraw-Hill.

National Academy of Sciences, 1973: Weather and Climate Modification: Problems and Progress. Washington, D.C.

Neyman, J., and E. L. Scott, 1965: Asymptotically optimal tests of composite hypotheses for randomized experiments. Journal of the American Statistical Association, 60, 699-721.

O'Reilly, F. J., and P. W. Mielke, 1980: Asymptotic normality of MRPP statistics from invariance principles of U-statistics. Communications in Statistics - Theory and Methods, A9, 629-637.

Puri, M. L., and P. K. Sen, 1971: Nonparametric Methods in Multivariate Analysis. New York, Wiley.

Simpson, J., W. L. Woodley, A. H. Miller, and G. F. Cotton, 1971: Precipitation results of two randomized pyrotechnic cumulus seeding experiments. Journal of Applied Meteorology, 10, 526-544.

Smith, E. J., E. E. Adderly, and F. D. Bethwaite, 1965: A cloud seeding experiment in New England, Australia. Journal of Applied Meteorology, 4, 433-441.

Smith, E. J., E. E. Adderly, and D. T. Walsh, 1963: Cloud-seeding experiment in the Snowy Mountains, Australia. Journal of Applied Meteorology, 2, 324-332.

Tamura, R., 1963: On a modification of certain rank tests. Annals of Mathematical Statistics, 34, 1101-1103.

Wu, S. C., J. S. Williams, and P. W. Mielke, 1972: Some designs and analyses for temporally independent experiments involving correlated bivariate responses. Biometrics, 28, 1043-1061.

12
Bayesian Inference

Robert L. Winkler

1. INTRODUCTION

Statistical inference deals with uncertainty, and probability can be thought of as the mathematical language of uncertainty. At any given time, the state of information about some uncertain quantity can be represented in probabilistic form, and when new information is obtained probabilities are revised accordingly. Bayes' theorem provides a formal mechanism to revise probabilities and provides the basis for the approach to statistical inference known as Bayesian inference.

In Bayesian inference, any variable about which there is uncertainty can be treated as a random variable. This means, for example, that parameters of statistical models (e.g., a proportion, a mean, a regression coefficient) can be viewed as random variables instead of as fixed, unknown quantities. As a result, probability distributions for such parameters can be considered and can be updated as new information is obtained. Indeed, the primary inferential statement about a parameter is a probability distribution for the parameter, and other types of inferences, such as point estimates, interval estimates, and tests of hypotheses, are based on this probability distribution. Therefore, the main concern in Bayesian inference is the determination of probability distributions for variables of interest and the revision of such distributions to reflect new information.

Bayes' theorem dates back over two centuries (Bayes, 1763), but its use as the basis for the Bayesian approach to statistics is relatively recent. Prior to the last two decades, virtually all statistical methodology was non-Bayesian in nature, based on what is often called the classical approach to statistics. Important foundational work by de Finetti (1937) and Savage (1954) helped to generate some interest in Bayesian statistics, and early books in the area (e.g., Schlaifer, 1959; Raiffa and Schlaifer, 1961) provided further motivation for Bayesian work. In the past two decades, the interest in Bayesian methods has increased considerably.

This chapter provides an introduction to Bayesian inferential procedures. Bayes' theorem is presented in Section 2, and Bayesian analyses with discrete and continuous prior distributions are discussed in Sections 3 and 4, respectively. The two inputs needed in a Bayesian analysis are a likelihood function and a prior distribution, and modeling issues concerning the assessment of these inputs are covered in Section 5. In Section 6, the Bayesian approach to problems of estimation, hypothesis testing, and prediction is discussed and compared with the classical approach. Finally, some applications of Bayesian inference in meteorology are discussed briefly in Section 7.

2. BAYES' THEOREM

In an inferential or decision-making problem it is desirable to make inferential statements or decisions conditional on any available information. Thus, conditional probabilities are of interest. Formally, a conditional probability can be defined in terms of the ratio of a joint probability and a marginal probability. If an event E is of interest, the conditional probability of E given the information y is

$$P(E|y) = P(E,y)/P(y). \qquad (1)$$

But $P(E,y)$ and $P(y)$ can be expressed in the forms

$$P(E,y) = P(E)P(y|E) \qquad (2)$$

and

$$P(y) = P(E)P(y|E) + P(\overline{E})P(y|\overline{E}), \qquad (3)$$

respectively, where $\overline{E}$ represents the complement of E. The substitution of (2) and (3) in (1) yields

$$P(E|y) = \frac{P(E)P(y|E)}{P(E)P(y|E) + P(\overline{E})P(y|\overline{E})}, \qquad (4)$$

which is Bayes' theorem for a single event E.

Bayes' theorem, then, is nothing more than a procedure for finding conditional probabilities. From Bayes' theorem, it can be seen that a conditional probability $P(E|y)$ can be thought of as an updated, or revised, version of the marginal probability $P(E)$. In (4), $P(E)$ and $P(\overline{E})$ are called prior probabilities because they are the probabilities of E and $\overline{E}$ before the information y is observed. Bayes' theorem provides the posterior probability $P(E|y)$, the probability of E after y is observed. Of course, the posterior probability of $\overline{E}$, $P(\overline{E}|y)$, can also be determined, either from Bayes' theorem or from $P(\overline{E}|y) = 1 - P(E|y)$. The new information y is incorporated into Bayes' theorem via $P(y|E)$ and $P(y|\overline{E})$, which are called likelihoods.

Thus, the probability revision process incorporates prior probabilities and likelihoods (representing the prior information and the new information, respectively) as inputs and provides output in the form of posterior probabilities.

A simple example serves to illustrate Bayes' theorem. Precipitation probabilities are issued routinely in weather forecasts in the United States, and such probabilities are revised as deemed necessary in the light of new information. Suppose that a forecast which includes a probability of 0.20 for "rain tomorrow in Denver" is issued one evening. That night, new information which suggests that an unanticipated storm is developing in the Rocky Mountains is received, and past data suggests that rainy days in Denver are preceded by this sort of information approximately 30 percent of the time, whereas nonrainy days are preceded by this sort of information only 5 percent of the time.

The prior probability of E, rain occurring in Denver on the day in question, is the original precipitation probability, 0.20, from the weather forecast. Furthermore, the past data concerning the incidence of y, unexpected stormy weather developing in the Rocky Mountains, provide the likelihoods $P(y|E) = 0.30$ and $P(y|\overline{E}) = 0.05$. Thus, from Bayes' theorem, the posterior probability of E is

$$P(E|y) = \frac{(0.20)(0.30)}{(0.20)(0.30) + (0.80)(0.05)} = \frac{0.06}{0.10} = 0.60.$$

The new information increases the probability of rain from 0.20 to 0.60. While rain is by no means a "sure thing," the increase in the probability of rain is sizable, as might be expected given the combination of (1) the low initial probability of rain and (2) the fact that the new information is six times as likely to occur before rainy days as it is to occur before non-rainy days.

Bayes' theorem can be applied repeatedly as new information is obtained. Suppose that a few hours after the information about the stormy weather in the mountains is received, strong winds are recorded in the Denver area. If y_1 represents the information about stormy weather and y_2 represents the information about strong winds, then the posterior probability of E following y_1 is now the prior probability of E with respect to y_2. In the second revision of Bayes' theorem, all probabilities are conditional upon y_1:

$$P(E|y_1,y_2) = \frac{P(E|y_1)P(y_2|E,y_1)}{P(E|y_1)P(y_2|E,y_1) + P(\overline{E}|y_1)P(y_2|\overline{E},y_1)}. \tag{5}$$

Given the stormy weather in the mountains, strong winds occur 60% of the time preceding rainy days and 30% of the time preceding non-rainy days. Therefore, $P(y_2|E,y_1) = 0.60$ and $P(y_2|\overline{E},y_1) = 0.30$, and the revised probability of rain is

$$P(E|y_1,y_2) = \frac{(0.60)(0.60)}{(0.60)(0.60) + (0.40)(0.30)} = \frac{0.36}{0.48} = 0.75.$$

As might be expected, y_2 causes another increase in the probability of rain, just as y_1 did.

When individuals are uncertain about whether or not a particular event will occur, they may revise their judgments about the likelihood of the event in response to information. In practice, for example, a weather forecaster may update a probability of precipitation forecast as new information becomes available over time. Indeed, it would seem foolish not to modify probabilities to reflect any information that bears on the event of interest. Bayes' theorem simply provides a formal way to represent this intuitively reasonable revision process.

3. BAYESIAN INFERENCE WITH DISCRETE PRIOR DISTRIBUTIONS

The weather forecasting example in Section 2 involves the revision of a probability for a single event, the occurrence of rain in Denver on a particular day. In statistics, parametric models are widely used, and interest centers upon the parameters of such models. Just as Bayes' theorem can be used to revise a probability for a single event, it can be used to revise a probability distribution for a parameter of a statistical model.

If the prior distribution of a parameter θ is discrete and the new information y comes from a discrete model, then Bayes' theorem can be expressed as follows:

$$P(\theta_i|y) = \frac{P(\theta_i)P(y|\theta_i)}{\sum\limits_{j=1}^{J} P(\theta_j)P(y|\theta_j)} \tag{6}$$

for $i = 1, 2, \ldots, J$, where $\theta_1, \theta_2, \ldots, \theta_J$ represent the possible values of θ. Here the prior probabilities are $P(\theta_1)$, $P(\theta_2)$, $\ldots$, $P(\theta_J)$, and the likelihoods are $P(y|\theta_1)$, $P(y|\theta_2)$, $\ldots$, $P(y|\theta_J)$. The posterior probabilities determined from Bayes' theorem are $P(\theta_1|y)$, $P(\theta_2|y)$, $\ldots$, $P(\theta_J|y)$.

3.1 Bernoulli Model

For example, suppose that forecasts of maximum temperature for a particular location are issued in terms of a range of values (e.g., "tomorrow's high temperature will be between 56°F and 60°F"). If extremely wide intervals are used, the forecast will not convey much information about maximum temperature. On the other hand, if very narrow intervals are used, the actual high temperature might fall outside the interval so often that the public might react negatively to the forecasts. Suppose that the intent is to have the high temperature fall in the interval about 60 percent of the time. Let x_i equal one if the high temperature falls in the forecast interval on day i and let x_i equal zero if it does not fall in the forecast interval. Moreover, suppose that the possible sequences of outcomes over a number of days are viewed as exchangeable. That is, for any number of days and any observed sequence of results, all sequences with the same number of day with the high temperature in the interval are viewed as equally likely. For a dichotomous process, exchangeability implies that the Bernoulli model is an appropriate model. The parameter of interest is thus a Bernoulli parameter, the probability p that the high temperature on any given day will fall within the limits of the range of values indicated in the forecast of high temperature for that day. The Bernoulli model can also be viewed in terms of outcomes on different days being independent with p remaining constant from day to day.

Consider three possible values of the Bernoulli parameter p: 0.5, which implies that the high temperature does not fall in the forecast interval quite as often as desired; 0.6, which implies that the forecasts meet the stated intent; and 0.7, which implies that the high temperature falls in the forecast interval slightly more often than would be liked. Moreover, assume that very little a priori information is available regarding p, and the prior probabilities are taken to be $P(p = 0.5) = P(p = 0.6) = P(p = 0.7) = 1/3$. A sample of 100 days is available, and the forecast interval "captures" the actual high temperature on 62 of these days. The likelihoods, which are the probabilities for this sample outcome conditional upon the different values of p, can be found from the binomial distribution, where r represents the number of times the observed temperature falls in the forecast interval and n is the number of days in the sample:

$$P(r = 62 \mid n = 100, p = 0.5) = \binom{100}{62}(0.5)^{62}(0.5)^{38} = 0.0045,$$

$$P(r = 62 \mid n = 100, p = 0.6) = \binom{100}{62}(0.6)^{62}(0.4)^{38} = 0.0754,$$

$$P(r = 62 \mid n = 100, p = 0.7) = \binom{100}{62}(0.7)^{62}(0.3)^{38} = 0.0191.$$

From Bayes' theorem, the posterior probabilities are as follows:

$$P(p = 0.5|r) = \frac{(1/3)(0.0045)}{(1/3)(0.0045) + (1/3)(0.0754) + (1/3)(0.0191)},$$

$$P(p = 0.6|r) = \frac{(1/3)(0.0754)}{(1/3)(0.0045) + (1/3)(0.0754) + (1/3)(0.0191)},$$

$$P(p = 0.7|r) = \frac{(1/3)(0.0191)}{(1/3)(0.0045) + (1/3)(0.0754) + (1/3)(0.0191)}.$$

The evidence is in favor of p = 0.6, with the sample results increasing the probability of p = 0.6 from 1/3 to 0.762. This result suggests that the forecasts seem to meet the initial intent of capturing the actual temperature in the interval approximately 60 percent of the time.

3.2 Poisson Model

For an example with a different statistical model of a data-generating process, consider the occurrence of severe storms (tornadoes or severe thunderstorms) in a particular rural area during the months of March, April, May, and June. It is assumed that the data-generating process can be represented as a Poisson process, with the Poisson parameter λ denoting the intensity of the process, or the average rate at which severe storms occur. This rate will be expressed in terms of storms per week. The Poisson model implies that the occurrence of severe storms in a particular time period (e.g., a day, a week) is independent of previous occurrences and that the intensity λ remains constant over time.

On the basis of past data regarding nearby locations, the values 0.4, 0.6, 0.8, and 1.0 are considered for λ, with prior probabilities $P(\lambda = 0.4) = 0.1$, $P(\lambda = 0.6) = 0.3$, $P(\lambda = 0.8) = 0.4$, and $P(\lambda = 1.0) = 0.2$. The area is then observed for a period of 15 weeks, and during this period exactly 13 severe storms are recorded. For a Poisson process with intensity λ per week, the expected number of severe storms in 15 weeks is simply 15λ, and the likelihoods in this example can be found from the Poisson distribution:

$$P(r = 13|\lambda = 0.4) = e^{-15(0.4)}[15(0.4)]^{13}/13! = 0.0052,$$

$$P(r = 13|\lambda = 0.6) = e^{-15(0.6)}[15(0.6)]^{13}/13! = 0.0504,$$

$$P(r = 13|\lambda = 0.8) = e^{-15(0.8)}[15(0.8)]^{13}/13! = 0.1056,$$

$$P(r = 13|\lambda = 1.0) = e^{-15(1.0)}[15(1.0)]^{13}/13! = 0.0956.$$

From Bayes' theorem, the posterior probabilities for the four possible values of λ are as follows:

$P(\lambda = 0.4|r)$

$$= \frac{(0.1)(0.0052)}{(0.1)(0.0052) + (0.3)(0.0504) + (0.4)(0.1056) + (0.2)(0.0956)}$$

$$= 0.007,$$

$P(\lambda = 0.6|r)$

$$= \frac{(0.3)(0.0504)}{(0.1)(0.0052) + (0.3)(0.0504) + (0.4)(0.1056) + (0.2)(0.0956)}$$

$$= 0.196,$$

$P(\lambda = 0.8|r)$

$$= \frac{(0.4)(0.1056)}{(0.1)(0.0052) + (0.3)(0.0504) + (0.4)(0.1056) + (0.2)(0.0956)}$$

$$= 0.549,$$

$P(\lambda = 1.0|r)$

$$= \frac{(0.2)(0.0956)}{(0.1)(0.0052) + (0.3)(0.0504) + (0.4)(0.1056) + (0.2)(0.0956)}$$

$$= 0.248.$$

The evidence practically eliminates $\lambda = 0.4$ from consideration, with the probability of $\lambda = 0.8$ increasing to 0.549, the probability of $\lambda = 1.0$ increasing slightly, and the probability of $\lambda = 0.6$ decreasing somewhat. If more data could be obtained, the probabilities could be revised again, of course.

4. BAYESIAN INFERENCE WITH CONTINUOUS PRIOR DISTRIBUTIONS

The Bernoulli and Poisson examples presented in Sections 3.1 and 3.2, respectively, are quite artificial in one important sense. It is highly unrealistic to assume in these cases that there are only a few possible values of θ, the parameter of interest. If p is the probability that the high temperature will fall in the forecast interval on a given day, the assumption that p must be either 0.5, 0.6, or 0.7 seems to be at best a rough approximation. Why are values such as 0.63, 0.548, and so on, excluded from consideration? Unless there is strong reason to believe that only 0.5, 0.6, and 0.7 are possible, the discrete treatment of p must be viewed as an approximation in which, for example, p = 0.6 might really represent something like 0.55 < p < 0.65. A similar argument can be made in the Poisson example.

The approximation provided by a discrete probability distribution for θ can be improved by increasing the number of values considered for θ. Ideally, however, the possibility of treating θ as a continuous random variable should be allowed. If the prior distribution of θ is represented by a density function $f(\theta)$, then the posterior distribution is also represented by a density function $f(\theta|y)$. The continuous analogues of (6) are

$$f(\theta|y) = \frac{f(\theta)P(y|\theta)}{\int_{-\infty}^{\infty}f(\theta)P(y|\theta)d\theta} \tag{7}$$

and

$$f(\theta|y) = \frac{f(\theta)f(y|\theta)}{\int_{-\infty}^{\infty}f(\theta)f(y|\theta)d\theta}. \tag{8}$$

Bayes' theorem with continuous prior and posterior distributions, then, is represented by (7) and (8), with (7) applying when the sampling distribution of the new information y is discrete and (8) applying when this sampling distribution is continuous.

Conceptually, (7) and (8) provide a convenient way to revise density functions in the light of sample information. In practice, however, it may prove quite difficult to apply these formulas. If the product of the prior density function and the likelihood function is not a fairly simple mathematical function, it may be a formidable task indeed to carry out the integration in the denominator of Bayes' theorem. For instance, in the Poisson example, suppose that the prior distribution of λ, the rate at which severe storms occur, is a normal distribution with mean 0.80 and standard deviation 0.20. The numerator of (7) is the product of this normal distribution and the Poisson probability that the number of storms is 13 given that the Poisson mean is 15λ, and the normalizing factor is provided by the denominator of (7), which is the integral of this product over the parameter space. The integral in the denominator actually is tractable in this case, but the posterior distribution is not of a well-known form and is not particularly easy to work with. Other examples can be found in which the integration in the denominator simply cannot be carried out analytically.

One way to avoid the problems that can be encountered because of the integration in the denominator of Bayes' theorem is to carry out the integration numerically. This means that any calculations involving posterior probabilities are also done numerically, and it amounts to a discrete approximation of the continuous prior and posterior distributions of θ. Another alternative is to restrict the prior distribution in such a way that the product of the prior density function and the likelihood function is relatively easy to work with. The restriction used most often is to require the prior distribution to be a member of the family of distributions that is conjugate to the likelihood function. A brief definition of a conjugate family of prior distributions is that a prior distribution is conjugate to the likelihood function when the prior density function is of the same mathematical form as the likelihood function insofar as factors involving the parameter θ are involved. The multiplication of the prior density and the likelihood function is then analogous to the multiplication of two likelihood functions from independent samples, and the combination of prior information and sample information via Bayes' theorem can be likened to the pooling of information from independent samples. When the prior distribution is a conjugate distribution, the posterior distribution is also a conjugate distribution, which means that any future revision of the posterior distribution will also be relatively easy to handle. For detailed developments of the notion of conjugate distributions in Bayesian analysis, see Raiffa and Schlaifer (1961), DeGroot (1970), and LaValle (1970).

4.1 Bernoulli Model

In the Bernoulli example involving the probability that the high temperature falls within the forecast interval, the likelihood function is

$$P(r = 62 \mid n = 100, p) = \binom{100}{62} p^{62} (1-p)^{38}.$$

A likelihood function can always be multiplied by a positive constant without affecting the inferences that are drawn from the likelihood function. Thus, the likelihood function in this example can be multiplied by the reciprocal of the combinatorial term $\binom{100}{62}$, leaving a likelihood function of the form $p^{62}(1 - p)^{38}$. In fact, if the likelihood were originally generated from the actual sequence of observations without bringing in the binomial distribution, the result would be $p^{62}(1 - p)^{38}$. In general, the likelihood function can be expressed as $p^{r}(1 - p)^{n-r}$, where n represents the number of days in the sample and r represents the number of times the observed high temperature falls in the forecast interval. Note that once the assumption of a Bernoulli process is invoked, the form of the likelihood function is determined, and only r and n are needed from the sample in order to specify the likelihood function. Other details, such as the exact sequence of observations inside and outside the forecast intervals, are not needed, although such details may be of interest in checking to see whether the original assumption that the process behaves like a Bernoulli process seems reasonable. As a result, r and n are called sufficient statistics in this example, and if n is predetermined by the individual who is collecting the data, r is the only statistic that must be computed from the actual results.

When the data-generating process is assumed to be a Bernoulli process, then a conjugate prior distribution must have a density function of the same form as the likelihood function $p^{r}(1 - p)^{n-r}$ and it must also be a proper density function. The family of distributions satisfying these requirements is the family of beta distributions. If the prior distribution of p is a beta distribution with density function

$$f(p) = \frac{(a+b+1)!}{a!b!} \, p^a(1-p)^b, \quad 0 \le p \le 1,$$

where a > -1 and b > -1, then the posterior distribution of p is a beta distribution with density function

$$f(p|y) = \frac{(a+b+n+1)!}{(a+r)!(b+n-r)!} \, p^{a+r}(1-p)^{b+n-r}, \quad 0 \le p \le 1.$$

That is, if the prior distribution is a beta distribution with parameters a and b, the posterior distribution is a beta distribution with parameters a* = a+r and b* = b+n-r. A beta distribution with parameters a and b has mean (a+1)/(a+b+2) and variance

$$(a+1)(b+1)/[(a+b+2)^2(a+b+3)],$$

and some tables of beta probabilities are available (e.g., Schlaifer, 1969).

In the Bernoulli example involving the probability that the high temperature falls in the forecast interval, suppose that the prior distribution of p is a beta distribution with a = 11 and b = 7. From beta tables, the quartiles of this distribution are 0.528, 0.603, and 0.676. The mean and standard deviation are 0.60 and 0.11. After the sample of n = 100 forecasts with r = 62 high temperatures within the forecast intervals, the posterior distribution of p is a beta distribution with a+r = 73 and b+n-r = 45. The quartiles of this distribution are 0.591, 0.617, and 0.643, the mean is 0.617, and the standard deviation is 0.044. The new information has increased the mean and the median of the distribution slightly and has reduced the dispersion of the distribution considerably.

4.2 Poisson Model

In the Poisson example involving the rate at which severe storms occur, the likelihood function is

$$P(r=13|\lambda) = e^{-15\lambda}(15\lambda)^{13}/13!,$$

which can be simplified to $\lambda^{13}e^{-15\lambda}$ by factoring out terms not involving λ. In general, the likelihood function for a sample from a Poisson process can be expressed as $\lambda^r e^{-t\lambda}$, where t represents the length of time (the number of weeks in this example) for which the process is observed and r represents the number of occurrences (severe storms in this example) during that time period. Thus, r and t are sufficient statistics, with t being predetermined in the example and r being observed when the sample of 15 weeks is taken.

Since the likelihood function for a sample from a Poisson process is of the form $\lambda^r e^{-t\lambda}$, a conjugate prior distribution in the Poisson case must have a density proportional to this same form. The appropriate family is the family of gamma distributions. If the prior distribution of λ is a gamma distribution with density function

$$f(\lambda) = e^{-a\lambda} a^{b+1} \lambda^b / \Gamma(b+1), \quad \lambda \geq 0,$$

where $a > 0$, $b > -1$, and Γ represents the gamma function, then the posterior distribution of λ is a gamma distribution with density function

$$f(\lambda|r) = e^{-(a+t)\lambda} (a+t)^{b+r+1} \lambda^{b+r} / \Gamma(b+r+1), \quad \lambda \geq 0.$$

As in the Bernoulli case with a conjugate beta prior distribution, the Bayesian revision in the Poisson case with a conjugate gamma distribution can be expressed in terms of the simple addition of prior parameters and sample statistics. The only difference between the prior and posterior gamma distributions is that the prior parameters a and b are replaced by a* = a+t and b* = b+r, respectively. The mean and variance of a gamma distribution with parameters a and b are $(b+1)/a$ and $(b+1)/a^2$, respectively. Gamma probabilities can be found from tables of chi-square probabilities (see, for example, Novick and Jackson, 1974).

On the basis of past information concerning the occurrence of severe storms in different areas and limited knowledge about the area now being considered, a forecaster assesses a gamma prior distribution for λ with parameters a = 4 and b = 2. This distribution has mean 0.75 and standard deviation 0.20. After the sample of t = 15 weeks with r = 13 severe storms, the posterior distribution of λ is a gamma distribution with a+t = 19 and b+r = 15. The average rate of storms per week during the sample period is r/t = 0.867, which is greater than the prior mean of λ, 0.75. As a result, the posterior mean, 0.842, is greater than the prior mean (but less than the sample average). The information provided by the sample reduces the standard deviation of λ from its prior value, 0.433, to a posterior standard deviation of 0.211.

4.3 Normal Model

The continuous model that is used most frequently in statistics to represent a data-generating process is the normal (or Gaussian) model. If a sample of size n is taken from a normally distributed population with known variance σ^2, the likelihood function can be expressed in the form

$$f(\bar{x}|\mu) = \frac{n^{1/2}}{\sigma(2\pi)^{1/2}} e^{-n(\bar{x}-\mu)^2/2\sigma^2},$$

where $\bar{x}$ is the sample mean. Here n and $\bar{x}$ are sufficient statistics. (The sample variance is not needed because the population variance σ^2 is assumed to be known.) The terms not involving μ can be factored out, leaving a likelihood function of the form $e^{-n(\bar{x}-\mu)^2/2\sigma^2}$.

A conjugate prior distribution for μ must be a proper distribution with a density proportional to $e^{-b(a-\mu)^2/2\sigma^2}$. Multiplying this expression by $(b/2\pi\sigma^2)^{1/2}$ yields a normal density function for μ with mean a and variance σ^2/b. When the prior density is of this form,

$$f(\mu) = \frac{b^{1/2}}{\sigma(2\pi)^{1/2}} e^{-b(\mu-a)^2/2\sigma^2},$$

where $b > 0$, the posterior density of μ is a normal density with mean $(ba+n\bar{x})/(b+n)$ and variance $\sigma^2/(b+n)$:

$$f(\mu|\bar{x}) = \frac{(b+n)^{1/2}}{\sigma(2\pi)^{1/2}} e^{-(b+n)\{[\mu-(ba+n\bar{x})]/(b+n)\}^2/2\sigma^2}.$$

The posterior mean is a weighted average of the prior mean a and the sample mean $\bar{x}$, and the posterior variance is smaller than the prior variance by a factor of $b/(b+n)$. Of course, probabilities involving μ can be found from prior and posterior distributions through the use of tables of standard normal probabilities.

Suppose that the errors in point forecasts of high temperature (i.e., the differences between the forecasts and the corresponding actual high temperatures) during the winter months at a particular location are assumed to be normally distributed with mean μ and variance 16. Furthermore, assume that the prior distribution of μ is itself a normal distribution with mean zero and standard deviation two. (The errors being considered here are not absolute errors, and it is expected that positive and negative errors will tend to cancel each other to provide a mean near zero.) Then $a = 0$ and $b = 16/4 = 4$. A sample of 20 forecasts yields an average forecast error of $\bar{x} = 0.54$. The posterior distribution of μ is a normal distribution with mean $(ba+n\bar{x})/(b+n) = 0.45$ and variance $16/(b+n) = 0.667$.

In most applications involving normally-distributed populations, it is probably unrealistic to assume that the population variance is known for certain. If the population variance is denoted by σ^2, then a realistic approach in most situations would be to assume that both parameters μ and σ^2 are unknown. Relaxing the known-variance assumption leads to a likelihood function of the form

$$\prod_{j=1}^{n} \frac{1}{\sigma(2\pi)^{1/2}} e^{-(x_j - \mu)^2/2\sigma^2},$$

which can be simplified to

$$\sigma^{-n} e^{-n(\overline{x}-\mu)^2/2\sigma^2} e^{-(n-1)s^2/2\sigma^2},$$

where

$$s^2 = \frac{1}{n-1} \sum_{j=1}^{n} (x_j - \overline{x})^2$$

represents the sample variance. Now that the population variance is no longer assumed known, the sample variance s^2 joins the sample size n and the sample mean $\overline{x}$ as sufficient statistics.

The conjugate family of distributions in the normal case with both mean and variance unknown is the family of normal-inverted-gamma distributions. A normal-inverted-gamma distribution consists of a conditional normal distribution for μ given σ^2 and a marginal inverted-gamma distribution for σ^2 (which implies a marginal gamma distribution for $1/\sigma^2$). For inferences about μ, the marginal distribution of μ is of interest, and when μ and σ^2 have a normal-inverted-gamma distribution, the marginal distribution of μ is a Student t distribution.

If the prior normal-inverted-gamma distribution of μ and σ^2 has density

$$f(\mu,\sigma^2) = f(\mu|\sigma^2)f(\sigma^2) = \frac{b^{1/2}}{\sigma(2\pi)^{1/2}} e^{-b(\mu-a)^2/2\sigma^2} \frac{(dc/2)^{d/2}}{\Gamma(d/2)\sigma^{d+2}} e^{-dc/2\sigma^2},$$

where $b > 0$, $c > 0$, and $d > 0$, then the conditional prior distribution of μ given σ^2 is a normal distribution with mean a and variance σ^2/b. The marginal prior distribution of μ is a Student t distribution with d degrees of freedom, mean a, and variance $[d/(d-2)]c/b$, with $d > 1$ required for the existence of the mean and $d > 2$ required for the existence of the variance. The uncertainty about σ^2 means that in the expression for the variance of μ, σ^2 is replaced by the expected value of σ^2. This expected value comes from the marginal prior distribution of σ^2, which is an inverted-gamma distribution with mean $[d/(d-2)]c$ and variance $2d^2c^2/(d-2)^2(d-4)$. Probabilities involving μ can be found from tables of the Student t distribution, and probabilities involving σ^2 can be related to inverse chi-square probabilities, for which some tables are available (Novick and Jackson, 1974; Isaacs, Christ, Novick, and Jackson, 1974).

Given the normal-inverted gamma density just described, and given a sample of size n with sample mean $\overline{x}$ and sample variance s^2, the posterior normal-inverted-gamma density for μ and σ^2 is of the same form as the prior density with the prior parameters a, b, c, and d replaced by

$$a* = (ba+n\overline{x})/(b+n),$$

$$b* = b+n,$$

$$c* = \{dc+ba^2 + (n-1)s^2 + n\overline{x}^2 - [(ba+n\overline{x})^2/(b+n)]\}/(d+n),$$

and

$$d* = d+n,$$

respectively. As in the known-variance case, the posterior mean of μ is a weighted average of the prior mean and the sample mean, and the sample size n is simply added to b, which might be thought of in terms of the "amount of prior information" concerning μ, and to d, which might be thought of in terms of the "amount of prior information" concerning σ^2. The revision of c is a bit more complicated, because new information about the variance of σ^2 is provided not just by the sample variance s^2, but also by any difference between the prior mean of μ, a, and the sample mean $\overline{x}$. When a and $\overline{x}$ are equal, the revision of c simplifies to $[dc + (n-1)s^2]/(d+n)$, which can be rewritten as a weighted average $(dc+nv)/(d+n)$ if

$$v = \frac{1}{n} \sum_{j=1}^{n} (x_j-\overline{x})^2 = [\frac{(n-1)}{n}] s^2$$

is used instead of s^2.

Returning to the distribution of errors in point forecasts of maximum temperature, suppose that the prior distribution for μ and σ^2

is normal-inverted-gamma with a = 0, b = 4, c = 16, and d = 10. The choices of a and b agree with the values used in the previous illustration of the known-variance case. The prior mean of μ, zero, is the same in the two cases, but the prior variance of μ is slightly larger in the unknown-variance case (5, as opposed to 4) because of the uncertainty about σ^2 and the resulting need to multiply by d/(d-2). Moreover, c = 16 appears to be consistent with the assumption of a variance of 16. If d were allowed to increase without bound, the results from the unknown-variance case would approach the results from the known-variance case. With d = 10, the prior mean of the variance σ^2 is 20, which is larger than 16 because of the factor d/(d-2), and the standard deviation of σ^2 is 11.55. Note that d > b, indicating that there is more prior information about σ^2 than about μ.

The forecaster observes the sample of 20 forecast errors, and, as before, calculates the sample mean, $\bar{x}$ = 0.54. With σ^2 unknown, it is also necessary to calculate the sample variance, and the result is s^2 = 14.6. Combining the prior information and the sample information yields a posterior normal-inverted-gamma distribution with parameters a* = 0.45, b* = 24, c* = 14.61, and d* = 30. The population mean now has an expected value of 0.45 and a variance of 0.65. As for σ^2, its posterior mean and standard deviation are 15.65 (a slight reduction from 20 due to the sample variance of 14.6) and 4.34, respectively.

This section has involved the revision of continuous prior distributions via Bayes' theorem. The notion of conjugate prior distributions has been developed briefly and illustrated. The presentation of this sort of material is somewhat mechanical in nature, with formulas presented and illustrated for the various cases considered. Revision formulas for other parameters such as differences in means, differences in variances, regression coefficients, and parameters of more complex models are available. Mechanical steps can be relegated to the computer, of course, and some computer packages for Bayesian revision have been developed (e.g., Novick, 1971; Schlaifer, 1971). It is important to look beyond the mechanics of probability revision to the underlying modeling issues. Thus, the generation of the necessary inputs to the revision process, the prior distribution and the likelihood function, is discussed briefly in the next section.

5. ASSESSMENT OF LIKELIHOOD FUNCTIONS AND PRIOR DISTRIBUTIONS

Once the likelihood function and the prior distribution have been determined, the generation of a posterior distribution is a conceptually straightforward step. The application of Bayes' theorem may be difficult in some instances, particularly if the prior distribution is not a member of the appropriate conjugate family. Sometimes it is necessary to resort to numerical methods. Nevertheless, once the inputs are determined, the die is cast. The opportunity to capture the important aspects of the real-world situation at hand comes when the inputs are determined. If the likelihood function is a poor representation of the actual data-generating process or if the prior distribution is a poor reflection of the prior information that is available, then even though the application of Bayes' theorem is correct, the results may not be very meaningful. Bayesian statistics certainly does not have a monopoly on this type of problem. Modeling is just as crucial in classical statistics as it is in Bayesian statistics. Indeed, in any modeling, quantitative or otherwise, the output of the model is only as good as the inputs. Using either the classical approach or the Bayesian approach, one can use a Bernoulli model and perform all sorts of calculations correctly with that model, but the effort will all be for naught (or worse than that, if the results mislead naive or even sophisticated readers) if the assumptions underlying the Bernoulli model are violated grossly. This is not to say that the assumptions of a model need to be met exactly; it would be surprising if they could be met exactly except in the simplest of real-world situations. What is needed is an understanding of the assumptions and some care in attempting to determine whether the assumptions provide a reasonable approximation and to investigate the effects on inferences of possible violations of the assumptions.

5.1 Assessment of Likelihoods

In the Bayesian approach to statistics, new information is incorporated into the analysis via the likelihood function. Likelihoods can be assessed directly without the use of any model, but the assessment and computational aspects of the analysis are generally simplified if a parametric model is used to represent the data-generating process. The step of choosing a parametric model is no different in a Bayesian analysis than it is in a classical analysis. Furthermore, although it may be possible to examine some data from the process to investigate the suitability of alternative models, the ultimate choice of a model is subjective in nature.

Consider a meteorologist studying the occurrence of rain at a particular location. The available information consists of a sequence of daily observations of "rain" or "no rain." The simplest model for this situation is that of a Bernoulli process, with the Bernoulli parameter p representing the probability of rain on any given day. However, the applicability of this model must be examined carefully. Because of persistence, the meteorologist may feel that the process is not independent. That is, the probability of rain on a certain day may not be independent of the previous weather. An alternative model allowing for some dependence is a first-order Markov model, with the probability of rain represented by p_1 if rain occurred on the previous day and p_2 if rain did not occur on the previous day. Persistence in the weather would be indicated in this model by $p_1 > p_2$, with the degree of persistence increasing as p_1 approaches one and p_2 approaches zero. Of course, persistence could have an effect for more than one day. A higher-order Markov model would take into account the relationship between, say, today's weather and the weather during the past several days.

In contemplating a model to represent the occurrence or non-occurrence of rain, the meteorologist would probably want to look at any available sequence of data to make sure that the data and the model are not seriously at odds. If, for example, rainy days seem to be followed by rainy days about the same proportion of the time as non-rainy days tend to be followed by rainy days, the meteorologist might feel that persistence is not an important factor at the location in question and might be willing to adopt a Bernoulli model. On the other hand, if rainy days tend to occur in clusters of several consecutive days, which seems more realistic, some sort of a Markov model might be judged to be appropriate. The meteorologist's investigation of the data could include informal "eyeballing" of the data as well as more formal measures and tests of properties such as independence and stationarity. For example, Yakowitz (1976) discusses Bayesian procedures for making inferences about the order of a Markov model and applies these procedures to hydrological data.

In general, it is advantageous to use the simplest model that provides a reasonable representation of the situation at hand. Thus, simplifying properties such as exchangeability are very important properties to consider. An informal definition of exchangeability is that a process is exchangeable if for any number of observations and sequence of outcomes, all sequences with the same outcomes are viewed as equally likely. If a process is exchangeable, then the observations may be treated as independent and identically distributed with some common distribution. When the observations are dichotomous, the Bernoulli model is appropriate. When the observations are continuous, then they have some common density function $f(x|\theta)$, and various models might be considered to represent f. A common choice for f is the normal density function, and various formal and informal procedures are available to test for normality.

The examples in this chapter have involved very simple models of data-generating processes, and the availability of "fancier" models should be emphasized. For example, second-order models might be suitable in certain situations. In the severe-storms example, for instance, the Poisson parameter λ might change from week to week. The occurrence of severe storms on a particular week might follow a Poisson process but the Poisson parameter itself might follow some second-order distribution (e.g., a gamma distribution). To make inferences about the distribution of λ, a second-order model is necessary because this distribution is two steps removed from the actual data. The intermediate step involves the sequence of weekly values of λ, and this sequence is unobservable. On a given week, the number of severe storms can be observed but the value of λ cannot be observed. Second-order models are among the models discussed in Lindley (1971). Other models, including models frequently used by classical statisticians, are discussed from the Bayesian viewpoint in Box and Tiao (1973).

Excellent examples of more complex models of data-generating processes can be found in the econometric literature, and the study of some of these models from the Bayesian viewpoint is presented in Zellner (1971) (see also Fienberg and Zellner, 1975). Such models range from linear regression models with identically distributed normal error terms to models involving autocorrelated errors, heteroscedastic errors, errors in the variables, nonlinear regression, autoregression, distributed lags, and simultaneous equations. As the model becomes more complicated, the likelihood function tends to become more complicated and the number of parameters generally increases. Thus, econometricians are pleased when it appears that a simple linear regression with identically distributed normal error terms is a suitable model. On the other hand, many of them are well aware of the need to carefully check the assumptions underlying this model and are willing to work with more complex models if assumptions such as linearity, normality, independence, or homoscedasticity are violated.

5.2 Assessment of Prior Distributions

When a parametric model is chosen to represent a data-generating process, interest generally focuses on the parameters of that model. Data from a sample or experiment provide information about the parameters, and this is brought into the formal analysis via the likelihood function. In the classical approach to statistics, inferences are often based directly on the likelihood function. Examples are maximum likelihood estimators and likelihood ratio tests. In the Bayesian approach to statistics, the likelihood function represents one input to the analysis, and the prior distribution represents another input. When a parametric model is used to represent the data-generating process, the choice of a model indicates the parameters for which a prior distribution is needed.

The prior distribution of a parameter is supposed to represent the information available about the parameter before the new information (i.e., the new sample or the new experiment) is observed. If the past information consists exclusively of past data from a previous sample, then the prior distribution is said to be data-based, and the combination of prior and sample information is directly analogous to the pooling of information from different samples. Often, however, much of the prior information will be of a more subjective nature, and the assessment of a prior distribution then involves the elicitation of subjective probabilities. The intent is to include all available information concerning the parameter of interest, and the formal

justification for the consideration of subjective probabilities is
provided by the axiomatic development of the theory of subjective
probability (e.g., see Savage, 1954, and de Finetti, 1970, 1972).

Various methods have been proposed and used for the assessment of
subjective probability distributions. In the discrete case, the
assessor could simply assess probabilities for the possible values of
the parameter, perhaps using concepts such as lotteries or betting
odds as mental aids in the assessment process. In the continuous
case, the assessor might assess cumulative probabilities at a number
of specific values of the parameter or assess some fractiles of the
distribution. These assessments can be graphed and used as the basis
for a cumulative distribution function. For discussions of assessment
techniques, see Winkler (1967) and Spetzler and Staël von Holstein
(1975). Of course, the assessment of subjective probabilities raises
psychological questions concerning judgmental processes, and a review
of the subjective probability literature from this viewpoint is provided in Hogarth (1975).

The assessment of prior distributions for use in Bayesian analysis raises some modeling questions in addition to questions concerning
the assessment process. Just as the goal in the choice of a parametric model to represent a data-generating process should be to capture the real-world situation as accurately as possible with as simple
a model as possible, the goal in the choice of a prior distribution
should be to represent the prior information as accurately as possible
with a prior distribution that simplifies the analysis as much as
possible. Simple discrete prior distributions such as those considered in the examples in Section 3 are relatively easy to work with,
but they often offer only very crude approximations of the prior
information. It is usually more realistic to assume that the parameter is continuous and to assess a continuous prior distribution. As
indicated in Section 4, however, the application of Bayes' theorem in
the continuous case might be very difficult unless the prior distribution is restricted in some manner. The restriction that is particularly appealing involves the use of conjugate prior distributions,
which lead to relatively tractable analyses. If a prior distribution
is assessed subjectively, then, one might attempt to find a member of
the conjugate family of distributions that provides a good approximation to the assessed distribution. The question of how good the
approximation should be in order to justify use of the conjugate
distributions is, of course, a modeling issue. An alternative
approach is to use an assessment technique that leads directly to a
conjugate prior distribution. Since the conjugate prior distribution
is of the same functional form (in terms involving the parameter of
interest) as the likelihood function, the parameters of the prior
distribution can be compared to sample statistics. If the prior
information can be viewed as equivalent to the information that would
be provided by a sample with certain sample statistics, then those
sample statistics can be taken as the prior parameters.

Ultimately, inferences about a parameter are based on its posterior distribution. Therefore, questions concerning the goodness of certain approximations in terms of the prior distribution should be viewed in terms of the impact of the approximation on the posterior distribution. Sometimes fairly substantial changes in the prior distribution lead to relatively minor changes in the posterior distribution, in which case it is said that the posterior distribution is relatively insensitive to changes in the prior distribution. One situation of particular interest is that in which the sample information overwhelms the prior information (see Edwards, Lindman, and Savage, 1963). This might occur, for example, when a large sample is taken and the prior information is quite limited. The prior distribution then has very little impact on the posterior distribution, and the prior information is said to be diffuse relative to the sample information. Since the prior information contributes very little to the analysis, it is not worthwhile to spend much time on its assessment. Although there is some disagreement as to the criteria that should be used to arrive at the exact form of a diffuse distribution, standard diffuse prior distributions have been developed. For the normal process with known variance, the conjugate family is the normal family, and a commonly-used representation of diffuseness is an improper normal prior distribution with infinite variance. The distribution is improper in the sense that the area under the density function is not one (the integral does not converge) because of the infinite variance. In the notation of Section 4, this is a normal prior distribution for μ with b = 0. The posterior mean and variance of μ are then simply $\bar{x}$ and σ^2/n. Even in cases where the prior distribution is not diffuse, a supplementary analysis with a diffuse prior distribution is often conducted to see what the results look like when based on the sample information alone.

The discussion in this section of modeling issues that arise in Bayesian statistics has merely scratched the surface. As the situation being modeled becomes more complex, the modeling process generally becomes more difficult. For example, when observations are multivariate, possible dependence among variables might be very important. Assumptions underlying potential models of the data-generating process can be hard to check, and these models usually have many parameters. In turn, as the number of parameters increases, relationships among these parameters must be considered, and the assessment of a prior distribution becomes more difficult. As the model becomes more complex, it may also be harder to investigate the sensitivity of inferences to deviations from the model. Nevertheless, the general approach to modeling is the same in complex situations as it is in relatively simple situations. Insofar as possible, the degree to which the model captures the real-world situation and the sensitivity of inferences or decisions to violations of the assumptions underlying the model should be investigated.

6. ESTIMATION, HYPOTHESIS TESTING, PREDICTION, AND DECISION MAKING

In Bayesian inference, the primary concern involves the application of Bayes' theorem to combine prior and sample information and form a posterior distribution. The posterior distribution itself is the primary inferential statement in the Bayesian approach. Sometimes it is convenient for reporting purposes to summarize this distribution in some manner. Moreover, the emphasis in classical statistics on estimation and hypothesis testing has led to the development of Bayesian counterparts to classical estimates and tests. In this section the use of posterior distributions in problems of estimation, hypothesis testing, prediction, and decision making is briefly discussed, with some comparison of Bayesian and classical approaches where appropriate. For a more detailed comparison of different approaches to statistical inference, see Barnett (1973).

6.1 Estimation

One type of inferential procedure is that of point estimation, in which a single point, or single value, is used to estimate an unknown parameter. The classical statistician, of course, bases point estimates entirely on the sample information. Thus, a sample statistic is used to estimate an unknown parameter; for instance, the sample mean $\bar{x}$ might be used to estimate the mean μ of a particular process, and the sample proportion r/n might be used to estimate the parameter p of a Bernoulli process. A number of properties (e.g., sufficiency, unbiasedness, efficiency, consistency) have been put forth as good properties for estimators to possess.

In the Bayesian approach, any point estimate should be based on the posterior distribution, and the determination of such a point estimate amounts to choosing a measure of location for the distribution. Possible choices include, but are not limited to, the mean, median, and mode of the posterior distribution. If a loss function is available for the point estimation problem, the choice of an estimate becomes a decision-making problem with the objective of minimizing the expected loss. A symmetric squared-error loss function, for example, would result in the posterior mean being the optimal estimate. Certain comparisons with classical procedures are sometimes possible. If the prior distribution is uniform (one possible choice for a diffuse distribution), then the posterior mode is equal to the maximum likelihood estimate.

Another form of estimation is interval estimation, in which the estimate is an interval of values rather than a single value. From the posterior distribution, the probability of any interval of values of the parameter can be determined. To find an interval containing 90 percent of the probability, one could consider the interval from the 0.05 fractile to the 0.95 fractile, the interval from the 0.01 fractile to the 0.91 fractile, and so on. Two common choices are (1) an interval that leaves the same amount of probability in each tail of the distribution and (2) a region (which need not be but usually is a single interval) for which the height of the density function for points outside the region is less than or equal to the density for points included in the region. The latter approach produces what is called a highest density region.

To illustrate Bayesian interval estimation, consider once again the known-variance case of the normal example from Section 4. Here the parameter of interest is μ and the posterior distribution of μ is a normal distribution with mean 0.45 and variance 0.667. An approximate 95 percent interval estimate for μ based on this posterior distribution has limits $0.45 \pm 1.96(0.667)^{1/2}$, or (-1.15, 2.05). By comparison, a classical 95 percent confidence interval in this example has limits $0.54 \pm 1.96(4)/(20)^{1/2}$, or (-1.21, 2.29). In the unknown-variance case, both Bayesian and classical interval estimates involve the use of tables of the Student t distribution.

In the normal example, the limits of the Bayesian 95 percent interval estimate differ from the limits of the classical 95 percent interval estimate. This difference is due to the inclusion of prior information in the Bayesian analysis. An even more important difference between the two intervals involves their interpretation. The Bayesian interval is based on the posterior distribution of μ and can be interpreted as a probability statement about μ: the probability is approximately 0.95 that μ lies in the interval from -1.15 to 2.05. In the classical approach, the parameter μ is viewed as unknown but fixed. It is not a random variable, and the classical interval estimate is not a probability statement about μ. The randomness is associated with the way in which the sample is drawn, and the interpretation of the classical interval estimate is that if repeated independent samples of the same size (n = 20) were drawn and a 95 percent confidence interval were found for each of the samples, then approximately 95 percent of these intervals would contain the true value of μ. The classical interpretation of confidence intervals is not as intuitively appealing as the Bayesian interpretation of interval estimates based on the posterior distribution, and many students and users of classical confidence intervals persist in interpreting these intervals as probability statements about μ rather than as probability statements about intervals genererated by repeated sampling. The natural way to think about an interval estimate is in terms of a probability statement about μ, and the Bayesian approach addresses the question of interest directly.

6.2 Hypothesis Testing

Instead of estimating a parameter, one may want to test a particular hypothesis concerning that parameter against some alternative hypothesis. In the Bayesian approach, the two hypotheses can be compared in terms of a posterior odds ratio, which is the ratio of the posterior probabilities of the hypotheses. If the two hypotheses of interest are labeled H_1 and H_2, the posterior odds ratio of H_1 to H_2 is $P(H_1|y)/P(H_2|y)$. A high odds ratio favors H_1, whereas a low odds ratio favors H_2. If one of the two hypotheses must be chosen, then it is necessary to determine a critical value of the odds ratio. Odds ratios above the critical value lead to the choice of H_1, while odds ratios below the critical value lead to the choice of H_2. As in any choice between two actions, two mistakes are possible: H_1 could be chosen when H_2 is in fact true, or H_2 could be chosen when H_1 is true. If these two mistakes are judged to be about equally serious, then the logical choice of a critical value for the odds ratio is one. When the posterior odds ratio is one, the two hypotheses are equally likely, so that a critical value of one amounts to a decision to choose the hypothesis that has the higher posterior probability. Of course, if one mistake is thought to be more serious than the other mistake, the critical value of the odds ratio should be adjusted accordingly.

Some comparisons can be made between Bayesian and classical hypothesis testing. Although classical tests are usually expressed in terms of critical regions for certain statistics (e.g., reject H_1 if $t \geq 1.7$), most tests used in practice are likelihood ratio tests. That is, the rejection region can be expressed in the form LR $\leq$ k, where k is a critical value of the likelihood ratio LR. The likelihood ratio is, as the name implies, simply a ratio of the likelihoods associated with the two hypotheses: LR $= P(y|H_1)/P(y|H_2)$ for discrete data-generating processes and LR $= f(y|H_1)/f(y|H_2)$ for continuous data-generating processes. But the likelihood ratio and posterior odds ratio are related, since Bayes' theorem can be expressed in terms of odds. The posterior odds ratio is equal to the product of the prior odds ratio (the ratio of the prior probabilities of H_1 and H_2) and the likelihood ratio. Thus, if the two hypotheses are equally likely a priori, the posterior odds ratio equals the likelihood ratio, and a Bayesian test is in fact a likelihood ratio test. When the prior odds ratio is not equal to one, of course, the posterior odds and the likelihood ratio are not equal, but the former might be thought of as an adjusted version of the latter.

As in interval estimation, the results of Bayesian and classical tests have different interpretations even when the numerical results happen to be similar. In classical hypothesis testing, the error probabilities are selected in advance (e.g., a significance level of 0.05 might be chosen), and they are interpreted as the probabilities of errors in repeated samples (e.g., if the procedure were carried out repeatedly and H_1 were true, H_1 would be rejected about 5 percent of the time). In Bayesian hypothesis testing, the error probabilities are posterior probabilities calculated after the sample is observed, and they are interpreted as probabilities about H_1 and H_2 given that particular sample. In the known-variance case of the normal example from Section 4.3, suppose that the hypotheses H_1: $\mu \geq 0$ and H_2: $\mu < 0$ are of interest. From the posterior distribution,

$$P(H_1|y) = P(\mu \geq 0|\overline{x} = 0.54) = 1 - \Phi[(0 - 0.45)/0.82] = 0.71$$

and

$$P(H_2|y) = P(\mu < 0|\overline{x} = 0.54) = 0.29,$$

where Φ represents the standard normal cumulative distribution function. Here H_1 appears to be more likely than H_2, with the posterior odds ratio equaling $0.71/0.29 = 2.45$.

Certain types of tests encountered frequently in classical analyses create some difficulties when studied from the Bayesian viewpoint. For example, consider the hypotheses H_1: $\mu = 0$ and H_2: $\mu \neq 0$. If the posterior distribution of μ is continuous, as it is when a normal prior distribution is assessed for μ, any single value such as zero has a posterior probability of zero. As a result, it makes no sense to even contemplate the hypothesis that $\mu = 0$. What is generally meant by this sort of hypothesis is that μ is close to zero, but the notion of "close" is left quite vague. The Bayesian approach makes it necessary to explicitly state the hypothesis of interest. If it is $-1 \leq \mu \leq 1$, then the posterior probability of the hypothesis can be computed. When the hypothesis is stated in the form $\mu = 0$, it makes sense only if the value zero is assigned a nonzero probability in the posterior distribution.

6.3 Prediction

The bulk of traditional statistical practice involves estimation and hypothesis testing, and the use of posterior distributions in problems of estimation and hypothesis testing has been discussed briefly. Another type of problem is that of predicting a future observation. In the Bayesian approach, a predictive distribution can be found for a future observation, and probability statements are based on this distribution. In the case of a discrete data-generating process and a discrete posterior distribution, for example, the predictive distribution for a future observation x is

$$P(x) = \sum_{j=1}^{J} P(\theta_j)P(x|\theta_j). \tag{9}$$

The probabilities $P(x|\theta_j)$ are obtained from the model chosen to represent the data-generating process. If we knew the actual value of θ, then this conditional probability distribution of x given θ could be used to make predictions about x. Since we do not know θ, conditional distributions of x given different values of θ are weighted by the posterior probabilities of different values of θ in order to determine the predictive distribution.

In the example in Section 3.2 with a discrete prior distribution for the Poisson parameter λ, the average rate of occurrence of severe storms, the posterior probabilities are $P(\lambda = 0.4|r) = 0.007$, $P(\lambda = 0.6|r) = 0.196$, $P(\lambda = 0.8|r) = 0.549$, and $P(\lambda = 1.0|r) = 0.248$. The conditional probabilities for x, the number of severe storms in the next week, are Poisson probabilities, and marginal, or predictive, probabilities for x are weighted averages of these Poisson probabilities. For example,

$$P(x = 0|\lambda = 0.4) = 0.6703,$$

$$P(x = 0|\lambda = 0.6) = 0.5488,$$

$$P(x = 0|\lambda = 0.8) = 0.4493,$$

$$P(x = 0|\lambda = 1.0) = 0.3679.$$

Thus,

$$P(x = 0) = (0.007)(0.6703) + (0.196)(0.5488)$$
$$+ (0.549)(0.4493) + (0.248)(0.3679)$$
$$= 0.4502.$$

In a similar manner, predictive probabilities for other value of x can be computed, and they are $P(x = 1) = 0.3550$, $P(x = 2) = 0.1443$, $P(x = 3) = 0.0402$, $P(x = 4) = 0.0086$, $P(x = 5) = 0.0015$, and $P(x = 6) = 0.0002$.

For continuous data-generating processes with continuous prior distributions, the predictive distribution for a future observation x has density

$$f(x) = \int_{-\infty}^{\infty} f(\theta)f(x|\theta)d\theta. \tag{10}$$

Certain combinations of likelihood functions and prior distributions (e.g., combinations in which the prior distribution is a member of the conjugate family) yield especially tractable results. For example, with a normal data-generating process with known variance, if the prior distribution of the mean μ is a normal distribution with mean

a and variance σ^2/b, then the predictive distribution for a future
observation from the process is normal with mean a and variance
$(b + 1)\sigma^2/b$. The uncertainty about the future observation can be
broken down into two sources of uncertainty: uncertainty about x due
simply to the variability of the process, and uncertainty about the
mean of the process. The former uncertainty contributes σ^2 to the
variance of x, whereas the latter uncertainty contributes σ^2/b to the
variance of x. Hence the total variance is $\sigma^2 + (\sigma^2/b)$, or
$(b + 1)\sigma^2/b$.

Probabilities for model parameters are not admitted in the
classical approach to statistics, and without such probabilities,
marginal probabilities for future observations cannot be determined.
In the Poisson example, for instance, a classical statistician can use
the Poisson distribution to make probability statements about a future
observation x, but these probabilities are conditional upon λ, the
Poisson parameter. Not having a distribution for λ, the classical
statistician cannot make unconditional probability statements about x.
As a result, the problem of predicting future observations from a data-
generating process has not received much attention from the classical
viewpoint. Yet in many real-world problems, the inferences or decisions of interest concern such future observations. Inferences about
the population mean error in categorical forecasts of maximum temperature may be of interest, but when a temperature-sensitive decision-
making problem is faced on a particular day, the variable of primary
interest is the actual forecast error on that day. Even among
Bayesian statisticians, problems of prediction have not received as
much attention as they deserve (however, see Aitchison and Dunsmore,
1975).

6.4 Decision Making

The primary focus in this chapter is on the inferential aspects
of Bayesian statistics. The use of Bayes' theorem to revise probabilities provides posterior probabilities, the main inferential statements of interest in the Bayesian approach. But these probabilities
can also be very important inputs in problems of decision making under
uncertainty. Indeed, the motivation for much of the recent interest
in Bayesian methods has been decision-theoretic in nature. The outcomes of decision-making problems often depend on particular events or
variables, and the decision maker is frequently quite uncertain about
the outcomes of the events or values of the variables. Probability
theory provides a framework for the quantification of uncertainty, and
Bayesian procedures enable the decision maker to update probabilities
as new information becomes available. Probability revision becomes
especially important in dynamic decision-making problems where sequences of decisions must be made over time and new information becomes

available over time. Furthermore, by anticipating the possible reactions to new sample information before it is actually obtained, the decision maker can determine the expected value of the sample information. For more details concerning the modeling of decision-making problems under uncertainty, see Chapter 13 of this book. The point of interest for the purposes of this chapter is that Bayesian procedures are of great value in the modeling of decision-making problems under uncertainty.

7. BAYESIAN INFERENCE IN METEOROLOGY

Bayesian inference is based on Bayes' theorem, which itself can be used to update probabilities in meteorological applications. In particular, probability forecasts can be revised as new pieces of information (possibly including other forecasts) become available. Moreover, Bayesian methods can be used to study the value of different types of forecasts (or other meteorological information) in weather-sensitive decision-making problems. For example, the value of climate information is investigated in Phillips and Keelin (1978) and Winkler, Murphy, and Katz (1980). This relates to the modeling of decision-making problems under uncertainty, which is discussed in the following chapter of this book.

More generally, "Bayesian inference" refers to an approach to problems of statistical inference. As such, Bayesian inference is applicable in the myriad of situations in which statistics is used in meteorology. Inferences about parameters of statistical models are based on posterior distributions of the parameters.

An example of Bayesian inference in meteorology is provided by Simpson, Olsen, and Eden (1975), who present a Bayesian analysis of a weather modification experiment designed to test dynamic seeding. Dynamic seeding is massive silver iodide seeding which is postulated to lead to invigorated cloud growth and hence increased precipitation. In the study by Simpson, Olsen, and Eden, the rain volume falling from a cloud is assumed to follow a gamma distribution, where the scale parameter is different for seeded (treatment) and unseeded (control) clouds but the shape parameter is the same for these two conditions. The posterior distribution for the "seeding factor," the ratio of the scale parameters for the seeded and unseeded clouds, is shown in Figure 1. Although this distribution is quite spread out, the posterior probability that $\theta \leq 1$ is less than 0.01. Thus, the authors conclude (p. 164) that "there is a positive seeding effect for single clouds and a tentative estimate of its magnitude is the modal value 2.4." The posterior distribution of the common shape parameter is centered at about 0.6, which is the value used for the shape parameter in previous work with this gamma model for rainfall amount.

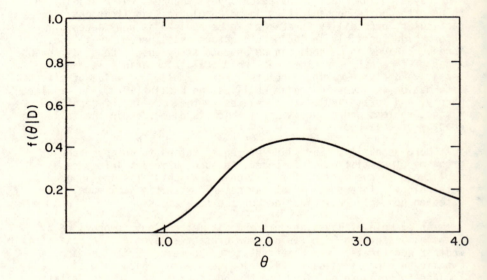

Figure 1. Posterior density for seeding factor θ, the ratio of the
scale parameters for seeded and unseeded clouds (from
Simpson et al., 1975).

A Bayesian analysis of a different type of parametric model involving time series data is provided by Rao (1980), who investigates the effects of urbanization on precipitation. Integrated Moving Average (IMA) models of annual precipitation data are used, and posterior distributions of the parameters of these IMA models are obtained. These posterior distributions suggest that changes in precipitation attributed to the effects of urbanization occur very gradually and that jumps in the data are small.

Bayesian methods can also be used in other types of statistical problems, such as classification. Rodgers, Siddalingaiah, Chang, and Wilheit (1979) study the ability to monitor remotely the coverage and movement of rain over land areas by using Nimbus 6 Electrically Scanning Microwave Radiometer (ESMR) measurements. The objective is to distinguish among land rain, dry ground, and wet soil. Three different classification procedures (Fisher, Bayesian, and nonparametric) are investigated, and the major difficulty is in distinguishing between rainfall over land and wet land surfaces. Overall, the Bayesian algorithm performs better than the other two methods.

These brief summaries of actual applications provide examples of Bayesian inference in meteorology. A posterior distribution is an intuitively appealing way to summarize the information concerning a parameter, as in Figure 1 for the weather modification study. Moreover, Bayesian methods can be used for various data-generating models (e.g., the IMA model in the rainfall study) and for different purposes (e.g., the classification algorithm in the study of remote monitoring of rainfall). As noted in Section 1, the interest in Bayesian methodology has increased considerably in the past two decades. The use of Bayesian procedures in practice should increase for meteorological applications as well as for applications in other fields.

REFERENCES

Aitchison, J., and I. R. Dunsmore, 1975: Statistical Prediction Analysis. Cambridge, Cambridge University Press.

Barnett, V., 1973: Comparative Statistical Inference. London, Wiley.

Bayes, T., 1763: An essay toward solving a problem in the doctrine of chance. Philosophical Transactions of the Royal Society, 53, 370-418. Reproduced with biography of Bayes in G. A. Barnard, 1958: Studies in the history of probability and statistics: IX. Biometrika, 45, 293-315.

Box, G. E. P., and G. C. Tiao, 1973: Bayesian Inference in Statistical Analysis. Reading, Mass., Addison-Wesley.

de Finetti, B., 1937: La prévision: Ses lois logiques, ses sources subjectives. Annales de L'Institut Henri Poincare, 7, 1-68. Translated by H. E. Kyburg, 1964: Foresight: Its logical laws, its subjective sources. In H. E. Kyburg and H. E. Smokler (eds.), Studies in Subjective Probability. New York, Wiley.

de Finetti, B., 1970: Teoria Delle Probabilita. Torino, Guilio Einaudi. Translated by A. Machi and A. Smith, Theory of Probability, Volumes 1 and 2, 1974 and 1975. London, Wiley.

de Finetti, B., 1972: Probability, Induction, and Statistics: The Art of Guessing. London, Wiley.

DeGroot, M. H., 1970: Optimal Statistical Decisions. New York, McGraw-Hill.

Edwards, W., H. Lindman, and L. J. Savage, 1963: Bayesian statistical inference for psychological research. Psychological Bulletin, 70, 193-242.

Fienberg, S. E., and A. Zellner (eds.), 1975: Studies in Bayesian Econometrics and Statistics in Honor of Leonard J. Savage. Amsterdam, North-Holland.

Hogarth, R., 1975: Cognitive processes and the assessment of subjective probability distributions. Journal of the American Statistical Association, 70, 271-289.

Isaacs, G. L., D. E. Christ, M. R. Novick, and P. H. Jackson, 1974: Tables for Bayesian Statisticians. Iowa City, University of Iowa.

LaValle, I. H., 1970: An Introduction to Probability, Decision, and Inference. New York, Holt, Rinehart and Winston.

Lindley, D. V., 1971: Bayesian Statistics: A Review. Philadelphia, Society for Industrial and Applied Mathematics.

Novick, M. R., 1971: Bayesian computer-assisted data analysis. Iowa City, American College Testing Program, Technical Bulletin 3.

Novick, M. R., and P. H. Jackson, 1974: Statistical Methods for Educational and Psychological Research. New York, McGraw-Hill.

Phillips, L. D., and T. W. Keelin, 1978: Bayesian modelling of improved climatological forecasts for large agricultural models. In M. H. Glantz, H. van Loon, and E. Armstrong (eds.), Proceedings, Multidisciplinary Research Related to the Atmospheric Sciences. Boulder, Colo., National Center for Atmospheric Research, 213-233.

Raiffa, H., and R. Schlaifer, 1961: Applied Statistical Decision Theory. Boston, Graduate School of Business Administration, Harvard University.

Rao, A. R., 1980: Stochastic analysis of annual rainfall affected by urbanization. Journal of Applied Meteorology, 19, 41-52.

Rodgers, E., N. Siddalingaiah, A. T. C. Chang, and T. Wilheit, 1979: A statistical technique for determining rainfall over land employing Nimbus 6 ESMR measurements. Journal of Applied Meteorology, 18, 978-991.

Savage, L. J., 1954: The Foundations of Statistics. New York, Wiley.

Schlaifer, R., 1959: Probability and Statistics for Business Decisions. New York, McGraw-Hill.

Schlaifer, R., 1969: Analysis of Decisions Under Uncertainty. New York, McGraw-Hill.

Schlaifer, R., 1971: Computer Programs for Elementary Decision Analysis. Boston, Graduate School of Business Administration, Harvard University.

Simpson, J., A. Olsen, and J. C. Eden, 1975: A Bayesian analysis of a multiplicative treatment effect in weather modification. Technometrics, 17, 161-166.

Spetzler, C. S., and C.-A. S. Staël von Holstein, 1975: Probability encoding in decision analysis. Management Science, 22, 340-358.

Winkler, R. L., 1967: The assessment of prior distributions in Bayesian analysis. Journal of the American Statistical Association, 62, 776-800.

Winkler, R. L., A. H. Murphy, and R. W. Katz, 1980: Assessing the value of climate information. In Proceedings of the Conference on Climate and Risk (Arlington, Va.). McLean, Va., MITRE Corp. 3-1-2-28.

Yakowitz, S. J., 1976: Small-sample hypothesis tests of Markov order with applications to simulated and hydrologic chains. Journal of the American Statistical Association, 71, 132-136.

Zellner, A., 1971: An Introduction to Bayesian Inference in Econometrics. New York, Wiley.

13
Decision Analysis

Robert L. Winkler and Allan H. Murphy

1. INTRODUCTION

Much statistical work focuses on problems of inference, as
indicated by the contents of this book. Given some information, but
not complete information, about certain events, variables, or model
parameters, the statistician attempts to make inferential statements
about their values. Under conditions of uncertainty, inferences may
be of interest, but frequently the interest goes beyond inferences.
Many important decisions are made in the face of uncertainty, and it
is often useful to model such decision-making problems. The modeling
effort can help the decision maker to understand the problem better
and can point the way toward a "good" decision.

In modeling a decision-making problem, a decision maker should
first consider the actions that are available and the events or
variables that are relevant in the sense that they may affect the
final outcome. The potential outcomes, or consequences that will be
received as a result of the decision maker's action and the events
that occur, must be considered, as must the decision maker's preferences concerning the various outcomes. Uncertainty is taken into
account in terms of forecasts of events or variables, with these
forecasts being expressed in probabilistic terms.

Once a decision-making problem is modeled, a decision criterion
can be applied to determine the decision that is optimal for the
given model and the chosen criterion. The criterion of primary
interest in this chapter is the maximization of expected utility,
although other criteria will be discussed briefly. The model can
also be used to investigate the advisability of purchasing additional
information (e.g., additional forecasts) before making a final decision, and the sensitivity of the results to changes in various inputs
(e.g., probabilities, outcomes) can be investigated.

Although not all decision-making models involve the use of
Bayesian techniques, it is often quite important in such models to
revise probabilities as new information is obtained. For example,

situations involving sequences of decisions to be made over time, with information being received between decisions, necessitate probability revision. As a result, the development of methods for modeling decision-making problems under uncertainty has to some degree paralleled the development of Bayesian methods. The book by Raiffa and Schlaifer (1961) provides an example of such parallel development. Modeling of decision-making problems under uncertainty has often been labeled statistical decision theory, or simply decision theory. In recent years, the term decision analysis, which is sometimes taken to suggest a more applied orientation than decision theory, but which deals essentially with the same problems and methods, has gained favor. Thus, decision analysis is used in this chapter to connote the modeling of decision-making problems under uncertainty.

The purpose of this chapter is to provide an introduction to decision analysis. For more extensive developments of decision analysis, see Raiffa (1968), Schlaifer (1969), Halter and Dean (1971), Lindley (1971), Winkler (1972), Brown et al. (1974), Moore and Thomas (1976), Anderson et al. (1977), LaValle (1978), Holloway (1979), and Keeney (1982).

The outline of this chapter is as follows. The elements of decision analysis (actions, events, probabilities, consequences, utilities) and the structuring of these elements in the form of decision tables and decision trees are presented in Section 2, and decision criteria are discussed in Section 3. The assessment of probabilities and utilities is covered in Sections 4 and 5, respectively. Section 6 considers the determination of the value of various types of information, sequential decisions are studied in Section 7, and a brief discussion of sensitivity analysis appears in Section 8. Some applications of decision analysis to weather-related decision-making problems are described in Section 9.

2. ELEMENTS OF DECISION ANALYSIS

Decision analysis necessitates the consideration of various elements: actions, events, probabilities, consequences, and utilities. The determination of such elements is the subject of this section. Also, the structuring of the elements in the form of decision tables and decision trees is discussed.

The most basic element of decision analysis is a set of possible actions, or decisions. Without alternative actions, there is no decision-making problem. Listing the possible actions seems like a simple task, and often it is. In some situations, however, a potential action is inadvertently ignored or opportunities to create additional actions are not considered. A careful analysis may suggest that certain actions not considered favorable initially are, in fact, quite appealing. At the beginning of the modeling effort, therefore, it is advisable to consider a broad set of actions which can be pared down later in the analysis.

The consequence eventually received by the decision maker will depend on the action which is chosen. This action is, of course, under the control of the decision maker. The consequence will also depend on certain events which are not under the decision maker's control. Thus, the decision maker should attempt to list these events. For example, an orchardist faced with possible frost damage on a particular night would be primarily concerned with that night's temperatures in deciding whether or not to attempt to protect his orchard against such frost damage (e.g., by using wind machines to circulate the air and/or heaters to warm the air). Of course, it is important that the events be defined carefully. Is the orchardist concerned only with the minimum temperature for the night? Perhaps the length of time the temperature is at or around that minimum makes a difference, and other factors such as dewpoint or wind speed could be relevant. In order to keep the model from becoming too complex, it is desirable to be parsimonious in the consideration of events. As in the case of actions, however, it is a good idea at least to contemplate a relatively wide range of events before paring the list down.

Once the actions and events are defined, the decision maker can focus on these actions and events and their relationships. In terms of the events, forecasts are needed. There is uncertainty concerning the events, and uncertainty can be represented formally by probabilities. Thus, the forecasts of interest are probability forecasts. If the occurrence of measurable precipitation is an event of interest, a PoP (probability of precipitation) forecast is needed. If the minimum temperature is of interest, a probability distribution for this variable is needed. If the only concern is whether the temperature will go below a certain value (e.g., if temperatures below that value will cause frost damage to fruit trees), then the forecast could consist of the probability that the temperature will dip below that value instead of the entire probability distribution of the minimum temperature.

Forecasts relate only to the events of interest, whereas the ultimate outcome, or consequence, to the decision maker depends both on the action that is chosen and on the events. For each possible combination of an action and an event, we would like to know the consequence. For example, if the orchardist does not protect against frost damage and the temperature remains above the critical level, then no damage is suffered and no cost of protection is incurred. If, on the other hand, the orchardist takes protective action unnecessarily (because the temperature remains high), then there is still no frost damage, but there is a cost associated with the protective action. With a low enough temperature, however, the lack of protection will lead to frost damage that could be avoided (again, at a certain cost) by taking protective action. The decision maker needs to assess all of the potential consequences, and it is possible that some of these consequences may themselves be uncertain events. For example, even if the temperature is very low and no protective action is taken, the orchardist may be uncertain as to the amount of frost damage, either because of lack of experience or because the amount of damage at a given temperature is inherently stochastic. Uncertainty concerning consequences is treated just like uncertainty concerning events, with probability forecasts being formulated to represent the uncertainty.

For modeling purposes, it is convenient to express the consequences of a decision-making problem in terms of a single numeraire. The simplest case occurs when the consequences are all monetary in nature or can be converted to monetary terms. Then the consequence is merely the amount of money involved. In some situations there are serious nonmonetary consequences that are difficult to convert into monetary terms. For example, in a decision as to whether to evacuate a coastal town that is threatened by a hurricane which is headed in the general direction of the town, the potential consequences include possibilities such as loss of life. It is difficult to assign a monetary value to potential fatalities, and it is probably preferable to consider the number of fatalities as a separate dimension of the consequences. In such situations, the consequences are expressed in quantitative terms but are multidimensional in nature. Decision analysis is capable of handling multidimensional consequences, but they do make the analysis a bit more complex.

The final element required for the decision analysis approach is a set of utilities corresponding to the consequences. Even if the consequences are all expressed in monetary terms, different people may have different preferences regarding money. One decision maker may be extremely risk averse, for instance, while another decision maker faced with a similar problem may be willing to take risks. These two decision makers could take two different courses of action and be perfectly reasonable in so doing. Attitude toward risk and other aspects of a decision maker's preferences can be captured for the purpose of modeling by utilities. When the consequences are all monetary, the decision maker's utility function for money is needed. With multidimensional consequences, a multidimensional (or multiattribute) utility function is of interest.

More details regarding the assessment of probabilities and utilities will be given in Sections 4 and 5. In the remainder of this section, we discuss how the various elements of decision analysis can be put together in a convenient structure. The structures most commonly used are decision tables (sometimes called payoff tables or utility tables) and decision trees. We will introduce decision tables and decision trees in the context of an example: a type of problem that is called a cost-loss ratio situation.

Suppose that a decision maker must decide whether or not to protect against the possible occurrence of adverse weather W. The two actions available to the decision maker are protect (P) or do not protect (P̄), and either the event W will occur or it will not occur (i.e., W̄ will occur). This problem is the simplest possible situation in the sense that there are only two actions and two events. It is assumed that no other actions are available and no other events are relevant enough to include in the analysis. Despite the simplicity, however, the situation is a reasonable approximation to realistic situations such as protecting fruit trees against frost damage or protecting concrete against rain damage.

If the decision maker fails to protect and adverse weather does not occur, no cost is incurred and no damage is suffered; the monetary consequence is zero. Failing to protect when adverse weather does occur causes damage in the amount L (the "loss" associated with the damage). The protection is 100% effective, but costs an amount C (the "cost" of protection). Thus, the consequences can be expressed in terms of C and L, and we assume that C < L (otherwise, it is clear that the decision maker should never protect). The cost table is given in Table 1. This table gives the cost (which in this case is either zero or positive) associated with each combination of an action and an event. Equivalently, we could express the problem in terms of a payoff table, as in Table 2, where the payoffs are simply negative costs. If utilities are assessed for the payoffs, the result will be a utility table, as in Table 3. This table, with the probabilities $p = Pr(W)$ and $1-p = Pr(\overline{W})$ included in the bottom margin, shows all of the elements of the problem in a concise format.

An alternative format is that of a decision tree. For the cost-loss ratio situation, a decision tree is presented in Figure 1. Reading from left to right, we first encounter branches corresponding to the two actions, which are followed by branches representing the two events. The probabilities p and 1-p are shown on the appropriate branches, and the payoffs are given at the ends of the branches. Note that the decision tree in Figure 1 is equivalent to the payoff table in Table 2. Similarly, by replacing payoffs with costs or utilities, we could generate decision trees corresponding to Tables 1 and 3. For this simple situation, tables and trees are equally easy to use. In more complex situations, especially situations with sequences of decisions and actions, the decision tree format makes it easier to see the interrelationships among successive actions than does the decision table format. Also, the determination of the value of various types of information can be expressed more clearly in terms of a tree diagram. Thus, we will use decision trees for the most part, particularly in Sections 6 and 7.

The elements described in this section, expressed in a table or a tree to show the structure of the problem, provide the decision analysis model. The decision maker's ultimate goal in this problem, of course, is to choose a course of action. Thus, we turn next to the selection of a decision criterion.

Event

		W	$\overline{W}$
Action	P	C	C
	$\overline{P}$	L	O

Table 1. Cost table for cost-loss ratio situation.

Event

		W	$\overline{W}$
Action	P	$-C$	$-C$
	$\overline{P}$	$-L$	O

Table 2. Payoff table for cost-loss ratio situation.

Event

		W	$\overline{W}$
Action	P	$U(-C)$	$U(-C)$
	$\overline{P}$	$U(-L)$	$U(0)$
Probability		p	$1-p$

Table 3. Utility table (with probabilities) for cost-loss ratio situation.

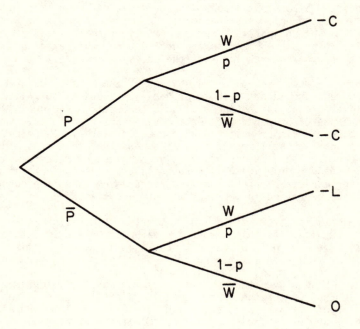

Figure 1. Decision tree (with payoffs) for cost-loss ratio
situation.

3. DECISION CRITERIA

In considering criteria for making decisions under uncertainty, we must keep in mind the fact that the decision is being made before the actual events have occurred. Thus, the decision maker cannot be sure about the outcomes under all of the actions; otherwise, the decision would be made under certainty. Even though an action may look quite good, with desirable consequences under most events, it may yield a bad outcome because an unfavorable event assigned a low probability ex ante does, in fact, occur. This bad outcome ex post does not necessarily mean that the original ex ante decision was a poor choice. Situations always look somewhat different with hindsight, and although a reevaluation of the model in light of what happened may be enlightening and may improve future modeling efforts, it is important to remember that a decision must be based on the information available at the time the decision is made.

There exists a school of thought which claims that decision-making models should not involve probabilities unless so-called "objective" probabilities are available (and some individuals would avoid the use of probabilities entirely). As a result, several non-probabilistic decision criteria have been developed. The decision analysis approach utilizes probabilities and therefore uses a probabilistic criterion. Before studying probabilistic criteria, however, we will consider briefly two nonprobabilistic criteria: maximin payoff and maximax payoff. These criteria assume that the consequences are expressed in terms of monetary payoffs.

The maximin payoff criterion instructs the decision maker to find the smallest possible payoff for each action and then to choose the action for which this smallest payoff is largest (maximize the minimum payoff, hence maximin payoff). For the cost-loss ratio situation, for example, the minimum payoffs are -C for P and -L for $\overline{P}$. Since -C is bigger than -L, the decision maker should always protect. In focusing on the worst possible payoff for each action, the maximin payoff criterion is extremely conservative, ignoring all of the other payoffs (as well as the probabilities, of course).

Exactly the opposite spirit is exemplified by the maximax criterion. Here the decision maker is to find the largest payoff for each action and then to choose the action for which this largest payoff is largest (maximize the maximum payoff). This approach ignores all but the best payoff for each action and appears to be potentially risky; throwing all caution to the winds in an attempt to attain the largest payoff in the entire problem. For the cost-loss ratio situation, the highest payoffs are -C for P and zero for $\overline{P}$, which means that the decision maker should never protect.

Other nonprobabilistic criteria are available, such as maximin loss, which involves opportunity losses instead of payoffs. All of these criteria suffer from a major weakness, in our opinion - they completely ignore all information about the likelihood of occurrence of the events of interest. In the cost-loss ratio situation, a minimax decision maker will protect and a maximax decision maker will not protect regardless of the forecast concerning adverse weather. In the remainder of this chapter, therefore, only probabilistic decision criteria will be considered.

Probabilities are generally introduced into the analysis by taking expected values. In terms of payoffs, this leads to the maximization of expected payoff as a decision criterion. For the cost-loss ratio situation, the expected payoff associated with "protect" is just $-C$, since that is the payoff whether adverse weather occurs or not. The expected payoff associated with "do not protect" is $p(-L) + (1-p)(0) = -pL$. Thus, protecting yields the higher expected payoff if $-C < -pL$, which simplifies to $p > C/L$. In order to make a decision, the expected-payoff maximizer need only compare the probability of the event of interest with the ratio of the cost of protection to the loss associated with the potential damage (hence the name cost-loss ratio situation). The decision maker should protect for high values of the probability forecast p and not protect for low values of p, with the dividing line between high values and low values depending upon the economic aspects of the situation.

Because we are talking about monetary payoffs, maximizing expected payoff is sometimes referred to as the EMV (expected monetary value) criterion. Note also that costs are just negative payoffs, which means that minimizing expected cost is equivalent to maximizing expected payoff. Finally, we are not going to discuss the concept of opportunity loss here, but it also happens that minimizing expected opportunity loss is equivalent to maximizing expected payoff.

The EMV criterion is appealing because it takes into consideration all possible payoffs associated with an action, weighting each payoff by the chance that it will actually be the amount received if that action is taken. What the EMV criterion does not do, however, is allow for the decision maker's attitude toward risk. The decision maker who maximizes expected payoff is behaving in a risk-neutral fashion. To take account of risk attitude in a formal way, it is necessary to work with utilities instead of payoffs. In the decision-analysis approach, therefore, the appropriate criterion is the maximization of expected utility. Justification for this criterion is provided by sets of "axioms of rational behavior" (e.g., Pratt et al., 1964) which imply that a decision maker must maximize expected utility in order to avoid violating any of the axioms.

For the cost-loss ratio situation, the expected utilities are

$$EU(P) = U(-C)$$

and

$$EU(\overline{P}) = pU(-L) + (1-p)U(0).$$

Thus, $EU(P) > EU(\overline{P})$ if and only if

$$p > \frac{U(-C) - U(0)}{U(-L) - U(0)}.$$

As we shall see in Section 5, some flexibility in scaling is allowed in the assessment of a utility function. If U is scaled such that $U(0) = 0$, then the expected utility criterion recommends taking protective action when

$$p > U(-C)/U(-L).$$

As in the case of the EMV-maximizer, the EU-maximizer compares p with a ratio. The ratio is no longer the ratio of C to L, but it is the ratio of the utilities associated with these two outcomes.

The appropriate criterion to use in decision analysis is the maximization of EU, but the EMV criterion need not be discarded. As noted earlier, the EMV criterion is consistent with risk neutrality. If the decision maker is risk neutral, which means that the utility function for money is linear, then the maximization of EMV is equivalent to the maximization of EU. This point will be discussed further in Section 5.

4. ASSESSMENT OF PROBABILITIES

Probability forecasts of the events of concern are an important input in decision analysis. In weather-related problems, a decision maker will generally consult a weather forecaster (unless an appropriate forecast is available in a public weather forecast). The forecaster may simply assess the probabilities directly, taking into account any available information concerning the events of interest. A considerable amount of evidence (e.g., Murphy, 1981) suggests that forecasters can assess precipitation probabilities in a reliable and skillful fashion. Also, subjective probability forecasts of other variables such as temperature and severe storms have been made on a more limited basis and have proved to be quite good forecasts (e.g., Winkler and Murphy, 1979; Murphy and Winkler, 1982). For additional details on the formulation and evaluation of probability forecasts, see Chapters 9 and 10.

The forecaster might find it valuable to consider certain consistency checks when formulating a probability forecast. For example, the probability of an event W and the probability that this event will not occur can both be assessed to assure that they seem reasonable to the forecaster and that they sum to one. Also, the forecaster may want to assess the probability of W and the odds in favor of W and then check these values for consistency.

Consistency checks are particularly valuable when more than a single probability forecast is desired. For example, the probability that measurable precipitation will occur somewhere in a forecast area must be at least as large as the probability that it will occur at a specific point in the area during the same forecast period (e.g., Winkler and Murphy, 1976). Also, certain relationships exist between, for example, six-hour precipitation probabilities and twelve-hour precipitation probabilities (e.g., Winkler and Murphy, 1968). Elementary probability theory indicates various relationships among probabilities, and it would be helpful for a forecaster to be aware of certain simple relationships of this nature (e.g., the probability of the union or intersection of events).

When an entire probability distribution, such as a distribution of minimum temperature, is required for a decision analysis, various assessment procedures can be used. The set of possible values of the variable might be divided into disjoint intervals and the variable treated in a discrete fashion. This procedure entails the assessment of probabilities for the intervals, with the restriction that the probabilities must sum to one. Or, instead of assessing probabilities for the individual intervals, the forecaster could assess cumulative probabilities such as the probability that the minimum temperature will be less than or equal to $0°C$.

If the variable is treated as a continuous variable, various assessment methods are available (e.g., Winkler, 1967). Given a sufficient understanding of continuous probability distributions, the forecaster might work directly with the entire probability density function or the cumulative distribution function. The density function is generally easier to visualize, but cumulative probabilities are easier to express in quantitative terms than densities. The forecaster could assess several points on the cumulative distribution function and then draw an entire cumulative function with the aid of these points. In the assessment of some points on the cumulative function, it is possible to begin by assessing cumulative probabilities corresponding to certain values of the variable or by assessing values of the variable corresponding to certain cumulative probabilities. For an example of the latter approach, consider the assessment of the median of the distribution, followed by the quartiles, the 0.125 and 0.875 fractiles, and so on. This approach, known as the method of successive subdivisions, has been used in probability forecasting of maximum and minimum temperature (Murphy and Winkler, 1974; Winkler and Murphy, 1979).

It may be useful to the assessment process to consider the possibility of using certain well-known probability distributions.

For example, a forecaster may feel that a normal distribution could provide a good approximation to his judgments concerning tonight's minimum temperature. Then the distribution can be completely specified by determining two parameters, the mean and the variance. Assessing the mean would be like formulating a categorical forecast for tonight's minimum temperature. Variances are more difficult to think about intuitively, but the variance of a normal distribution can be found directly by assessing a cumulative probability which can be used, in conjunction with the mean, to find the standard deviation and hence the variance.

The discussion in this section has focused on the formulation of subjective probability forecasts. The ultimate assessment of probabilities to be used in a decision analysis is up to the decision maker, who may choose to delegate this authority to an expert (e.g., a weather forecaster). Of course, no restrictions are placed on the methods that can be used by the decision maker or expert to formulate the required probability forecasts. One alternative is to use models and/or statistical methods to generate the probability forecasts. Statistical and probabilistic weather forecasting are covered in Chapters 8 and 9, respectively, so we will not discuss these procedures in detail here.

5. ASSESSMENT OF UTILITIES

A decision maker's preferences concerning different consequences can be represented in terms of utilities. In this section we discuss the assessment of utilities. Most of the discussion will involve the case in which all consequences have been expressed in monetary terms, although the more general problem of the assessment of utilities for multivariate consequences will be mentioned at the end of the section.

When all consequences are expressed in monetary terms, then the decision maker needs to assess a utility function for money. If the potential monetary payoffs in a decision-making problem range from -$800 to +$920, the decision maker might choose limits of -$1000 and +$1000 and attempt to assess a utility function for monetary payoffs in the interval [-$1000, +$1000]. To determine relationships among utilities, the decision maker can compare simple lotteries such as the following:

Lottery I: Receive $500 for certain.

Lottery II: Receive $1000 with probability 0.5, lose $1000 with probability 0.5.

Lottery II might be thought of as a bet of $1000 on the flip of a coin.

Suppose the decision maker contemplates the lottery and states a preference for Lottery I over Lottery II. The expected utilities of the lotteries are

$$EU(\text{Lottery I}) = U(\$500)$$

and

$$EU(\text{Lottery II}) = 0.5\ U(\$1000) + 0.5\ U(-\$1000).$$

Thus, a preference for Lottery I implies that

$$U(\$500) > 0.5\ U(\$1000) + 0.5\ U(-\$1000).$$

In this manner, the decision maker can generate utility inequalities by considering different lotteries.

A more general form of Lottery II is to receive $1000 with probability p and to lose $1000 with probability 1-p. For small values of p, the decision maker may prefer Lottery I to Lottery II. For very large values of p (e.g., try p = 0.99), Lottery II might be preferable. After considering various values of p, the decision maker should be able to find a value that makes him indifferent between the two lotteries. Suppose that this indifference value of p is 0.96. Indifference implies equal utilities, so we have

$$U(\$500) = .96\ U(\$1000) + .04\ U(-\$1000).$$

A linear transformation of U results in another utility function that is strategically equivalent to U. Thus, we have flexibility in choosing the scale for a utility function. In the example, suppose that the decision maker sets U($1000) = 1 and U(-$1000) = 0. Then the above indifference between lotteries implies that U($500) = 0.96. By changing the $500 in Lottery I to another value, say $0, and attempting to find the indifference value of p once again, the decision maker can assess U($0). In a similar fashion, utilities can be assessed for a number of monetary payoffs. Alternatively, p can be preselected and the decision maker can find a fixed amount of money which leaves him indifferent between Lottery I and Lottery II.

After several utilities have been assessed, they can be plotted on a graph and the decision maker can attempt to fit a curve to the points, as in Figure 2. This curve is the decision maker's assessed utility function for money, and we can investigate the decision maker's attitude toward risk by examining the shape of the curve. If the curve is concave, as in Figure 2, the decision maker is a risk avoider. If, on the other hand, the curve is convex, as in Figure 3, the decision maker is a risk taker. In between these two types of behavior is the risk-neutral decision maker, who has a utility function that is linear in money. It should be noted that many individuals do not fall strictly into any of these three categories. For

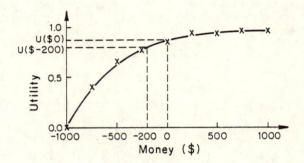

Figure 2. Utility function of a risk avoider.

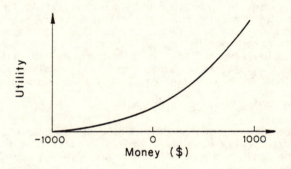

Figure 3. Utility function of a risk taker.

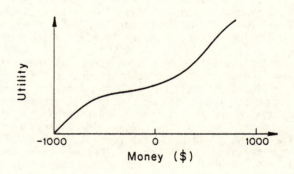

Figure 4. Utility function of a risk taker for monetary payoffs near $0 and a risk avoider for larger amounts.

example, Figure 4 shows the utility function of someone who is a risk taker for monetary payoffs near zero and a risk avoider for larger amounts.

The decision maker's utility function is incorporated into a decision analysis by converting all monetary payoffs into utilities. The expected utilities can then be found for the various actions. For example, consider a decision maker faced with a cost-loss ratio situation with C = \$200 and L = \$1000. If the decision maker's utility function is given by Figure 2, then U(-\$1000) = 0, U(-\$200) = 0.82, and U(\$0) = 0.89. Thus,

$$EU(P) = 0.82$$

and

$$EU(\overline{P}) = p(0) + (1-p)(0.89).$$

The decision maker should protect if

$$0.82 > (1-p)(0.89),$$

which is equivalent to

$$p > 0.08.$$

By contrast, a risk-neutral decision maker would protect if

$$p > C/L = 0.20.$$

The breakeven p is lower for the risk-averse decision maker because of a greater desire to avoid the \$1000 loss.

For convenience, certain models are sometimes used as approximations to utility functions for money. The most common example of such an approximation is a linear model, which simplifies matters considerably by making it possible to work with expected payoffs. Linear utility (i.e., risk neutrality) is often thought to be a reasonable assumption for decisions where the payoffs (positive or negative) are not too large. In any given situation, the assessment of a few utilities will generally reveal whether a linear model provides a reasonable approximation. If the linear model is not reasonable, other models such as exponential, logarithmic, and quadratic functions might be considered. One advantage of such models is that utilities can be found by simply using a mathematical formula instead of reading values from a graph. Also, the decision maker's attitude toward risk can be investigated in terms of certain measures such as the Pratt-Arrow measure of risk aversion (Pratt, 1964), $r = -U''/U'$, where the primes denote differentiation. A positive (negative, zero) r indicates risk aversion (risk taking, risk neutrality), and the decision maker's risk attitude for different wealth levels can be examined by looking at r as a function of wealth.

As noted in Section 2, decision analysis is capable of handling multidimensional consequences, but a utility function for these consequences is needed. Here utility is defined on a multidimensional space, and the assessment of a utility function is more complex than in the case of a utility function for a single numeraire such as money. Much of the work in the area of multiattribute preference has concentrated on sets of preference assumptions that imply certain simplified forms for the utility function. In particular, it is convenient if a multiattribute utility function can be decomposed into a function of single-attribute utility functions so that each single-attribute function can be assessed separately and tradeoffs between attributes need be considered only in combining the single-attribute functions into a single multiattribute utility function. The structuring of preferences and the assessment of multiattribute utilities are covered in detail by Keeney and Raiffa (1976), which is highly recommended to those seeking further information about this topic.

6. VALUE OF INFORMATION

In decision making under uncertainty, the term "uncertainty" indicates that the decision maker is not sure which event or events will occur (e.g., whether or not adverse weather will occur in the cost-loss ratio situation). In the face of uncertainty, additional information may help to reduce the uncertainty and thereby enable the decision maker to make a better decision. In the sense that it may improve the decision, information has some value to the decision maker.

In decision analysis, the value of a certain type of information in a particular decision-making situation is defined as the maximum amount that the decision maker should be willing to pay for the information. This amount will depend, of course, on the details of the problem, on the decision maker's initial state of uncertainty, and on how "good" the information is expected to be. To simplify matters slightly, we assume in this section that the decision maker is risk neutral and is thus willing to make decisions on the basis of the EMV criterion.

Consider the cost-loss ratio situation shown in Table 2, and suppose that the decision maker's probability is p that the adverse event W will occur. As indicated in Section 3,

$$EMV(P) = -C$$

and

$$EMV(\overline{P}) = -pL.$$

Therefore, the decision maker should protect if $p > C/L$ and not protect if $p < C/L$. If p happens to equal C/L, the decision maker is indifferent between protecting and not protecting.

Suppose that the decision maker could find out for certain whether W or W̄ will occur. This type of information, which is seldom available, is called perfect information. If the perfect information indicates that W will occur, the optimal decision is to protect, and the payoff is -C. If, on the other hand, the perfect information indicates that W̄ will occur, the best choice is not to protect, and the payoff is zero. That is, before the information is obtained, the decision maker knows that with the information the payoff will either be -C or zero. Moreover, since the decision maker's probability of W is p, the probability is also p that the best post-information payoff will be -C, and the probability is 1-p that the best post-information payoff will be zero. Hence, before the information is obtained, the expected value of the post-information payoff is

$$\text{EMV(perfect information)} = \text{E(post-information payoff)},$$

$$= p(-C) + (1-p)(0),$$

$$= -pC.$$

Note that EMV(perfect information) is ex ante in the sense that it is calculated before the information is actually obtained. Obviously, the decision maker, in deciding whether or not to buy information, has not yet seen the information and must therefore use an ex ante analysis.

Without the information, the decision maker's expected payoff is -C if $p > C/L$ and -pL if $p < C/L$. The expected value of perfect information (EVPI) is the difference between EMV(perfect information) and the current (pre-information) EMV:

$$\text{EVPI(perfect information)} = \begin{cases} -pC - (-C) = (1-p)C & \text{if } p > C/L, \\ -pC - (-pL) = p(L-C) & \text{if } p < C/L. \end{cases}$$

This expression is graphed in Figure 5 as a function of p, the initial probability of W, when C = \$200 and L = \$1000. Here C/L = 0.2, and the EVPI is highest when $p = C/L$ (i.e., when the decision maker is just indifferent between the two actions). As p moves away from C/L, the decision maker feels more confident in the initial decision, and the EVPI decreases. When $p = 1$ or $p = 0$, of course, the decision maker already feels certain about what will happen, and the information has no value whatsoever. It is important to keep in mind that the EVPI refers to the value of the information to a specific decision maker with particular values of C, L, and p. Different decision makers may have different economic parameters (i.e., different values of C and L) or different initial information (i.e., different values of p). As a result, the same information may be worth different amounts to different people.

It is unusual to be able to obtain perfect information. Except for relatively short lead times, perfect forecasts are generally

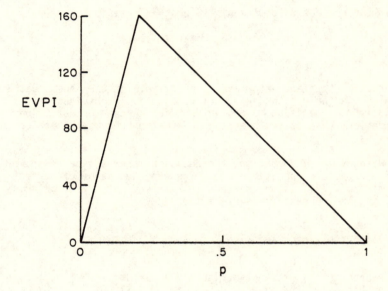

Figure 5. EVPI as a function of p in the cost-loss ratio situation
with C = $200 and L = $1000.

unattainable. The EVPI is still of some interest, however, because it provides an upper bound on the value of less-than-perfect informa- tion. If EVPI = $80 for a given problem, for example, the decision maker knows that information of any sort can never be worth more than $80.

With the same general approach used to find EVPI, we can deter- mine the value of different types of information. It is necessary to anticipate what the information might tell us (e.g., what forecasts might be obtained) and how likely it is that we will receive various pieces of information (e.g., numerical values of forecasts). For instance, consider the cost-loss ratio situation with C = $200, L = $1000, and p = 0.4. From Figure 5, EVPI = $120. Instead of perfect information, consider a forecast that will indicate that the proba- bility of W is either 0.1 or 0.7, with the two forecast values being equally likely. We might think of the initial p = 0.4 as being based on climatology (the only initial information available to the deci- sion maker), with the forecast of 0.1 or 0.7 being provided by an experienced weather forecaster. The limitation to only two forecast values, while unrealistic, is used to simplify the situation a bit for illustrative purposes.

Given the initial p = 0.4, the decision maker would protect and would have an expected payoff of -C = -$200. A forecast of 0.7 for W would leave the decision and the expected payoff unchanged. However, a forecast of 0.1 would cause the decision maker to change the deci- sion to $\overline{P}$, with an EMV of -pL = -(0.1)($1000) = -$100. Thus, the post-information payoff will either be -$200 (if the forecast is 0.7) or -$100 (if the forecast is 0.1). The two forecast values are equally likely, and the expected value of the post-information payoff is thus

$$EMV(forecast) = E(post\text{-}information\ payoff)$$

$$= 0.5(-\$200) + 0.5(-\$100)$$

$$= -\$150.$$

The forecast can be thought of as sample information, and the expected value of sample information (EVSI) is simply the difference between EMV(forecast) and the current (pre-information) EMV:

$$EVSI(forecast) = -\$150 - (-\$200) = \$50.$$

The forecast is worth $50 to the decision maker, as compared with EVPI = $120.

For the sake of comparison, suppose that the decision maker could hire a different forecaster who is not as capable at recogniz- ing different meteorological conditions and at being able to provide a forecast that departs appreciably from climatology. In particular,

with this forecaster the forecast will either be 0.3 or 0.5, with the
two values being equally likely. If the forecast is 0.5, the decision maker will protect and the expected payoff will be -$200. If
the forecast is 0.3, the decision maker will still protect, with an
expected payoff of -$200. As a result,

$$EMV(forecast) = 0.5(-\$200) + 0.5(-\$200) = -\$200,$$

and

$$EVSI(forecast) = -\$200 - (-\$200) = \$0.$$

The decision maker should not be willing to pay anything for this
forecaster's services! No matter what the forecaster says, the decision will not be changed. When we speak of the value of information,
we mean value in terms of the decision-making problem. Information
that will never lead to a change in the decision simply has no value
in terms of the problem.

The expected value of a forecast tells us how much the forecast
is worth in terms of expected improvement in EMV for a decision-
making problem. If the forecast has a certain cost, this cost must
be subtracted from the EVSI to arrive at the expected net gain (ENG)
from the forecast. In deciding between different sources of information (e.g., different forecasters), the decision maker should compare
their expected net gains and choose the forecaster (or forecast
system) providing the highest ENG.

7. SEQUENTIAL DECISIONS

Heretofore, in this chapter, situations in which a single, isolated decision is to be made have been considered. However, many
real-world decision-making problems involving the use of meteorological information consist of sequences of decisions to be made over
time. In such cases, the action to be taken at any point in time -
and the value of the information used in choosing this action - may
depend on previous actions and events and on possible future actions
and events. Such decision-making situations can become very complex
to model and analyze. Here, a very simple sequential decision-making
problem is considered, namely an extension of the cost-loss ratio
situation.

Suppose that a decision maker in the cost-loss ratio situation
must decide whether or not to take protective action against adverse
weather on N occasions. As before, the cost of taking protective
action is C (on each occasion), and no loss is incurred if protective
action is not taken and adverse weather does not occur. However, it
will be assumed that the loss L, which is suffered if protective
action is not taken and adverse weather occurs, can be incurred at
most once on the N occasions. If the decision maker suffers the loss

L on a particular occasion, then the sequence of decisions is assumed to terminate (i.e., no additional costs or losses are incurred). For example, if an orchardist's entire crop is destroyed by low temperatures, there is no need to consider protective action for the rest of the season because there is nothing left to protect.

The multiple-occasion cost-loss ratio situation is a sequential decision-making problem (as opposed to a situation involving repeated application of the static cost-loss ratio model discussed previously in this chapter). It is sequential in the sense that the decision on a particular occasion can influence future decisions. As a result, the anticipation of these future decisions may affect the choice of a current action.

For simplicity, we consider here the case of only two occasions (i.e., N = 2), which will be referred to as two days. The sequential cost-loss ratio situation with N = 2 is illustrated in the form of a decision tree in Figure 6. In this tree, decision points (or nodes) are denoted by squares and event nodes are denoted by circles. At each decision node the decision maker must decide whether to protect (P) or not to protect ($\overline{P}$), and at each event node the uncertainty concerning the events is resolved by the occurrence of adverse weather (W) or no adverse weather ($\overline{W}$). As before, the probability of adverse weather is denoted by p. Moreover, we assume here that the probability of adverse weather p is the same on both days (for example, p might be based on climatological information), that the weather on Day 2 is independent of the weather on Day 1, and that the decision maker wants to maximize the total expected payoff over the two-day period.

A sequential decision-making problem is analyzed by starting with the final-period decision and working back to the initial decision. This process is known as "backward induction," or "averaging out and folding back." For example, in the situation represented in Figure 6, the decision for the second day is analyzed first. But since Day 2 is the final day, this decision is a single-period decision which is uninfluenced by any future decisions. Therefore, the optimal action on Day 2 is dictated by the result obtained for the static cost-loss ratio model: protect if p > C/L. Knowing that this strategy will be optimal on Day 2, we can move back to the initial decision on Day 1.

First, suppose that p > C/L, which means that the decision will always be to protect on Day 2 (unless, of course, there is no choice on Day 2 because the loss L was incurred on Day 1). As a result, if the decision maker protects on Day 1 the overall expected payoff will be -2C (a cost of C will be incurred on each day). If the decision maker does not protect on Day 1 the overall expected payoff will be -pL - (1-p)C [with the -pL corresponding to the probability p of suffering the loss L on Day 1 and the -(1-p)C corresponding to the probability (1-p) of no adverse weather occurring on the first day and then

514

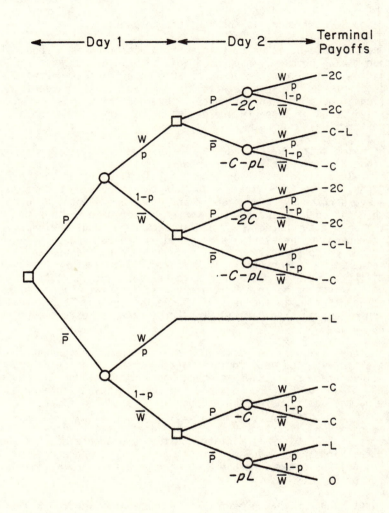

Figure 6.   Decision tree for two-day sequential cost-loss ratio situation.

protecting on Day 2]. Thus, the decision maker should protect on Day 1 if

$$-2C > -pL - (1-p)C,$$

which simplifies to

$$p > C/(L-C).$$

Note how the anticipation of the second-day decision changes the critical value of p on the first day from C/L to C/(L-C), a larger value. If C/L < p < C/(L-C), then the decision maker should not protect on Day 1 but plan to protect on Day 2 (if $\overline{W}$ occurs on Day 1, of course). If p > C/(L-C), the probability of adverse weather is high enough to warrant taking protective action on both days.

Next, consider the case in which p < C/L. On Day 2, the decision should be not to protect. Moving back to Day 1, the decision maker finds that protecting on Day 1 leads to an overall expected payoff of -C - pL (the -C for protecting on the first day and the -pL for the probability of suffering the loss L on the second day, when no protection will be used). Not protecting on Day 1 yields an overall expected payoff of -pL - (1-p)pL [the -pL corresponding to the probability p of suffering the loss on Day 1 and the -(1-p)pL corresponding to the probability p(1-p) of surviving the first day but suffering the loss on the second day]. On Day 1, then, the decision maker should protect if

$$-C - pL > -pL - (1-p)pL,$$

which reduces to

$$p > C/(1-p)L.$$

But it has been assumed in this case that p < C/L, which implies that

$$p < C/L < C/(1-p)L.$$

Therefore, the decision maker should not protect on either day if p < C/L.

An alternative way to analyze sequential problems is to evaluate overall strategies, where a strategy consists of a sequence of daily choices. For example, in the two-day sequential cost-loss ratio situation, there are four possible overall strategies: P on both days, P on Day 1 but $\overline{P}$ on Day 2, $\overline{P}$ on Day 1 but P on Day 2 if L is not incurred on Day 1, and $\overline{P}$ on both days. These four strategies are listed in Table 4 along with their expected payoffs and the values of p for which each strategy is optimal. Each expected payoff is determined by taking the terminal payoffs associated with the branches corresponding to the strategy and weighting these payoffs by the probabilities of reaching these particular end points. The values of

Table 4. Expected payoffs and values of p for which respective strategies are optimal in two-day sequential cost-loss ratio situation.

| Strategy | Action | | EMV | Values of p for which strategy is optimal |
	Day 1	Day 2		
1	P	P	$-2C$	$C/(L-C) < p \leq 1$[a]
2	P	$\overline{P}$	$-C - pL$	None[b]
3	$\overline{P}$	P	$-pL - (1-p)C$	$C/L < p < C/(L-C)$
4	$\overline{P}$	$\overline{P}$	$-[1-(1-p)^2]L$	$0 \leq p < C/L$

[a] Since $C/(L-C) > 1$ if $C/L > 1/2$, the set of values of p for which strategy 1 is optimal is empty if $C/L > 1/2$.

[b] Since $EMV(3) \leq EMV(2)$ for all values of p, strategy 2 is never optimal.

p for which each strategy is optimal agree with the results obtained in the sequential analysis (e.g., under no circumstances should the decision maker protect on Day 1 but not on Day 2).

For a brief example, suppose that C = $200 and L = $1000. Then the decision maker should protect on Day 1 if p > 0.25, as compared with protecting if p > 0.20 under the non-sequential model. Furthermore, as N increases, the breakeven value of p increases for the decision on Day 1. When there are many periods to consider, the decision maker must be careful about committing too much money to pay for protection early in the season.

Many real-world decision-making problems are sequential in nature, and models of such problems must reflect this consideration. Decision analysis is able to handle sequential decisions, including simple sequential decisions such as that considered in this section and more complex decisions involving more periods, nonindependence of successive events, probability revision as decisions are made and events observed, and so on. Sequential models generally yield different optimal courses of action than static models, and the value of information is also affected by sequential considerations.

8. SENSITIVITY ANALYSIS

The results (e.g., expected payoffs or utilities, recommended course of action, value of information) obtained from a model of a decision-making problem depend on the various inputs (e.g., payoffs, probabilities, utilities) used to generate these results. As the inputs are changed, the results may also change. The more the results vary for a given change in inputs, the more sensitive the model is said to be. Sensitivity analysis, which involves the investigation of such variations, is an important step in the application of the modeling procedures of decision analysis.

For example, consider the simple static cost-loss ratio situation. If C = $200, L = $1000, and p = 0.15, then the results are quite sensitive to relatively small variations in the inputs. The optimal strategy is not to protect, but a decrease of just over $50 in C or an increase of just over 0.05 in p would make protecting the action to choose. On the other hand, if C = $200, L = $1000, and p = 0.70, then very large shifts in the inputs would be required to change the optimal decision. The sensitivity of the expected value of perfect information to changes in p when C = $200 and L = $1000 can be seen from Figure 5.

Sensitivity analysis can be used to investigate the impact of numerical changes in inputs, as in the case of varying C, L, or p in the cost-loss ratio situation. It can also be used to study the impact of more general changes in the model, such as the consideration of sequential aspects. Whether conducted formally or informally, sensitivity analysis is an important step in modeling decision-making problems. It can pinpoint the crucial elements of a problem and therefore indicate the elements which should be given special attention in additional modeling efforts.

9. SOME APPLICATIONS OF DECISION ANALYSIS IN METEOROLOGY

Studies based on decision-analytic concepts and methods have been applied to a variety of meteorological problems. Most applications have involved studies of the use and value of meteorological information (generally forecasts), but a few studies during the last decade have examined decision-making situations related to both weather modification and climatic impacts. Some of this work is described briefly in this section.

The early work in this area consisted primarily of studies of the use and value of weather forecasts in simple decision-making situations such as the cost-loss ratio situation (e.g., Thompson, 1952; Thompson and Brier, 1955). These studies are based on an ex post approach and are concerned principally with the relative merits of climatological probabilities and categorical (i.e., deterministic) forecasts. In contrast to the ex ante approach used in decision analysis, an ex post approach involves the determination of the value of information after the forecasts and relevant observations have become available. More recently, the ex post approach has been used to compare the value of probabilistic forecasts with that of other types of forecasts in the cost-loss ratio situation (Thompson, 1962; Murphy, 1977) and to study relationships between the quality and value of probabilistic forecasts (Murphy, 1972; Murphy and Thompson, 1977).

The ex post approach has been used in a weather forecasting context in conjunction with various extensions of the cost-loss ratio model and with more general decision-making models by a number of individuals over the last twenty-five years. This work includes studies by Gringorten (1958, 1959), Borgman (1960), Glahn (1964), Anderson (1973), and Kernan (1975), among others. Some of these studies investigate decision-making situations with more than two actions or events as well as problems in which considerations of risk or utility are important. Moreover, the studies examine the use and value of weather forecasts in a variety of different contexts (e.g., agriculture, industry, transportation).

Among the first papers to adopt a Bayesian decision-theoretic, or decision-analytic, approach to the value and use of meteorological information are works by Nelson and Winter (1960, 1964), Epstein (1962), and Lave (1963). Nelson and Winter consider the simple "protect - don't protect" problem within a general decision-making framework, examine the optimal use of weather forecasts in this framework, and describe the formulation of optimal information structures. Their work has served as an important stimulus to meteorologists, economists, and others to conduct more refined studies of the value and use of weather information. Epstein describes a Bayesian approach to decision making under uncertainty in applied meteorology and investigates the implications of adopting various decision criteria in a hypothetical situation involving different prior distributions. Lave uses the model developed by Nelson and Winter and studies the value of improved weather forecasts to the raisin industry in California.

More recently, a decision-analytic approach has been adopted by several investigators in studies of the use and value of both short-range and long-range (or climate) forecasts. For example, Baquet et al. (1976) and Katz et al. (1982) have examined the so-called "fruit-frost" situation in which an orchardist must decide whether or not to protect his orchard against freezing temperatures. It should perhaps be noted that the model developed by Katz et al. is a dynamic model that takes into account the relationships between decisions and events on successive occasions. Howe and Cochrane (1976) formulate a decision-making model incorporating both long-run (strategic) and short-run (tactical) decisions and apply this model to the problem of a municipal authority's response to urban snow storms (the snow removal problem). With regard to long-range forecasts, Agnew and Anderson (1977) and Phillips et al. (1978) employ decision-analytic methods to study the value of potential improvements in seasonal forecasts for segments of the agricultural sector in the U.S. In addition, Winkler et al. (1983) use decision analysis to investigate the relationships between the quality and value of climate forecasts in a hypothetical agricultural decision-making situation. Finally, since this chapter has devoted considerable attention to the cost-loss ratio situation, it should be noted that Murphy et al. (1982) present a dynamic model of this decision-making situation (i.e., a N-day extension of the sequential decision-making problem considered in Section 7).

Another potential area of application of decision analysis is that of planned weather modification, in which decisions must be made concerning whether or not to seed clouds to increase precipitation or decrease hail and to seed storms to reduce their intensity. Decision analysis has been applied to the problem of deciding whether or not to seed hurricanes in a pioneering study by Howard et al. (1972). At that time, meteorologists generally agreed that the destructive force of hurricanes could be mitigated by seeding them with silver iodide crystals. The authors' analysis is based on both expert judgments provided by meteorologists and the results of previous seeding experiments and other studies related to hurricane wind speeds. The outcomes are measured in economic terms and include losses due to property damage and government responsibility costs. Howard et al. concluded, on the basis of the information available at that time, that the decision to seed is preferable to the decision not to seed and that information obtained from additional seeding experiments would be of positive value. This study provides an excellent illustration of the application of decision analysis in a meteorological context (including examples of determining the value of additional information).

A third area in which decision analysis could play an important role is that of advertent or inadvertent impacts of man's activities on climate (and, in turn, the impacts of any climatic change on man). Examples of such impacts include the effects of emissions from super-sonic aircraft in the stratosphere (the "ozone problem") and the effects of increases in carbon dioxide content in the atmosphere due to burning of fossil fuels (the "CO_2 problem"). Decisions related to these effects have been studied using decision-analytic method-ology in recent years (National Academy of Sciences, 1975; Laurmann, 1980). These decision-making problems obviously are very complex, involving great uncertainties concerning the events as well as out-comes consisting of many different dimensions or attributes.

These brief descriptions of actual applications provide examples of the use of decision analysis in meteorology. The examples include a variety of decision-making problems involving weather and climate information, weather modification, climatic impacts, and forecast evaluation. Interest in decision analysis among meteorologists and others has increased considerably in recent years, and it is antici-pated that this interest will be reflected in further and more refined applications of this methodology to meteorological problems in the future.

ACKNOWLEDGMENT

This work was supported in part by the National Science Founda-tion (Division of Atmospheric Sciences) under grant ATM80-04680.

REFERENCES

Agnew, C. E., and R. J. Anderson, 1977: The economic benefits of improved climate forecasting. Princeton, N.J., Mathematica, Inc., MATHTECH, Report (to NOAA), 159 pp.

Anderson, J. R., J. L. Dillon, and B. Hardaker, 1977: Agricultural Decision Analysis. Ames, Iowa State University Press.

Anderson, L. G., 1973: The economics of extended-term weather fore-casting. Monthly Weather Review, 101, 115-125.

Baquet, A. E., A. N. Halter, and F. S. Conklin, 1976: The value of frost forecasting: a Bayesian appraisal. American Journal of Agricultural Economics, 58, 511-520.

Borgman, L. E., 1960: Weather-forecast profitability from a client's viewpoint. Bulletin of the American Meteorological Society, 41, 347-356.

Brown, R. V., A. S. Kahr, and C. Peterson, 1974: Decision Analysis for the Manager. New York, Holt, Rinehart and Winston.

Epstein, E. S., 1962: A Bayesian approach to decision making in applied meteorology. Journal of Applied Meteorology, 1, 169-177.

Glahn, H. R., 1964: The use of decision theory in meteorology, with an application to aviation weather. Monthly Weather Review, 92, 383-388.

Gringorten, I. I., 1958: On the comparison of one or more sets of probability forecasts. Journal of Meteorology, 15, 283-287.

Gringorten, I. I., 1959: Probability estimates of the weather in relation to operational decisions. Journal of Meteorology, 16, 663-671.

Halter, A. N., and G. W. Dean, 1971: Decisions under Uncertainty with Research Applications. Cincinnati, Ohio, South-Western Press.

Holloway, C. A., 1979: Decision Making Under Uncertainty: Models and Choices. Englewood Cliffs, N.J., Prentice-Hall.

Howard, R. A., J. E. Matheson, and D. W. North, 1972: The decision to seed hurricanes. Science, 176, 1191-1202.

Howe, C. W., and H. C. Cochrane, 1976: A decision model for adjusting to natural hazard events with application to urban snow storms. The Review of Economics and Statistics, 58, 50-58.

Katz, R. W., A. H. Murphy, and R. L. Winkler, 1982: Assessing the value of frost forecasts to orchardists: a dynamic decision-making approach. Journal of Applied Meteorology, 21, 518-531.

Keeney, R. L., 1982: Decision analysis: an overview. Operations Research, 30, 803-838.

Keeney, R. L., and H. Raiffa, 1976: Decisions with Multiple Objectives: Preferences and Value Tradeoffs. New York, N.Y., John Wiley and Sons.

Kernan, G. L., 1975: The cost-loss decision model and air pollution forecasting. Journal of Applied Meteorology, 14, 8-16.

522

Laurmann, J. A., 1980: Assessing the importance of CO_2-induced climatic changes using risk-benefit analysis. In Interactions of Energy and Climate (W. Bach, J. Pankrath, and J. Williams, Editors). Dordrecht-Holland, D. Reidel, pp. 437-460.

LaValle, I. H., 1978: Fundamentals of Decision Analysis. New York, Holt, Rinehart and Winston.

Lave, L. B., 1963: The value of better weather information to the raisin industry. Econometrica, 31, 151-164.

Lindley, D. V., 1971: Making Decisions. London, Wiley-Interscience.

Moore, P. G., and H. Thomas, 1976: The Anatomy of Decisions. Harmondsworth, England, Penguin Books Ltd.

Murphy, A. H., 1972: Ordinal relationships between measures of the "accuracy" and "value" of probability forecasts: preliminary results. Tellus, 24, 531-542.

Murphy, A. H., 1977: The value of climatological, categorical and probabilistic forecasts in the cost-loss ratio situation. Monthly Weather Review, 105, 803-816.

Murphy, A. H., 1981: Subjective quantification of uncertainty in weather forecasts in the United States. Meteorologische Rundschau, 34, 65-77.

Murphy, A. H., R. W. Katz, R. L. Winkler, and W.-R. Hsu, 1982: A dynamic model for repetitive decisions in the cost-loss ratio situation. Corvallis, Oregon State University, Department of Atmospheric Sciences, and Bloomington, Indiana University, Graduate School of Business, manuscript, in preparation. To be submitted for publication.

Murphy, A. H., and J. C. Thompson, 1977: On the nature of the non-existence of ordinal relationships between measures of accuracy and value of probability forecasts: an example. Journal of Applied Meteorology, 16, 1015-1021.

Murphy, A. H., and R. L. Winkler, 1974: Credible interval temperature forecasts: some experimental results. Monthly Weather Review, 102, 784-794.

Murphy, A. H., and R. L. Winkler, 1982: Subjective probabilistic tornado forecasts: some experimental results. Monthly Weather Review, 110, 1288-1297.

National Academy of Sciences, 1975: Environmental Impacts of Stratospheric Flight: Biological and Climatic Effects of Aircraft Emissions in the Stratosphere. Washington, D.C., National Academy of Sciences, 348 pp.

Nelson, R. R., and S. G. Winter, Jr., 1960: Weather information and economic decisions: a preliminary report. Santa Monica, Calif., RAND Corporation, Memorandum RM-2620-NASA, 121 pp.

Nelson, R. R., and S. G. Winter, Jr., 1964: A case study in the economics of information and coordination: the weather forecasting system. Quarterly Journal of Economics, 78, 420-441.

Phillips, L. D., C. R. Peterson, and T. W. Keelin, 1978: The value of improved forecasts of climate for agricultural decision making. McLean, Va., Decisions and Designs, Inc., Final Report PR 77-7-44, 148 pp.

Pratt, J. W., 1964: Risk aversion in the small and in the large. Econometrica, 32, 122-136.

Pratt, J. W., H. Raiffa, and R. Schlaifer, 1964: The foundations of decision under uncertainty: an elementary exposition. Journal of the American Statistical Association, 59, 353-375.

Raiffa, H., 1968: Decision Analysis. Reading, Mass., Addison-Wesley.

Raiffa, H., and R. Schlaifer, 1961: Applied Statistical Decision Theory. Boston, Mass., Harvard Business School, Division of Research.

Schlaifer, R., 1969: Analysis of Decisions under Uncertainty. New York, McGraw-Hill.

Thompson, J. C., 1952: On the operational deficiencies in categorical weather forecasts. Bulletin of the American Meteorological Society, 33, 223-226.

Thompson, J. C., 1962: Economic gains from scientific advances and operational improvements in meteorological prediction. Journal of Applied Meteorology, 1, 13-17.

Thompson, J. C., and G. W. Brier, 1955: The economic utility of weather forecasts. Monthly Weather Review, 83, 249-254.

Winkler, R. L., 1967: The assessment of prior distributions in Bayesian analysis. Journal of the American Statistical Association, 62, 776-800.

Winkler, R. L., 1972: An Introduction to Bayesian Inference and Decision. New York, Holt, Rinehart and Winston.

Winkler, R. L., and A. H. Murphy, 1968: Evaluation of subjective precipitation probability forecasts. Proceedings of First National Conference on Statistical Meteorology (Hartford). Boston, Mass., American Meteorological Society, pp. 148-157.

Winkler, R. L., and A. H. Murphy, 1976: Point and area precipitation probability forecasts: some experimental results. Monthly Weather Review, 104, 86-95.

Winkler, R. L., and A. H. Murphy, 1979: The use of probabilities in forecasts of maximum and minimum temperatures. Meteorological Magazine, 108, 317-329.

Winkler, R. L., A. H. Murphy, and R. W. Katz, 1983: The value of climate information: a decision-analytic approach. Journal of Climatology, 3, in press.

About the Contributors

HARALD DAAN is head of the Bureau for Techniques Development and Evaluation of the Royal Netherlands Meteorological Institute, De Bilt (3730 AE The Netherlands). His interests include improvement of operational methods used in weather forecasting as well as forecast evaluation and the use of meteorological information.

K. RUBEN GABRIEL is a professor of statistics and biostatistics at the University of Rochester, Rochester, NY (14618). His research interests include multivariate statistics, multiple comparisons, and randomization, as well as their applications to meteorological data.

HARRY R. GLAHN is director of the Techniques Development Laboratory, National Weather Service, NOAA, Silver Spring, MD (20910). He has worked in the areas of objective weather map analysis, numerical weather prediction, weather forecast verification, and the application of statistical models to weather forecasting.

THOMAS E. GRAEDEL is a member of the technical staff in the Chemical Kinetics Research Department at Bell Laboratories, Murray Hill, NJ (07974). His research interests include the atmospheric chemistry of gases and condensed water and the atmospheric corrosion of metals and alloys.

RICHARD H. JONES is a professor of biometrics at the University of Colorado School of Medicine, Denver (80262). His primary research interests include time series analysis, particularly problems relating to unequally spaced data, and statistical computing.

RICHARD W. KATZ is a scientist in the Environmental and Societal Impacts Group at the National Center for Atmospheric Research, Boulder, CO (80307). His research interests include probabilistic models and their application to atmospheric sciences.

BEAT KLEINER is a group supervisor at Bell Laboratories, Short Hills, NJ (07974). His primary research interests involve data analysis, graphical methods in statistics, and time series analysis.

DONALD W. MARQUARDT is consultant manager of the Applied Statistics Group at the E.I. du Pont de Nemours & Co., Wilmington, DE (19898). He has published extensively in regression analysis, nonlinear estimation, time series, experimentation with mixtures, and the practice of consulting.

PAUL W. MIELKE, JR. is a professor in the Departments of Statistics and Atmospheric Science at Colorado State University, Fort Collins (80523). His research interests include permutation and nonparametric procedures, evaluation of weather modification projects, and the development and application of statistical and probabilistic techniques in meteorology, climatology, and other disciplines.

ALLAN H. MURPHY is a professor and the director of the Statistics and Climatic Impacts Laboratory in the Department of Atmospheric Sciences at Oregon State University, Corvallis (97331). His primary research interests involve the fields of statistical and applied meteorology, including probabilistic and statistical weather forecasting, forecast evaluation, and the value and use of meteorological information.

RONALD D. SNEE is consultant supervisor, Applied Statistics Group, Engineering Department L3167, at the E.I. du Pont de Nemours & Co., Wilmington, DE (19898). He has worked in the areas of regression analysis, analysis of variance, graphical display of data, design and analysis of mixture experiments and formulation studies, and the training of statisticians.

ROBERT L. WINKLER is IBM Research Professor, Fuqua School of Business, Duke University, Durham, NC (27706). His primary research interests involve probability forecasting, Bayesian inference, and statistical decision theory.

Index

Other Titles of Interest from Westview Press

World Climate Change: The Role of International Law and Institutions, edited by Ved P. Nanda

*Climate Change and Society: Consequences of Increasing Atmospheric Carbon Dioxide, William W. Kellogg and Robert Schware

Weather Modification: Technology and Law, edited by Ray Jay Davis and Lewis O. Grant

**Climate and Environment: The Atmospheric Impact on Man, John F. Griffiths

Applied Climatology: A Study of Atmospheric Resources, John E. Hobbs

Climate's Impact on Food Supplies: Strategies and Technologies for Climate-Defensive Food Production, Lloyd E. Slater and Susan K. Levin

*Available in hardcover and paperback.
**Available in paperback only.